An Introduction to

ATMOSPHERIC
PHYSICS
Second Edition

An Introduction to

ATMOSPHERIC PHYSICS

Second Edition

ROBERT G. FLEAGLE
JOOST A. BUSINGER

Department of Atmospheric Sciences
University of Washington
Seattle, Washington

 ACADEMIC PRESS, INC.

(Harcourt Brace Jovanovich, Publishers)

Orlando   San Diego   San Francisco   New York   London
Toronto   Montreal   Sydney   Tokyo   São Paulo

ACADEMIC PRESS, INC.
Orlando, Florida 32887

United Kingdom Edition published by
ACADEMIC PRESS, INC. (LONDON) LTD.
24/28 Oval Road, London NW1 7DX

Library of Congress Cataloging in Publication Data

Fleagle, Robert Guthrie, Date
    An introduction to atmospheric physics.

    (International geophysics series ;   )
    Bibliography: p.
    Includes index.
    1. Atmosphere. I. Businger, Joost Alois, joint
author. II. Title. III. Series.
QC880.F53 1980   551.5   80–766
ISBN 0–12–260355–9

PRINTED IN THE UNITED STATES OF AMERICA

84 85 86 87   9 8 7 6 5 4 3 2

Contents

CHAPTER III   PROPERTIES AND BEHAVIOR
OF CLOUD PARTICLES

Part I: Growth

Part II: Electrical Charge Generation and Its Effects

CHAPTER IV   ATMOSPHERIC MOTIONS

CHAPTER V   SOLAR AND TERRESTRIAL RADIATION

Part I: Principles of Radiative Transfer

Part II: Radiation outside the Atmosphere

Part III: Effects of Absorption and Emission

Part IV: Photochemical Processes

CHAPTER VI TRANSFER PROCESSES

CHAPTER VII ATMOSPHERIC SIGNAL PHENOMENA

APPENDIX I MATHEMATICAL TOPICS

APPENDIX II PHYSICAL TOPICS

Bibliography

Index

Preface to Second Edition

Since publication of the first edition in 1963, developments in science and in the relation of science to society have occurred that make this revised edition appropriate and, in fact, somewhat overdue. The principles of physics, of course, have not changed, and the atmosphere has changed little. What is different is that in some respects we are now better able to relate atmospheric processes and properties to physical principles. Research results have extended our understanding and brought the various topics of atmospheric physics closer together. At the same time the vital linkages of atmospheric processes to habitability of our planet have become increasingly apparent and have stimulated greater interest in atmospheric physics.

A new chapter on atmospheric motions has been included, reflecting (a) the essential coupling of motions with the state and physical processes of the atmosphere and (b) the recognition that other books published in the past decade make unnecessary the separate volume on motions that had been planned originally. Accordingly, in the new Chapter IV we concentrate on the fundamentals of atmospheric motions that interact with the topics of energy transfer and signal phenomena discussed in other chapters. Ionospheric and magnetospheric physics have developed in a specialized manner and to a high level of sophistication, so that these topics deserve separate treatment from the remainder of atmospheric physics. Also, the ionized upper atmosphere in most respects acts essentially independently of the neutral lower atmosphere. For these reasons, we have with some reluctance omitted ionospheric and magnetospheric physics from this revised edition. With these changes, the text now provides a reasonably concise but complete course in atmospheric physics that is suitable for upper division physics students, as well as for students of the atmospheric sciences.

The final two chapters on energy transfer and signal phenomena have been rewritten and extended, and other chapters have been revised in accordance with insights provided by recent research. The interrelationships of boundary layer structure and energy transfer, the roles of radiation and atmospheric motions in atmospheric composition and in climate, and the use of remote sensing in atmospheric measurement are some of the subjects that receive new emphasis. We have taken advantage of the opportunity to bring other material up to date, to make a variety of corrections, and to improve clarity. We have responded to the most frequent criticism of the earlier edition by including solutions to most of the problems.

In the process of revision, many individuals have responded to our questions and requests, and to each of them we are grateful. In particular, we appreciate the reviews of separate sections and chapters by Peter V. Hobbs, James R. Holton, Conway B. Leovy, C. Gordon Little, and George A. Parks. The task of revision was greatly aided by Phyllis Brien's careful preparation and accurate review of the manuscript.

April 1980

Preface to First Edition

This book is addressed to those who wish to understand the relationship between atmospheric phenomena and the nature of matter as expressed in the principles of physics. The interesting atmospheric phenomena are more than applications of gravitation, of thermodynamics, of hydro-dynamics, or of electrodynamics; and mastery of the results of controlled experiment and of the related theory alone does not imply an understand-ing of atmospheric phenomena. This distinction arises because the extent and the complexity of the atmosphere permit effects and interactions that are entirely negligible in the laboratory or are deliberately excluded from it. The objective of laboratory physics is, by isolating the relevant variables, to reveal the fundamental properties of matter; whereas the objective of atmospheric physics, or of any observational science, is to understand those phenomena that are characteristic of the whole system. For these reasons the exposition of atmospheric physics requires substantial exten-sions of classical physics. It also requires that understanding be based on a coherent "way of seeing" the ensemble of atmospheric phenomena. Only then is understanding likely to stimulate still more general insights.

In this book the physical properties of the atmosphere are discussed. Atmospheric motions, which are part of atmospheric physics and which logically follow discussion of physical properties, have not been included because they require more advanced mathematical methods than those used here. We hope to treat atmospheric motions in a later volume.

The content of the book has been used as a text for a year's course for upper division and beginning graduate students in the atmospheric sciences. In the course of time, it has filtered through the minds of more than a dozen groups of these critics. It also should be of useful interest to students of other branches of geophysics and to students of physics who want to extend their horizons beyond the laboratory. And, finally, we believe that the tough-minded amateur of science may find pleasure in seeking in these pages a deeper appreciation of natural phenomena.

Although an understanding of the calculus and of the principles of physics is assumed, these fundamentals are restated where they are rele-vant; and the book is self-contained for the most part. Compromise has been necessary in a few cases, particularly in Chapter VII, where full development would have required extended discussion of advanced and specialized material.

The text is not intended for use as a reference; original sources must be sought from the annotated general references listed at the end of each chapter. Exceptions have been made in the case of publications that are so recent that they have not been included in standard references and in a few cases where important papers are not widely known. In the case of new or controversial material, an effort has been made to cut through irrelevant detail and to present a clear, coherent account. We have felt no obligation to completeness in discussing details of research or in recounting conflicting views. Where it has not been possible for us to make sound judgments, we have tried to summarize the problem in balanced fashion, but we have no illusions that we have always been right in these matters. We accept the inevitable fact that errors remain imbedded in the book, and we challenge the reader to find them. In this way, even our frailties may make a positive contribution.

We are indebted for numerous suggestions to our colleagues who have taken their turns at teaching the introductory courses in Atmospheric Physics: Professors F. I. Badgley, K. J. K. Buettner, D. O. Staley, and Mr. H. S. Muench. Parts of the manuscript have profited from the critical comments of Professors E. F. Danielsen, B. Haurwitz, J. S. Kim, J. E. McDonald, B. J. Mason, H. A. Panofsky, T. W. Ruijgrok, and F. L. Scarf. Last, and most important of all, we acknowledge the silent counsel of the authors whose names appear in the bibliography and in the footnotes. We are keenly aware of the wisdom of the perceptive observer who wrote:

> Von einem gelehrten Buche abgeschrieben ist ein Plagiat,
> Von zwei gelehrten Büchern abgeschrieben ist ein Essay,
> Von drei gelehrten Büchern abgeschrieben ist eine Dissertation,
> Von vier gelehrten Büchern abgeschrieben ist ein fünftes gelehrtes Buch.

March 1963

CHAPTER 1

Gravitational Effects

"We dance round in a ring and suppose,
But the Secret sits in the middle and knows."   ROBERT FROST

We live at the bottom of an ocean of air which presses on us with a force of about 10^5 newtons for each square meter of surface (10^5 pascals or 10^3 millibars). Our senses give us limited impressions of the atmosphere immediately around us: its temperature, velocity, the precipitation of particles from it, perhaps its humidity, its visual clarity, and its smell, but little of its pressure or its other properties. The atmosphere sustains our life in a variety of ways, but we of course take it for granted except occasionally when storms or drought, or some other examples of severe weather, threaten us, or when we notice a particularly striking or beautiful phenomenon.

Viewed from a fixed point in space at a distance of several earth radii, our planet appears as a smooth spheroid strongly illuminated by a very distant sun. As seen from above the northern hemisphere the earth rotates about its axis in a counterclockwise sense once in 23 hr 56 min and revolves about the sun in the same sense once a year in a nearly circular orbit. The axis of rotation is inclined to the plane of the earth's orbit at an angle of 66° 33'.

Photographs made from earth satellites provide an excellent description of the atmosphere on a nearly continuous basis. Satellites in nearly polar orbits observe the earth in north–south bands. They cover nearly the entire earth twice each 24-hr day, once during the day and once during the night as shown in the composite pictures in Fig. 1.1. The satellites regularly take pictures sensitive to infrared and others sensitive to visible light, and they also measure the vertical distributions of temperature and humidity by sensing radiation emitted from the atmosphere. "Synchronous" satellites at a height of about 36,000 km over fixed points on the equator photograph the field below them (extending to about 70° north and 70° south) once each 30 min. A sequence of these pictures can describe the motions of clouds in exquisite detail. Satellite cloud pictures are regularly seen on television and published in magazines, so that each of us is now able to see the earth and its atmosphere far more completely than anyone had ever seen it prior to the development of meteorological satellites which began in the 1960s. These pictures show that the atmosphere always contains regions of cloud extending over thousands of kilometers but imbedded in a layer no greater than 6–10 km in thickness lying just above the earth's surface. Each of these

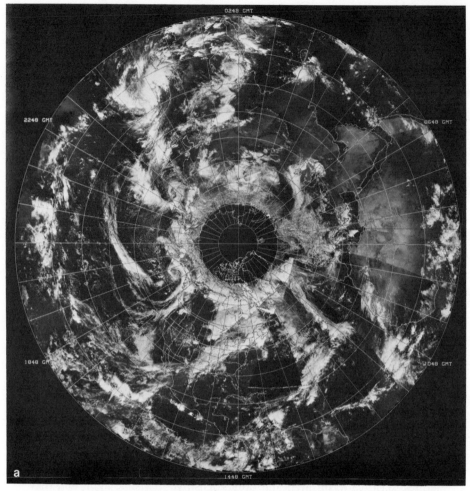

FIG. 1.1a.   Northern hemisphere cloud distribution observed in visible light by polar orbiting satellite NOAA-3, 0636 GMT November 4 to 0745 GMT November 5, 1974. The picture is formed by computer from 13 successive north–south passes of the satellite at the times indicated along the equator.

cloud areas undergoes a characteristic life cycle. Within each large area there are smaller scale cloud systems exhibiting complex structures, movements, and life cycles. Systematic relations can be readily detected among the size and structure of the cloud areas and their latitude, the season of the year, and the underlying earth's surface.

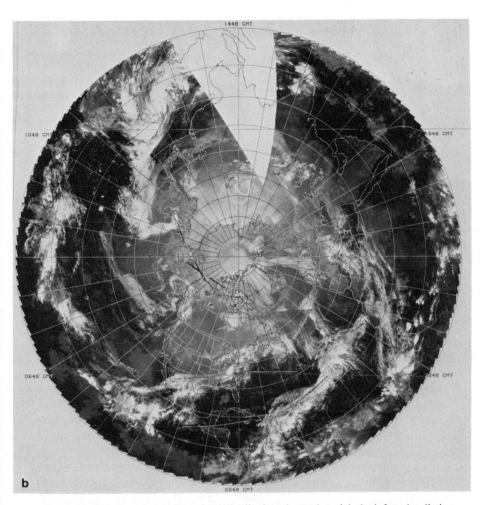

FIG. 1.1b. Northern hemisphere cloud distribution observed at night by infrared radiation by NOAA-3, 2040 GMT November 4 to 2001 GMT November 5, 1974. The picture is formed by computer from 13 successive passes of the satellite at the times indicated along the equator (by courtesy of National Environmental Satellite Service, National Oceanic and Atmospheric Administration).

Half of the atmosphere is always confined to the 6-km layer in which most of the clouds are found, and 99% is confined to the lowest 30 km. The relative vertical and horizontal dimensions of the atmosphere are suggested in Fig. 1.2, which shows the atmosphere as photographed by astronaut

F_{IG}. 1.2.   The atmosphere as seen by John Glenn from a height of about 200 km over the Indian Ocean after sunset 20 February 1962. The photograph covers about 10° of latitude (by courtesy of Project Mercury Weather Support Group, U.S. Weather Bureau).

John Glenn in 1962 on his first flight around the earth. In this photograph the atmosphere is visible only to a height of about 40 km. Although the atmosphere is heavily concentrated close to the earth, air density sufficient to produce observable effects is present at a height of 1000 km, and a very small proportion of the atmosphere extends even farther into space. Therefore, for applications in which mass is the relevant factor, the atmosphere properly is considered a thin shell; for other applications the shell concept is inadequate.

   How is this curious vertical distribution of mass in the atmosphere to be explained? Deeper understanding of this problem requires consideration of complex energy transformations which will be discussed at a later stage; we shall then have occasion to refer to this problem of the vertical distribution of density in the atmosphere. At this stage, however, we shall be concerned with understanding some of the effects of the earth's gravitation on the atmosphere. Before that, some of the elementary concepts and principles of mechanics will be reviewed briefly. Readers who are familiar with these principles should be able safely to skip the next three sections.

1.1   Fundamental Concepts

   The fundamental concepts of physics, those defined only in terms of intuitive experience, may be considered to be space, time, and force. We utilize these concepts, which arise from direct sensory impressions, with the faith that they have the same meaning to all who use them. The choice of fundamental concepts is somewhat arbitrary; it is possible to choose other quantities as fundamental, and to define space, time, and force in terms of these alternate choices.

The events which are important in ordinary or Newtonian mechanics are considered to occur within a frame of reference having three spatial coordinates. These define the position of the event in terms of its distance from an arbitrary reference point; the distance may be measured along mutually perpendicular straight lines (Cartesian coordinates) or along any of a number of lines associated with other coordinate systems. Time is expressed in terms of the simultaneous occurrence of an arbitrary familiar event, for instance, the rotation of the earth about its axis or the characteristic period of an electromagnetic wave emitted by an atom. Force is measured by its effect on bodies or masses.

The primary system of units in this book is the SI system (Système International d'Unités). It is described in Appendix II.A. Note that length, time, and mass are the base quantities of mechanics in the SI system.

1.2 Law of Universal Gravitation

Sir Isaac Newton first recognized that the motions of the planets as well as many terrestrial phenomena (among which we may include the free-fall of apples) rest on a universal statement which relates the separation of two bodies (the separation must be large compared to the linear dimensions of either body), their masses, and an attractive force between them. This law of universal gravitation states that every point mass of matter in the universe attracts every other point mass with a force directly proportional to the product of the two masses and inversely proportional to the square of their separation. It may be written in vector form

$$\mathbf{F} = -G\frac{Mm}{r^2}\frac{\mathbf{r}}{r} \tag{1.1}$$

where $\mathbf{F}$ represents the force exerted on the mass m at a distance $\mathbf{r}$ from the mass M, and G represents the universal gravitational constant. Vector notation and elementary vector operations are summarized in Appendix I.B. The scalar equivalent of Eq. (1.1), $F = GMm/r^2$, gives less information than the vector form, and in this case the direction of the gravitational force must be specified by an additional statement. The universal gravitational constant has the value 6.67×10^{-11} N m^2 kg^{-2}.

The law of universal gravitation is fundamental in the sense that it rests on insight into the structure of the universe and cannot be derived by logical deduction from more fundamental principles. It is an example of the inverse square relation which makes its appearance over and over in fundamental principles and applications. Relativity theory introduces a generalization of the law of universal gravitation, but the change from Eq. (1.1) is so small that it may be neglected for most geophysical purposes.

1.3   Newton's Laws of Motion

Publication in 1686 of Newton's "Philosophiae Naturalis Principia Mathematica," or the "Principia" as it has come to be known, may be recognized as the most important single event in the history of science. For in one published work Newton stated both the law of universal gravitation and the laws of motion; together these laws provided the base for all later developments in mechanics until the twentieth century.

Newton's first law, which states that a body at rest remains at rest and a body in motion remains in motion until acted on by a force, is actually a form of the more general second law. The second law states that *in an unaccelerated coordinate system the resultant force acting on a body equals the time rate of change of momentum of the body*. It may be written in vector form

$$\mathbf{F} = \frac{d}{dt}(m\mathbf{v})$$

(1.2)

where $\mathbf{v}$ represents velocity and t, time. Equation (1.2) is the central equation in development of the physics of atmospheric motions. It, like the law of universal gravitation, is fundamental in the sense that it cannot be derived from more fundamental statements; it states a property of the universe. Newton's second law can be transformed into the equation of conservation of angular momentum. This is left as Problem 1 at the end of the chapter.

Forces occur in pairs. If I push against the wall, I feel the wall pushing against my hand; the two forces are equal in magnitude and opposite in direction. If I push against a movable body, for example, in throwing a ball, I experience a force which is indistinguishable from that exerted by the wall in the first example. Newton's third law of motion expresses these observations in the form: *for every force exerted on a body A by body B there is an equal and opposite force exerted on body B by body A*. This may be expressed by

$$\boxed{\mathbf{F}_{A,B} = -\mathbf{F}_{B,A}}$$ (1.3)

Equation (1.3) is a third fundamental equation.

Notice that the forces referred to in discussing the second law and the law of universal gravitation are forces acting *on* the body under consideration. Only the forces acting *on* a body can produce an effect on the body.

1.4 The Earth's Gravitational Field

The gravitational force with which we are most familiar is that exerted by the earth on a much smaller object: a book, a stone, or our bodies. We call this (with a correction to be explained in Sections 1.5 and 1.8) the weight of the object. If we consider the earth to be a sphere of mass M, in which density depends only on distance from the center, it may be shown that a small external mass m at a distance r from the center of the earth is attracted toward the center of the earth by a force given by Eq. (1.1). This result may be verified by solving Problem 2. A gravitational field is said to be associated with the earth; a mass experiences a force at any point in this field.

For a small mass within the earth the attractive force turns out to depend only on the portion of the earth's mass within a sphere whose radius is the distance of the small mass from the center of the earth. Thus the gravitational effect of the mass of the outer shell cancels exactly. This result is found by solving Problem 3.

The gravitational force exerted on a mass m by the earth may be written conveniently in the form

$$\mathbf{F} = m\mathbf{g}^*$$ (1.4)

where

$$\mathbf{g}^* \equiv -G\frac{M}{r^2}\frac{\mathbf{r}}{r}$$ (1.5)

The vector $\mathbf{g}^*$ is called the gravitational force per unit mass or the gravitational acceleration, and it is directed vertically downward toward the center of the earth. The slight flattening of the earth at the poles makes this statement less than exact, but still adequate for nearly all atmospheric problems. In this discussion the contributions to the gravitational field made by the moon, sun, and other terrestrial bodies are not considered; these effects are applied in Section 1.10 to the discussion of gravitational tides. Problem **4** provides an application of Eq. (1.5).

It is useful to express Eq. (1.5) in terms of the radius of the earth R and height above the earth z. The scalar form is

$$g^* \equiv \frac{GM}{R^2(1 + z/R)^2} \equiv \frac{g_0^*}{(1 + z/R)^2} \tag{1.6}$$

The identity on the right-hand side of (1.6) represents the definition of g_0^*, the gravitational force per unit mass at sea level. Its value as computed from Eq. (1.6) varies by 0.066 N kg^{-1} between the poles, where R is 6356.9 km, and the equator, where R is 6378.4 km. But Eq. (1.6) is not quite correct for the triaxial spheroidal shape of the earth; it is easy to recognize that Eq. (1.6) overestimates g_0^* at the equator because the "extra" mass is distributed symmetrically about the plane of the equator. As a result of these effects g_0^* is 9.832 N kg^{-1} or m s^{-2} at the poles and 9.814 N kg^{-1} at the equator. There are also small local anomalies in g_0^* associated with mountains and the inhomogeneous distribution of density within the earth's crust. For applications in which z is very much less than R, Eq. (1.6) may be expanded in a binomial or a Taylor series about $z = 0$; if only the first two terms are retained, the result is

$$g^* = g_0^*(1 - 2z/R) \tag{1.7}$$

Equation (1.7) shows that near the earth's surface g^* decreases linearly with height. Even at 400 km above the earth, where g^* is about 11% less than g_0^*, Eq. (1.7) is accurate to within about 1%. At increasing heights g^* decreases less rapidly with height than the linear rate given by Eq. (1.7). A brief account of the Taylor series is given in Appendix I.C.

1.5   The Force of Gravity

The force acting on a static unit mass or the acceleration which we measure in a laboratory fixed on the earth's surface is not exactly that given by Eq. (1.5) because the coordinate system within which the measurements are made is itself an accelerating system. It is most convenient in this case to consider that an "apparent" centrifugal force acts on a body rotating with the earth, even though it is clear that, as viewed from a nonrotating coordinate system in space, no "real" centrifugal force acts on the rotating body. It is the coordinate system fixed on the rotating earth which is accelerated, but by adding the apparent centrifugal force, we can use Newton's second law even in the rotating system.

The apparent centrifugal force per unit mass is easily found to be given by $\Omega^2\mathbf{r}$, where Ω represents the *angular frequency* (often called angular velocity) of the earth and $\mathbf{r}$ is the radius vector drawn perpendicularly from the earth's axis to the point in rotation as shown in Fig. 1.3. The vector sum of $\mathbf{g}^*$ and $\Omega^2\mathbf{r}$ represents the force per unit mass or acceleration measured by plumb bob or by a falling weight. It is called the *acceleration of gravity* and is expressed by the vector $\mathbf{g}$. The *force of gravity* is then defined by

$$\mathbf{F}_g \equiv m\mathbf{g} \tag{1.8}$$

The direction of $\mathbf{F}_g$ and $\mathbf{g}$ is of course normal to horizontal surfaces (sea level, in the absence of currents of air or water, for example). The angle between $\mathbf{g}^*$ and $\mathbf{g}$ clearly vanishes at pole and equator, and it is very small even at 45° of latitude. Although small, this angle accounts for the flattening of the earth at the poles and the bulge at the equator; for the earth is a plastic body and takes on the form which tends to minimize internal stresses. The relation between $\mathbf{g}$ and $\mathbf{g}^*$ is represented in vector form in Section 4.3. Problem 5 requires calculation of the angle between $\mathbf{g}^*$ and $\mathbf{g}$.

The magnitude of g is easily calculated if we assume that $\mathbf{g}^*$ and $\mathbf{g}$ have the same direction. The component of $\Omega^2\mathbf{r}$ in the direction of $\mathbf{g}^*$ is $\Omega^2(R+z)$ $\cos^2 \phi$, from which it follows that

$$g = g^*\left[1 - \left(\frac{\Omega^2(R+z)\cos^2\phi}{g^*}\right)\right] \tag{1.9}$$

By combining Eqs. (1.6) and (1.9), the mass of the earth may be calculated from a measured value of g. Numerical evaluation is left to Problem 6.

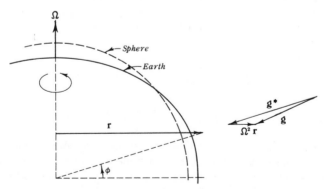

FIG. 1.3.   Cross section through the earth showing the vector sum of the gravitational ($\mathbf{g}^*$) and centrifugal ($\Omega^2\mathbf{r}$) forces acting on unit mass fixed with respect to the rotating earth and the resulting distortion of the earth's figure. Latitude is represented by ϕ and the earth's angular frequency by Ω.

The numerical value of $\Omega^2 R/g^*$ is 3.44×10^{-3}, from which it follows that the effect of the earth's rotation is to make the value of g at the equator $0.0337 \, \text{N kg}^{-1}$ less than at the poles. As pointed out earlier, the oblateness of the earth makes g^* dependent on latitude; the result is that the measured value of g is $9.832 \, \text{N kg}^{-1}$ at the pole, 9.806 at $45°$ latitude, and 9.780 at the equator.

1.6 Geopotential

In order for a body to be raised from sea level, work must be done against the force of gravity. If we recall that *work* is defined by the line integral $\int \mathbf{F} \cdot d\mathbf{s}$ where $d\mathbf{s}$ represents a differential displacement, then the work done per unit mass in displacing a body in the field of gravity is expressed by

$$\Phi \equiv \int_0^z \frac{F_g}{m} \, dz = \int_0^z g \, dz \tag{1.10}$$

Equation (1.10) defines the *geopotential*, the potential energy per unit mass. The importance of the geopotential derives from the fact that at any point in the field, no matter how complicated, the geopotential is unique, that is, it is a function of position and does not depend on the path followed by the mass (m) in reaching the height z. Consequently, the line integral of the force of gravity around any closed path must equal zero; this is the condition required for existence of a potential function. The potential and its relation to vector fields is discussed in Appendix I.G. Because of this property of the field of gravity, the energy required to put a satellite into orbit or to travel to any specified point in the solar system may be calculated easily. Geopotential may be used in place of height to specify position in the vertical direction. The geopotential (Φ) due to the earth alone is found by substituting Eqs. (1.6) and (1.9) in Eq. (1.10) with the result

$$\Phi = \frac{GMz}{R(R + z)} - \Omega^2 \cos^2 \phi \left(R + \frac{z}{2} \right) z \tag{1.11}$$

If potential energy is calculated with respect to a nonrotating coordinate system, only the first term on the right-hand side of Eq. (1.11) is used; this is called *gravitational potential* and is useful in problems in which rotation of the earth is irrelevant. Problems 2 and 3 can be solved more easily using the gravitational potential than using the gravitational force as discussed in Section 1.4.

1.7 Satellite Orbits

If Eq. (1.1) is substituted into Eq. (1.2), there emerges the differential equation which governs the motion of bodies in a gravitational field

$$\frac{d\mathbf{v}}{dt} = -\frac{GM}{r^2}\frac{\mathbf{r}}{r}$$

For a small mass in the gravitational field of a specified large mass, the solution is fairly simple; the result is that the small mass moves along a path which is described as a conic section, that is, it may be elliptic, parabolic, or hyperbolic. The shape of the path is determined by the initial velocity and the initial position of the small mass. For certain critical conditions the path is parabolic. If the initial speed exceeds the critical value, the small body moves along a hyperbola; an example is the path a satellite projected from the earth out into the solar system never to return to the neighborhood of the moon or earth. For an initial speed less than critical, the path is elliptic; examples are the orbits of earth satellites, the moon's orbit around the earth, and the earth's orbit about the sun. We may understand these motions intuitively if we imagine the vector sum of the gravitational force exerted by the large mass and the apparent centrifugal force. For exact balance of these forces, the orbit is circular; if the two are not balanced, acceleration occurs. The earth accelerates as it "falls" toward the sun (gravitational force exceeds centrifugal) from July to January and decelerates as it "rises" against the gravitational pull from January to July. In the case of the earth, accelerations are small because the eccentricity of the orbit is only 0.017. In the case of some satellites projected from the earth large eccentricities have been established.

If we properly can assume balance between the apparent centrifugal force and the gravitational force, certain important results are easily obtained. This balance may be expressed by

$$\frac{GM}{(R+z)^2} = \omega^2(R+z) \tag{1.12}$$

where ω represents the angular frequency of the satellite with respect to the fixed stars. From Eq. (1.12) we may observe that as the height of the satellite decreases, the angular frequency increases. An exercise is provided in Problem 7. If the earth's orbit is assumed to be circular, Eq. (1.12) can be used to calculate the mass of the sun. Applied to the earth–moon system the equation permits calculation of the mass of the earth, as required in Problem 8.

A satellite in the plane of the equator may rotate about the earth with the same angular frequency with which the earth itself rotates. The satellite remains above a fixed point, so that it can view a portion of the earth continuously. The height of such "synchronous" satellites can be found from Eq. (1.12) by replacing ω by the earth's rotational frequency Ω. The solution for the height of the synchronous orbit is

$$z_s = R\left[\left(\frac{g_0^*}{R\Omega^2}\right)^{1/3} - 1\right] \qquad (1.13)$$

Upon substituting numerical values, z_s is found to be about 35,900 km.

The total energy per unit mass needed to put a satellite into a polar orbit may be developed by adding the gravitational potential to the kinetic energy per unit mass $[\frac{1}{2}\omega^2(R + z)^2]$. It follows that the total energy per unit mass is expressed by

$$T = \frac{GM}{R + z}\left(\frac{z}{R} + \frac{1}{2}\right) \qquad (1.14)$$

The energy per unit mass required to enable a body to escape from the earth's gravitational field (that is, to follow a parabolic or hyperbolic path) is easily found from the equation for gravitational potential [Eq. (1.11) without the last term on the right]. We see immediately that the *escape energy* per unit mass is g_0^*R. The absolute velocity (velocity with respect to the solar system) which must be imparted to a free projectile (or a molecule) in order for it to escape from the gravitational field of the earth is found by equating the kinetic energy and the escape energy. The *escape velocity* is then expressed by

$$v_{es} = (2g_0^*R)^{1/2} \qquad (1.15)$$

The escape velocity for the earth is about 11 km s^{-1} and for the moon is only about 2.5 km s^{-1}. The absolute velocity of a point fixed on the surface of the earth at the equator is 0.47 km s^{-1}, so that there is economy in launching satellites from low latitude in the direction of the earth's rotation (toward the east).

1.8   Hydrostatic Equation

Although all planetary bodies (planets, moons, asteroids) possess gravitational fields, not all possess atmospheres, for there is constant escape of gas molecules. Rate of escape depends on strength of the gravitational and magnetic fields and on the velocities of the molecules near the outer limit of the atmosphere. Uncharged molecules which move upward with speeds in excess of the escape velocity and which fail to collide with other molecules or to become ionized leave the planet's gravitational field and are lost to the atmosphere. Ionized molecules moving in the earth's magnetic field are also strongly influenced by an induced electromagnetic force. In the case of the moon virtually the whole atmosphere has escaped, but in the case of the earth it is not known whether the total mass of the atmosphere is increasing through release of gas from the solid earth and by collection of solar and interstellar matter or decreasing through escape of gas from the upper atmosphere.

The presence of the atmosphere as a shell surrounding the earth can now be recognized as a direct consequence of the earth's gravitational field. Each molecule of air is attracted toward the center of mass of the earth by the force of gravity and is restrained from falling to the earth by the upward force exerted by collision with a molecule below it. This collision produces a downward force on the lower molecule, and this force is balanced, in the mean, by collision with a still lower molecule. Therefore, molecules above a horizontal reference surface exert a downward force on the molecules below the surface, a force which is called the *weight* of the gas above the surface. Because weight is proportional to the force of gravity, the weight of the atmosphere includes the effect of the earth's rotation but does not include inertial effects which arise from accelerations measured with respect to the rotating earth. Under static conditions (accelerations negligible) the weight of a vertical column of unit base cross section extending from the earth to the top of the atmosphere is equal to the atmospheric pressure, and the weight therefore may be measured with a barometer. The vertical column is illustrated in Fig. 1.4. Since g varies only slightly within the layer which contains virtually the entire atmosphere, the mass of air contained in the vertical column under static conditions is very nearly proportional to the barometric pressure. At sea level there is about 1 kg above each square centimeter of the earth's surface, and this mass rarely changes by more than $\pm 3\%$.

The force of gravity exerted on a unit volume of air at any point in the column shown in Fig. 1.4 is expressed by ρg, where ρ represents the mass per unit volume (density). Consequently, the pressure at any height z is expressed by

$$p = \int_z^\infty \rho g \, dz \tag{1.16}$$

One of the most important equations in atmospheric physics, the *hydrostatic equation,* is now derived by differentiating Eq. (1.16) with respect to height, holding x, y, and t constant. This yields

$$-\frac{\partial p}{\partial z} = \rho g \qquad (1.17)$$

The hydrostatic equation states that pressure decreases upward at a rate equal to the product of density and force of gravity per unit mass; this result is the same for incompressible and for compressible fluids. It represents a generalization of Archimedes' principle which states that a body immersed in a fluid experiences an upward force equal to the weight of the displaced fluid. Problem 9 provides an illustrative exercise. The density of a compressible fluid like air depends on the pressure, so that we cannot integrate Eqs. (1.16) or (1.17) without knowing density as a function of pressure. Integration of the hydrostatic equation will be discussed in the following chapter; here we shall only point out that density of air decreases with decreasing pressure, so that the rate of pressure decrease itself decreases with height entirely apart from the decrease of g with height. We may understand now how it is possible that the compressible atmosphere extends to very great distances from the earth.

The vertical walls shown in Fig. 1.4 are, of course, only mental aids to understanding; they do not confine the air in any way. It is possible to use

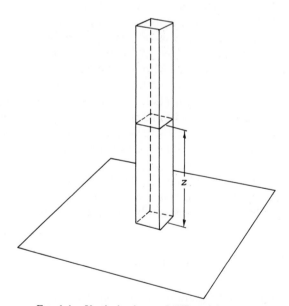

FIG. 1.4.   Vertical column of differential cross section.

this mental aid because under static conditions each vertical column sustains each adjacent vertical column by horizontal pressure. If mass is added to one column, say by an airplane entering the column, the increased pressure within the lower part of the column leads to unbalanced horizontal forces and to accelerations. These lead, in turn, to increased pressure in adjacent columns, and the net result is that the weight of the airplane is distributed over a very large area of ground surface. This is, of course, a most fortunate property of the atmosphere.

The hydrostatic equation may be written in other forms which are particularly revealing for certain applications to be discussed later. First, if Eq. (1.17) is written in the form

$$-\frac{1}{\rho}\frac{\partial p}{\partial z} = g \tag{1.18}$$

the right-hand side represents the force of gravity per unit mass. We have required in the derivation of the hydrostatic equation that accelerations be negligible, so that forces must be balanced. The force which balances the force of gravity in Eq. (1.18) is called the *pressure force* per unit mass.

A second form of the hydrostatic equation may be written if we recognize from Eq. (1.10) that $d\Phi = g\,dz$; therefore, Eq. (1.17) may be written

$$-\frac{\partial p}{\partial \Phi} = \rho \tag{1.19}$$

In this transformation the geopotential has replaced the height as the independent variable; in so doing the number of variables has been reduced by one. Geopotential is used as the vertical coordinate in most atmospheric applications in which energy plays an important role, particularly in the field of large scale motions.

The hydrostatic equation holds under the condition that the pressure and gravity forces balance. In the general case they may not be exactly balanced, and in this case by substituting the pressure and gravity forces in the vertical component of Eq. (1.2) the vertical equation of motion takes the form

$$\frac{dv_z}{dt} = -\frac{1}{\rho}\frac{\partial p}{\partial z} - g \tag{1.20}$$

where v_z represents the vertical component of velocity. If other forces, for example the frictional force, are to be considered, they may be added to form a still more general equation. Obviously, the hydrostatic equation properly may be regarded as the special form of the vertical equation of motion which applies if the acceleration is negligible. A change of vertical component of velocity of $0.1\ \mathrm{m\,s^{-1}}$ in $1\ \mathrm{s}$ makes dv_z/dt amount to about

1% of the pressure and gravity forces. This magnitude of acceleration occurs occasionally in thunderstorms, tornadoes, and severe turbulence, but in most of the atmosphere most of the time the vertical acceleration is much less than 0.1 m s^{-2}, so that we may conclude that the hydrostatic equation is accurate except in unusual circumstances.

1.9   Distribution of Sea Level Pressure

With occasional exceptions sea level pressure varies with time and place between about 98.0 and 104.0 kPa (980 and 1040 mb†), and the mean value for the world is about 101.3 kPa. Therefore, from Eq. (1.16) we find that the total mass of the atmosphere is 5×10^{18} kg. The mean distribution of pressure at sea level for the month of January is shown in Fig. 1.5. The outstanding features are the semipermanent systems of high and low pressure. In general, high pressure predominates at about 30° north and south latitude; low pressure predominates at high latitudes and in the tropics, and the lowest pressures are found in the regions of Iceland and the Aleutian Islands. Figures 1.5 also reveals that in the winter hemisphere pressure is higher over land than over sea, whereas in the summer hemisphere pressure is higher over the sea than over land. In the northern hemisphere low pressure regions over the North Pacific and North Atlantic are dominant features; these are called, respectively, the Aleutian and Icelandic Lows.

Gradients of pressure in the horizontal are associated with horizontal pressure forces. The form of the horizontal pressure force per unit mass may be developed using Fig. 1.6 although it is also easily inferred from the prior discussion of the vertical pressure force. The pressure force acting on the left-hand face of the volume element is directed along the y axis, and it is expressed by $p \, dx \, dz$ where p represents pressure at the center of the face. The pressure on the right-hand face is expressed by the first two terms of a Taylor expansion in the form, $p + (\partial p/\partial y) \, dy$, and the pressure force acting in the y direction is given by $[p + (\partial p/\partial y) \, dy] \, dx \, dz$. Therefore, the net pressure force in the y direction is

$$p \, dx \, dz - \left(p + \frac{\partial p}{\partial y} \, dy \right) dx \, dz = -\frac{\partial p}{\partial y} \, dy \, dx \, dz$$

and, upon dividing by the mass of the differential element ($\rho \, dx \, dy \, dz$), the pressure force per unit mass in the y direction is $-(1/\rho)(\partial p/\partial y)$. Similarly, the pressure force per unit mass in the x direction is expressed by $-(1/\rho)$ $(\partial p/\partial x)$. These results show that pressure force is directed from high to low pressure.

---

† The symbol mb is widely used as an abbreviation for millibar(s), especially in meteorological literature.

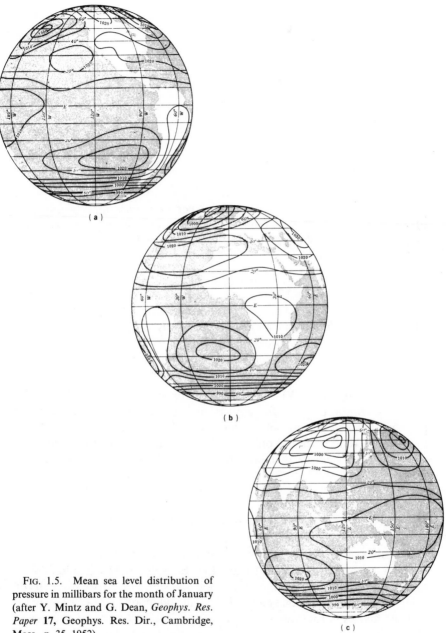

FIG. 1.5. Mean sea level distribution of pressure in millibars for the month of January (after Y. Mintz and G. Dean, *Geophys. Res. Paper* **17**, Geophys. Res. Dir., Cambridge, Mass., p. 35, 1952).

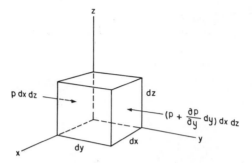

FIG. 1.6.   Pressure forces acting normal to opposite faces of a differential element of volume.

The three-dimensional pressure force per unit mass may be expressed by multiplying the three components respectively by their corresponding unit vectors (**i** in the x direction, **j** in the y direction, and **k** in the z direction) and adding as follows

$$-\frac{1}{\rho}\nabla p = -\frac{1}{\rho}\left(\mathbf{i}\frac{\partial p}{\partial x} + \mathbf{j}\frac{\partial p}{\partial y} + \mathbf{k}\frac{\partial p}{\partial z}\right) \tag{1.21}$$

The horizontal component may be expressed by

$$-\frac{1}{\rho}\nabla_{\mathrm{H}} p = -\frac{1}{\rho}\left(\mathbf{i}\frac{\partial p}{\partial x} + \mathbf{j}\frac{\partial p}{\partial y}\right) \tag{1.22}$$

The gradient operator (∇) is discussed in Appendix I.B.

The horizontal pressure force plays an important role in the generation and maintenance of atmospheric motions, which are discussed in Chapter IV. Here we shall point out only that the horizontal pressure gradients indicated in Fig. 1.5 imply that persistent large scale wind systems must also exist.

1.10   Gravitational Tides

Newton's third law requires that the gravitational forces which are exerted by the moon and sun on the earth and the atmosphere be equal and opposite to those exerted by the earth on the moon and sun. For clarity we shall consider only the earth–moon system. The earth and moon form a system of two coupled masses in rotation about their common center of mass. Because the mass of the earth is about 80 times that of the moon, the center of mass is at a point about 1/80 of the distance between the centers of the earth and moon or about 1600 km below the earth's surface. The earth and moon rotate about this point once in 27.3 days as illustrated in Fig. 1.7a. The diurnal rotation of the earth about its axis may be ignored because its effects are incorporated into the force of gravity as discussed earlier in this chapter. We imagine therefore that the earth moves around the centers of mass without rotation about its axis. The motion may be compared to the motion of a coin placed between the thumb and index finger while the hand makes a circular movement.

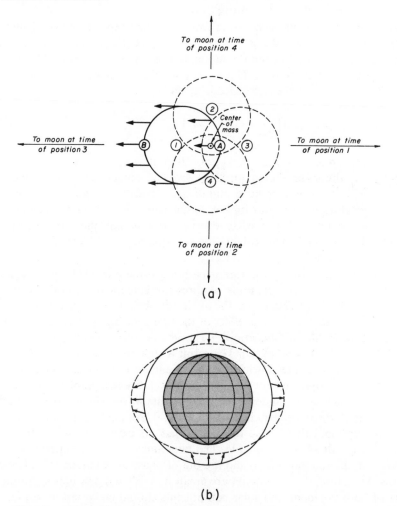

FIG. 1.7. (a) Successive positions of the earth with respect to the center of mass of the earth and moon. Parallel arrows represent centrifugal force when the earth is in position 1. (b) The resultants of the centrifugal and gravitational forces (the tide generating force) and the consequent distortion of the atmospheric shell (not to scale).

Every unit mass in the earth and the atmosphere experiences an equal and parallel centrifugal force due to the 27.3-day rotation of the moon about the earth, and the total of these forces must be equal and opposite to the total gravitational force exerted by the moon. However, gravitational force is inversely proportional to the square of the distance from the moon, so that gravitational force is greater on the side facing the moon than it is on the side away from the moon. The resultant of the two forces, called the *tide generating force*, is illustrated in Fig. 1.7b.

To formulate the tide generating force, we may replace the centrifugal force per unit mass by the gravitational force acting at the distance (D) of the center of the earth from the moon, that is, by GM/D^2 where M represents the mass of the moon. The gravitational force acting at point A is $GM/(D - R)^2$ so that the difference is given by

$$GM\left(\frac{1}{D^2} - \frac{1}{(D - R)^2}\right) \approx -\frac{2GMR}{D^3} \qquad (1.23)$$

so long as R is much less than D. At point B on the opposite side of the earth the result is the same except for the sign. The numerical value is about 10^{-6} N kg^{-1}. These forces at A and B act vertically. At any other point the tide generating force and its horizontal component may be calculated in a similar manner; they must be everywhere less than the vertical forces calculated at A and B. This calculation is required in Problem 10.

In order to deduce the tidal velocity field, equations of motion must be developed by substituting the tide generating force into Newton's second law. To integrate these equations requires mathematical methods beyond the scope of this book. However, the qualitative behavior of the atmosphere can be recognized readily. The effect of the tide generating force is to accelerate air toward the sublunar point (A) and toward the point on the earth opposite the sublunar point (B). The result is that bulges in the atmosphere tend to form on the sides toward the moon and away from the moon while depressions tend to form between the bulges. Relative to a point on the rotating earth, these bulges and depressions constitute waves which travel around the earth each day (the semidiurnal lunar tide). An analogous semidiurnal solar gravitational tide is also to be expected; its magnitude is about 40% of the magnitude of the lunar tide. The two components are superimposed at times of full and new moon and are out of phase at the intervening lunar phases. Therefore, the amplitudes are greatest at full and new moon. Amplitudes of both the lunar and solar components should be largest in low latitudes and least in high latitudes.

The tide generating force acting on the ocean produces the familiar ocean tide. Ocean tidal currents and the height of the tide are influenced strongly by the boundaries of the ocean basins, and there are therefore large local differences in the ocean tide. The tide generating force acting on the elastic body of the earth produces earth tides of which we are not usually aware, but which can be detected by sensitive instruments.

To detect the atmospheric tide, records of sea level atmospheric pressure must be averaged over long periods in order to eliminate the large pressure changes associated with weather and to separate the lunar and solar effects. In this way observations have revealed that the amplitude of the semidiurnal lunar wave in low latitudes is somewhat less than 10 Pa (0.1 mb). The observed semidiurnal solar wave is considerably larger, about 1.5×10^2 Pa (1.5 mb), even though the tide generating force of the sun is less than that of the moon.

Lord Kelvin recognized this paradox in the 1870s, and he consequently attributed the solar semidiurnal wave not to gravitational effects at all, but to thermal excitation of the atmosphere through absorption of solar radiation. However, development of this idea in a quantitatively satisfactory manner proved to be elusive and difficult. It was not until the late 1960s that R. S. Lindzen showed that absorption of solar radiation by ozone between 20 and 80 km above the earth, together with absorption by water vapor in the lower atmosphere, results in atmospheric motions which are associated with the semidiurnal pressure changes observed at the ground.

List of Symbols

		First used in Section
D	Separation of centers of earth and moon	1.10
$\mathbf{F}$	Vector representing force	1.2
g	Scalar force of gravity per unit mass	1.5
$\mathbf{g}$	Vector force of gravity per unit mass	1.5
g^*	Scalar gravitational force per unit mass	1.4
$\mathbf{g}^*$	Vector gravitational force per unit mass	1.4
g_0^*	Scalar gravitational force per unit mass at sea level	1.4
G	Universal gravitational constant	1.2
M, m	Masses	1.2
p	Pressure	1.8
r	Scalar separation of two points	1.2
$\mathbf{r}$	Vector separation of two points	1.2
R	Radius of earth	1.4
t	Time	1.3
T	Total kinetic plus potential energy per unit mass	1.7
v_x	Velocity component toward the east	1.8
v_z	Vertical component of velocity	1.8
v_{es}	Scalar escape velocity	1.7

$\mathbf{v}$	Vector velocity	1.3
x, y	Horizontal coordinates	1.8
z	Height above sea level	1.4
z_s	Height of synchronous orbit	1.7
ρ	Density	1.8
ϕ	Latitude	1.5
Φ	Geopotential	1.6
Ω	Scalar angular frequency of the earth	1.5
ω	Scalar angular frequency of a satellite	1.7

Problems

1.   Show that Eq. (1.2) may be transformed to $\tau = d\mathbf{p}/dt$ where τ represents the torque (defined by $\mathbf{r} \times \mathbf{F}$) and $\mathbf{p}$ represents the angular momentum (defined by $m\mathbf{r} \times \mathbf{v}$).

2.   Show by integration over the volume of a sphere whose density ρ depends only on distance from the center that the gravitational force between the sphere of mass M and an external point mass m is given by Eq. (1.1) where r represents the distance of mass m from the center of the sphere.

3.   Find the gravitational force between the sphere of Problem 2 and the point mass m if the point mass is located within the sphere of radius R.

4.   Find the height above the earth at which the component in the direction to the center of the earth of the gravitational force due to the earth and the moon vanishes

(a) along a line joining the centers of the earth and moon,

(b) along a line at 30° to the line joining the centers of the Earth and moon.

5.   Find the angle between $\mathbf{g}^*$ and $\mathbf{g}$ at the surface of the earth as a function of latitude. What is its maximum numerical value, and at what latitude does it occur?

6.   Calculate the mass of the earth if the measured value of the force of gravity per unit mass at sea level is 9.80 N kg^{-1} at latitude 30°.

7.   Find the periods of revolution for satellites in circular orbits at heights of 200 and 600 km above the earth. What are the apparent periods as seen from a point on the earth if the satellites move from west to east around the equator?

8.   Calculate the mass of the earth assuming that the moon moves in a circular path about the earth once in 27.3 days at a distance of 384,400 km from the earth. Compare the result with Problem 6.

9.   Find the net upward force exerted by the atmosphere on a spherical balloon of 1 m radius if the air density is 1.25 kg m^{-3}, the force of gravity per unit mass is 9.80 N kg^{-1} and the mass of the balloon and its gas is 1.00 kg.

10.   Calculate the components of the lunar tide generating force, and find the point on the earth where the tide generating force is horizontal.

Solutions

1.   Vector multiplication of Eq. (1.2) by $\mathbf{r}$ yields

$$\mathbf{r} \times \mathbf{F} = \frac{d}{dt}(\mathbf{r} \times m\mathbf{V}) - \frac{d\mathbf{r}}{dt} \times m\mathbf{V}$$

Since $\mathbf{V} \equiv d\mathbf{r}/dt$, the second term on the right vanishes, leaving

$$\tau = \frac{d\mathbf{p}}{dt}$$

2. Consider a differential element of mass situated on the ring shown in the accompanying

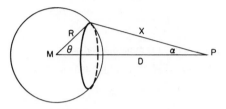

figure. It exerts a gravitational force on a mass m at point P given by

$$\frac{Gm\rho R^2}{x^2} \sin\theta \, d\phi \, d\theta \, dR$$

where ϕ represents the azimuthal angle measured from the center of M. The component in the direction along r is found by multiplying by $\cos\alpha$.

If we integrate over ϕ, all components normal to the r direction cancel and the force due to the mass of the ring is

$$2\pi Gm\rho R^2 \, dR \, \frac{\sin\theta \cos\alpha}{x^2} = 2\pi Gm\rho R^2 \, dR \, \frac{(D - R\cos\theta)\sin\theta \, d\theta}{[r^2 + R^2 - 2DR\cos\theta]^{3/2}}$$

The right side may be integrated over the surface of the sphere by setting $y \equiv \cos\theta$ to yield

$$\frac{4\pi Gm\rho}{D^2} R^2 \, dR$$

This represents the gravitational force exerted by a shell of differential thickness. Upon integrating this from 0 to R gives for the gravitational force exerted by the spherical mass M

$$\frac{4}{3} \pi R^3 \rho \, \frac{Gm}{D^2} = \frac{GmM}{D^2}$$

Therefore, the gravitational force exerted by the sphere is that which would be exerted by an equal mass concentrated at the center of the sphere.

3. The result of Problem 2 can be used to express the gravitational force due to the mass within the sphere defined by the position of the point mass. Then the force due to the outer shell can be evaluated by integrating over a ring and then the shell as in Problem 2.

Another way to evaluate the contribution from the outer shell is to consider a cone with apex at the point mass and solid angle $d\omega$. This cone subtends an area of the shell given by $r_1^2 \, d\omega$, and the gravitational attraction due to this portion of the shell is

$$\frac{Gr_1^2 \rho \, d\omega \, dR}{r_1^2} = G\rho \, d\omega \, dR$$

The opposite side of the shell exerts a gravitational force given in a similar way by $G\rho \, d\omega \, dR$. These forces are in opposite directions. Upon integrating over all solid angles, the net force due to the outer shell remains zero.

4. (a) The balance of gravitational forces due to the earth and moon is expressed by

$$\frac{GM_e}{(R+h)^2} = \frac{GM_m}{(D-R-h)^2}$$

where D represents the separation of the centers of the earth and moon. Upon solving for h,

$$h = \frac{D - R(1-a) \pm Da^{1/2}}{1-a}$$

where a represents M_m/M_e. Using the numerical values

$$R = 6400 \text{ km} \qquad D = 384,400 \text{ km} \qquad a = 1/81$$

the values of h are

$$h = 330,500 \quad \text{or} \quad 416,000 \text{ km}$$

The second value is on the side of the moon opposite the earth and does not apply since both gravitational forces are in the same direction. (b) A mathematical relationship can be written equating the earth's gravitational force at height h to the component of the moon's gravitational force at the same point in the direction toward the center of the earth. This leads to a cubic equation which has no solution for h. This can be shown easily by comparing the absolute values of the earth's and moon's gravitational forces at the point on the 30° line which is closest to the moon. Since the magnitude of the force exerted by the earth exceeds that exerted by the moon, the components in the direction of the earth cannot be equal.

5. From Fig. 1.3 we may express the angle between $\mathbf{g}^*$ and $\mathbf{g}$ by

$$\alpha = \frac{\Omega^2 R \cos\phi \sin\phi}{g^*}$$

because $\Omega^2 R/g^* \ll 1$. The maximum value of α is found by setting the derivative of α with respect to ϕ equal to zero.

$$\frac{d\alpha}{d\phi} = \frac{\Omega^2 R}{g^*}(\cos^2\phi - \sin^2\phi) = 0$$

Therefore,

$$\tan^2\phi = 1 \qquad \phi = 45°$$

At this latitude $\alpha = \frac{1}{2}(\Omega^2 R/g^*) = 0.0015$ rad.

6. At latitude $\phi = 30°$ and $z = 0$, Eq. (1.9) yields

$$g = GM/R^2 - \tfrac{3}{4}\Omega^2 R$$

Solving for the mass of the earth gives

$$M = 6.0 \times 10^{24} \text{ kg}$$

7. By equating the gravitational and centrifugal forces per unit mass the absolute angular frequency of a satellite is expressed by

$$\omega = \frac{(GM)^{1/2}}{(R+h)^{3/2}}$$

The corresponding absolute periods are

$$\tau_a = \frac{2\pi(R + h)^{3/2}}{(GM)^{1/2}}$$

$$= 5.35 \times 10^3 \text{ s} \quad \text{for} \quad h = 200 \text{ km}$$
$$ 5.82 \times 10^3 \text{ s} \quad \text{for} \quad h = 600 \text{ km}$$

For satellites rotating from west to east in the plane of the equator, the angular frequency of the earth's rotation is subtracted and

$$\tau_r = 5.71 \times 10^3 \text{ s} \quad \text{for} \quad h = 200 \text{ km}$$
$$ 6.25 \times 10^3 \text{ s} \quad \text{for} \quad h = 600 \text{ km}$$

8. By equating the gravitational and centrifugal forces acting on the moon the earth's mass may be expressed by

$$M = \frac{\omega^2 r^3}{G} = \frac{4\pi^2 r^3}{\tau^2 G}$$

where r is the distance of the moon center from the earth's center. Substitution of numerical values gives

$$M = 6.1 \times 10^{24} \text{ kg}$$

which is close to the mass found in Problem 6. The differences may be accounted for by local variations in measured values of g as well as by the limited accuracy of given numerical values.

9. The net upward force given by Archimedes' principle is

$$F_g = gV\rho - gm$$

where V represents volume of the balloon and m the mass of the balloon and the gas it contains. Substitution of numerical values gives

$$F_g = 41.5 \text{ N}$$

10. The component of the tide generating force parallel to the D direction can be expressed by

$$GM\left(\frac{1}{D^2} - \frac{1}{(D - R\cos\theta)^2}\right) \approx \frac{2GMR\cos\theta}{D^3}$$

where θ represents the angle between D and the zenith at the point considered and θ is also considered equal to the zenith angle of the moon. The component perpendicular to D can be expressed by

$$\frac{GM\sin\alpha}{(D - R\cos\theta)^2}$$

where α is the angle at the center of the moon between D and the line to the point considered on the earth. It can be recognized that

$$\sin\alpha = \frac{R\sin\theta}{D - R\cos\theta}$$

so that the component of the tide generating force normal to D is

$$\frac{GMR \sin \theta}{(D - R \cos \theta)^3} \sim \frac{GMR \sin \theta}{D^3}$$

For the tide generating force to be horizontal the ratio of the normal component to the component along D is given by

$$\tan \theta = \frac{2 \cos \theta}{\sin \theta}$$

Therefore,

$$\sin^2 \theta = 2 \cos^2 \theta$$

$$\cos \theta = \tfrac{1}{3} \sqrt{3} \quad \text{and} \quad \theta = 54° \, 44'$$

General References†

Chapman and Lindzen, *Atmospheric Tides*, summarizes the history of investigations of atmospheric tides, presents the recently developed quantitative theory, and compares theory with observations.

Haltiner and Martin, *Dynamical and Physical Meteorology*, and other texts in meteorological theory treat the hydrostatic equation somewhat more generally than is attempted here.

Joos, *Theoretical Physics*, 2nd ed., is one of many classical physics texts which provide a good concise account of the theory of planetary (satellite) motion.

Sears, *Principles of Physics*, Volume 1: *Mechanics Heat and Sound*, gives an excellent introduction to mechanical concepts and to Newton's laws. There are other recent introductory physics texts which are equally good.

† General references which provide necessary background or which substantially extend the material presented are listed at the end of each chapter. Complete references for these books are listed in the Bibliography.

CHAPTER II

Properties of Atmospheric Gases

"Work expands so as to fill the time (and space)
available for its completion." CYRIL NORTHCOTE PARKINSON

In the previous chapter we have found it useful to look upon the earth and its atmosphere from the outside; the external point of view has helped to focus attention on certain important macroscopic properties of the atmosphere. In this chapter we shall investigate other properties for which an internal view is more useful. We can imagine the internal view as made by a particularly versatile and powerful microscope although we realize that our understanding comes from many observations of varied sorts, both direct and indirect.

First, we recognize that the gaseous ocean surrounding the earth consists of countless particles, or molecules, which move about in random motion. The chaos of these motions may be reduced to tractable order by considering the statistical properties of a very large number of molecules. We then may replace the concept of innumerable minute molecules in random motion by the concept of a continuous fluid medium with continuous properties; however, wherever it is useful to do so, these continuous properties can be interpreted in statistical terms. Statistical treatment of molecular behavior falls under the branch of physics known as *statistical mechanics*, and the study of the related continuous macroscopic properties falls under the branch known as *thermodynamics*.

2.1 Molecular Behavior of Gases

The atomic structure of matter was introduced as a philosophical speculation by the Greek philosophers Empedocles and Democritus, and Dalton in 1802 introduced this concept as a scientific hypothesis.

The characteristic behavior of gases involved in chemical reactions inspired Avogadro to suggest in 1811 that equal volumes of different gases at the same pressure and temperature contain equal numbers of molecules. It has been possible to verify this hypothesis and to use it for determining the relative masses of the various atoms. The carbon 12 nucleus has been assigned a *molecular mass* value of 12 exactly, and this is used as a standard with which the masses of all atoms are compared.† A quantity of any substance

† Adopted by the General Assembly of the International Union of Physics and Applied Physics, September 1960, Ottawa, Canada.

whose mass in grams is equal to its molecular mass is called a *mole*. The volume occupied by a mole of gas at standard atmospheric pressure and $0°$ C is 22.4 liters [$(2.241383 \pm 0.000070) \times 10^{-2}$ m^3 mole^{-1}] and is the same for all gases. The number of molecules contained by a mole is therefore a constant and is called *Avogadro's number* (N_0).

Avogadro's number has been determined by several methods. The method which is easiest to understand and which gives a very accurate determination is based upon observations of the electrolysis of solutions. When an electric field is created between electrodes placed in a solution, the positive ions in the solution migrate toward the cathode and the negative ions migrate toward the anode. If each ion carries just one fundamental quantity of charge, the charge of the electron, the number of ions which transfer their charge to the electrodes is equal to the total transferred charge divided by the charge of the electron. By measuring the mass deposited on the electrode, the number of ions deposited per mole (Avogadro's number) may be calculated. An accurate value of this number is

$$N_0 = (6.022045 \pm 0.000031) \times 10^{23} \text{ mole}^{-1}$$

The empirical evidence that any substance consists of a very large number of molecules has led to the development of several statistical theories designed to explain the macroscopic properties of matter. These theories utilize idealized models of the molecules. A particularly simple model will be discussed in Section 2.3.

2.2   Composition of Air

The atmosphere is composed of a group of nearly "permanent" gases, a group of gases of variable concentration, and various solid and liquid particles. If water vapor, carbon dioxide, and ozone are removed, the remaining gases have virtually constant proportions up to a height of about 90 km. Table 2.1 shows the concentrations of these permanent constituents of air.

From Table 2.1 it may be calculated that the molecular mass of air is 28.97. It is apparent that nitrogen, oxygen, and argon account for 99.997% of the permanent gases in the air. Although all these gases are considered invariant, very small changes in space or time may be observed. The uniformity of the proportions is produced by mixing associated with atmospheric motions. Above a height of 90 km the proportion of the lighter gases increases with height as diffusion becomes more important relative to mixing. Diffusive equilibrium and the process of diffusion are discussed in Sections 2.6 and 2.16.

TABLE 2.1

PERMANENT CONSTITUENTS OF AIR[a]

Constituent	Formula	Molecular mass	% by volume
Nitrogen	N_2	28.0134	78.084
Oxygen	O_2	31.9988	20.9476
Argon	Ar	39.948	0.934
Neon	Ne	20.183	18.18×10^{-4}
Helium	He	4.0026	5.24×10^{-4}
Krypton	Kr	83.80	1.14×10^{-4}
Xenon	Xe	131.30	0.087×10^{-4}
Hydrogen	H_2	2.01594	0.5×10^{-4}
Methane	CH_4	16.04303	2×10^{-4}

[a] From *U. S. Standard Atmosphere, 1976*, NOAA, NASA, USAF, Washington, D.C., 1976.

TABLE 2.2

VARIABLE CONSTITUENTS OF AIR

Constituent	Formula	Molecular mass	% by volume
Water vapor	H_2O	18.0160	0–7
Carbon dioxide	CO_2	44.00995	0.01–0.1 (near the ground) average $0.033 + 1 \times 10^{-4} t$[a]
Ozone	O_3	47.9982	0–0.01
Sulfur dioxide	SO_2	64.064	0–0.0001
Nitrogen dioxide	NO_2	46.0055	0–0.000002

[a] t in years starting at 1976.

The major variable gases in air, taken judiciously from a variety of sources, are listed in Table 2.2. Variations in CO_2 are caused by absorption and release by the oceans, by photosynthesis, and by combustion, especially of fossil fuels; these variations have a significant effect on atmospheric absorption and emission of infrared radiation. This subject is further discussed in Section 6.15. Variations in water vapor and ozone are important in processes which are discussed in later sections. Other gases not included in Table 2.2 are chiefly products of combustion and are found in the atmosphere in variable concentration.

The solid and liquid particles which are suspended in the atmosphere (the *aerosol*) play an important role in the physics of clouds. The distribution and properties of these particles are discussed in Sections 3.5 and 3.9.

2.3 Elementary Kinetic Theory

A simple theory of dilute gases based on classical mechanics may be developed by assuming that the gas consists of molecules which have their mass concentrated in very small spherical volumes and which collide with each other and with the walls surrounding them without loss of kinetic energy. The molecules are assumed to exert no forces on each other except when they collide. With this model it is possible quite accurately to account for the behavior of the permanent gases under atmospheric pressure and temperature. The great power of kinetic theory, which goes far beyond the elementary theory developed here, may be attributed to the fact that, whereas thermodynamics is concerned with equilibrium states only, kinetic theory recognizes no such limitation.

Consider N molecules in a rectangular volume V. The molecules are separated by large distances compared with their own diameters. Between collisions they move in straight lines with constant speed. Since the walls are considered perfectly smooth, there is no change of tangential velocity in a collision with the walls. If it is assumed that the molecules are distributed uniformly, the number of molecules in any volume element is

$$dN = n \, dV \tag{2.1}$$

where n represents the number of molecules per unit volume. It is also assumed that the distribution of molecular velocities is the same in all directions. It will be shown in Section 2.15 that these last two assumptions are valid for the condition of maximum probability. Solution of Problem 1 yields the smallest volume increment for which the assumption of a uniform distribution still is true.

To find an expression for the number of molecules striking a wall per unit area and per unit time, consider the small element dA of the wall sketched in Fig. 2.1 and construct the normal to the element and a reference plane through the normal. We now ask how many molecules traveling in the particular direction θ, ϕ and with specified speed v strike the surface in time dt. This means that we consider all cases between θ and $\theta + d\theta$, ϕ and $\phi + d\phi$, v and $v + dv$. Now construct the cylinder of length $v \, dt$ whose axis lies in the direction θ, ϕ as shown in the figure. All the molecules with the specified speed and direction in the cylinder hit the wall in time dt, and also all the molecules with this speed and direction that hit the wall are contained by the cylinder.

The next question is: How many of these molecules are there in the cylinder? Equation (2.1) shows that the total number is $nv \, dt \, dA \cos \theta$. If dn_v represents the total number of molecules with speeds between v and $v + dv$ per unit volume, the number of molecules with the required speed

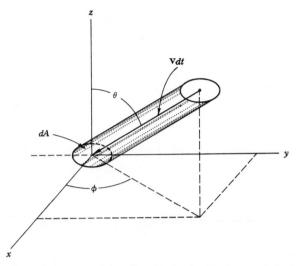

FIG. 2.1. Volume element containing all molecules that hit the area dA in the time interval dt coming from the direction ϕ, θ with speed v.

in the cylinder is $dn_v \, v \, dt \, dA \cos \theta$. The directions of these molecules are uniformly distributed over all angles as shown in Appendix I.I. The number with speed v in the direction ϕ, θ may be expressed by

$$v \, dn_v \, dt \, dA \cos \theta \, \frac{d\omega_{\theta\phi}}{4\pi}$$

where $d\omega_{\theta\phi}$ is the increment of solid angle from where the molecules come. Using Eq. (I.1) from Appendix I.I and considering that the number is the same for each cylinder with an angle θ, integration over ϕ yields the number of molecules hitting the wall in area dA and in time dt from all directions defined by the ring between θ and $\theta + d\theta$. The result of this integration is

$$\tfrac{1}{2} v \, dn_v \sin \theta \cos \theta \, d\theta \, dt \, dA \tag{2.2}$$

The number of molecules hitting the area dA in time dt is found by integrating from $\theta = 0$ to $\theta = \pi/2$ to be

$$\tfrac{1}{4} v \, dn_v \, dt \, dA$$

Consequently, the molecules with speeds between v and $v + dv$ experience

$$\tfrac{1}{4} \, v \, dn_v$$

collisions with the wall per unit area per unit time. The total number of collisions per unit area and per unit time is obtained by integrating over the entire range of speeds. If the average speed is defined by

$$\bar{v} \equiv \frac{1}{n} \int_0^n v \, dn_v \tag{2.3}$$

then the total number of collisions per unit area per unit time may be written in the form

$$\tfrac{1}{4} n \bar{v}$$

In Problem 2 an expression is developed for the total number of collisions per unit area per unit time per unit solid angle for molecules of all velocities.

2.4   Equation of State of an Ideal Gas

A molecule which collides with the wall experiences a change in momentum. If m represents the mass of the molecule and θ the angle of its direction with the vertical, the momentum change illustrated in Fig. 2.2 is given by

$$mv \cos \theta - (-mv \cos \theta) = 2mv \cos \theta$$

Therefore, it follows from Eq. (2.2) that the change in momentum due to all collisions taking place per unit area per unit time coming from molecules with directions defined by the ring between θ and $\theta + d\theta$ is expressed by

$$mv^2 \, dn_v \sin \theta \cos^2 \theta \, d\theta$$

and, upon integrating between 0 and $\pi/2$, the change in momentum from collisions coming from all directions is given by

$$\tfrac{1}{3} mv^2 \, dn_v$$

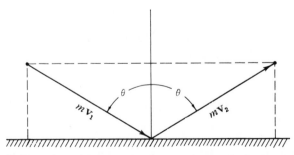

FIG. 2.2.   The momentum vectors of a moving molecule before and after making a perfectly elastic collision with a plane wall.

Finally, the change of momentum per unit time may be equated to the force exerted on the surface dA. The force is then expressed by

$$dF = \tfrac{1}{3}m\left(\int_0^n v^2 \, dn_v\right) dA$$

and the pressure exerted on the surface is

$$p \equiv \frac{dF}{dA} = \tfrac{1}{3}m \int_0^n v^2 \, dn_v$$

This may be written in the form

$$p = \tfrac{1}{3}mn\overline{v^2} \qquad (2.4)$$

where the average value of the square of the speed is defined by

$$\overline{v^2} \equiv \frac{1}{n} \int_0^n v^2 \, dn_v \qquad (2.5)$$

Upon eliminating n from Eq. (2.4)

$$pV = \tfrac{1}{3}mN\overline{v^2} \qquad (2.6)$$

Equation (2.6) may be written in the form

$$p\alpha_m = \tfrac{1}{3}mN_0\overline{v^2} \qquad (2.7)$$

where α_m represents the volume containing one mole of the gas, the *molar specific volume*.

We now define the *temperature* as proportional to the translational kinetic energy of the molecules, and write

$$\tfrac{1}{2}m\overline{v^2} \equiv \tfrac{3}{2}kT \qquad (2.8)$$

where the factor $\tfrac{3}{2}k$ is the constant of proportionality and k is known as the *Boltzmann constant*. Upon substituting Eq. (2.8) into (2.7)

$$p\alpha_m = kN_0T$$

which is the *equation of state* for an *ideal gas*. This equation is usually written in the form

$$p\alpha_m = RT \qquad (2.9)$$

where R, the *gas constant*, replaces kN_0. The Boltzmann constant (k) may be interpreted as the gas constant for a single molecule. Avogadro's number (N_0), the Boltzmann constant (k), and the gas constant (R) are universal, that is, they are the same for any ideal gas.

It is also useful to express the equation of state in terms of *specific volume* (α) rather than the molar specific volume (α_m). By definition

$$\alpha \equiv \alpha_m / M$$

where M represents molecular mass. Therefore, Eq. (2.9) may be written

$$p\alpha = (R/M)T \equiv R_m T \qquad (2.10)$$

where R_m is the specific gas constant for a gas with molecular mass M.

The numerical value of R may be calculated using the results of experiment. A specific mass of a gas brought into contact with melting ice at standard atmospheric pressure always reaches the same volume after a long exposure. Similar behavior is observed if a volume of gas is brought into contact with boiling water at standard atmospheric pressure. In each case the average kinetic energy of the gas molecules approaches the average kinetic energy of the contacting molecules, and it is recognized that the systems in contact have uniform temperature and are in equilibrium. It further has been observed that whenever two different masses of gas (or *systems*) are brought into equilibrium with melting ice, they are also in equilibrium with each other. This is a general thermodynamic principle called the *zeroth law of thermodynamics*.

Now Eq. (2.9) predicts that at constant pressure a volume of an ideal gas varies linearly with T and that the constant of proportionality is the same for all ideal gases. By comparing volumes of various gases at the steam point of water (s) with their volumes at the ice point (i), it is found that the ratio V_s/V_i is constant for all the permanent gases. If we arbitrarily assign to the difference in temperature between the ice and steam points the value 100°, the equation of state gives

$$\frac{V_s}{V_i} = \frac{T_s}{T_0} = \frac{T_0 + 100}{T_0}$$

Solving for T_0 it is found that the temperature of the ice point on this scale, referred to as the absolute or Kelvin scale, is

$$T_0 = 273.155 \text{ K}$$

By subtracting 273.155 K from the absolute temperature, the Celsius temperature is defined. Thus on the Celsius scale the ice and steam points are at 0° C and 100° C, respectively. It is now possible to determine from Eq. (2.9) that R is given by

$$R = (8.31441 \pm 0.00026) \text{ J mole}^{-1} \text{ K}^{-1}$$

and for dry air

$$R_m = 0.2871 \times 10^3 \text{ J kg}^{-1} \text{ K}^{-1}$$

It follows that $k = (1.380662 \pm 0.000044) \times 10^{-23}$ J K^{-1}. A definition of temperature which is independent of kinetic theory has been given by Kelvin using the second law of thermodynamics.

The root mean square (rms) speed of the molecules may now be calculated from Eq. (2.8) for any temperature if the mass of the molecules and the Boltzmann constant are known. In this way the rms speed for nitrogen molecules in air at 280 K is found to be 500 m s^{-1}. This speed is proportional to the square root of the temperature and inversely proportional to the square root of the mass. In Problem 3 it is required to relate this information to the escape velocity.

The kinetic theory developed so far only gives an adequate equation of state for the special group of permanent gases. All other substances apparently obey more complicated equations of state. This is not surprising because the molecular model used as the basis of this theory is particularly simple. In Section 2.17 a more complicated model will be described in relation to real gases.

The equation of state is a corner stone of thermodynamics. The development just completed emphasizes that the equation of state refers to a system in *equilibrium*, that is, to a system in which every statistical sample of molecules exhibits the same average condition. The pressure, specific volume and temperature are called the *state variables*. It is clear that two state variables define the state of a system, and the endeavor of thermodynamics is to describe the characteristic properties of matter during various processes entirely in terms of the state variables. It is evident that only a special class of processes can be described, processes that are so gradual that at each instant the state of the system can be defined. In other words, the process evolves from one equilibrium state to the next. This type of process has the characteristic that it is *reversible*. Processes that involve nonequilibrium conditions of the system are *irreversible* processes and, in general, cannot be described by thermodynamics. Although the limitations of thermodynamics in describing the properties of matter are severe, the number of atmospheric processes that can be described by thermodynamics should not be underestimated.

2.5 The Velocity Distribution of Molecules

In section 2.4 we have found it possible to derive the equation of state utilizing only the average kinetic energy of the molecules. Certain properties of a gas in equilibrium depend not only on the mean molecular energy but also on the distribution of energies (the proportion of molecules with each particular energy) and the associated velocity distribution.

The distribution of a single velocity component, say v_x, can be represented in normalized form by

$$\phi(v_x) \equiv \frac{1}{n}\frac{dn}{dv_x} \tag{2.11}$$

Integration yields

$$\int_{-\infty}^{\infty} \phi(v_x)\,dv_x = 1 \tag{2.12}$$

Similarly, velocity distributions can be defined in the y and z directions. It is intuitively clear that for a gas in equilibrium the velocity distribution must be independent of the direction in which the components are chosen; therefore,

$$\phi(v_x) = \phi(v_y) = \phi(v_z) \qquad \text{for} \quad v_x = v_y = v_z$$

Furthermore, ϕ should be an even function of the velocity components, so that

$$\phi(-v_x) = \phi(v_x), \text{ etc.}$$

For easy visualization only a two-dimensional velocity distribution is considered; later, the result will be generalized to three dimensions.

It follows from the point just made that all the molecules with an x velocity component lying between v_x and v_{x+dx} have the same distribution in the y direction as all the molecules in the system. Using Eq. (2.11) the number of molecules with velocity components between v_x and v_{x+dx} and between v_y and v_{y+dy}, as shown in Fig. 2.3, may be expressed by

$$d^2n = n\phi(v_x)\phi(v_y)\,dv_x\,dv_y$$

But the distribution is independent of the coordinate system, so that in the system defined by the axis connecting O and P in Fig. 2.3 the same number may be expressed by

$$d^2n = n\phi(v_2)\phi(0)\,dv_x\,dv_y$$

where $v_2^2 \equiv v_x^2 + v_y^2$. Because the intersections of the function with vertical planes, as shown in Fig. 2.3, are geometrically similar, it follows that

$$\phi(v_2)\phi(0) = \phi(v_x)\phi(v_y) \tag{2.13}$$

This functional equation can be solved by first differentiating with respect to v_x and next differentiating with respect to v_y. From the two relations so

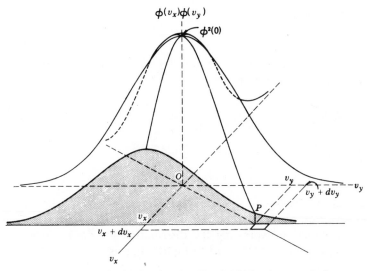

FIG. 2.3.   The molecular distribution function [Eq. (2.13)] for the two velocity components v_x and v_y.

obtained there results

$$\frac{1}{v_x}\frac{\phi'(v_x)}{\phi(v_x)} = \frac{1}{v_y}\frac{\phi'(v_y)}{\phi(v_y)}$$

where the prime denotes the derivative. The left-hand side of this equation is independent of v_y and the right-hand side is independent of v_x. This is only possible when each side is equal to the same constant, say $-\lambda$. Therefore,

$$\frac{1}{v_x}\frac{\phi'(v_x)}{\phi(v_x)} = -\lambda$$

Integration of this equation yields

$$\phi(v_x) = A \exp[-(\lambda/2)v_x^2] \qquad (2.14a)$$

Similarly,

$$\phi(v_y) = A \exp[-(\lambda/2)v_y^2] \qquad (2.14b)$$

and

$$\phi(v_x^2 + v_y^2)^{1/2} = A \exp[-(\lambda/2)(v_x^2 + v_y^2)] \qquad (2.14c)$$

The constant of integration is found by substituting Eqs. (2.14) into

(2.13), which yields

$$A = \phi(0)$$

The value of $\phi(0)$ is found by substituting one of Eqs. (2.14) into (2.12) and carrying out the integration. Therefore

$$\phi(0) \int_{-\infty}^{\infty} \exp\left(-\frac{\lambda}{2}v_x^2\right) dv_x = \phi(0) \int_{-\infty}^{\infty} \exp\left(-\frac{\lambda}{2}x^2\right) dx = 1$$

and because $\int_{-\infty}^{\infty} \exp[-(\lambda x^2/2)] \, dx = (2\pi/\lambda)^{1/2}$

$$\phi(0) = (\lambda/2\pi)^{1/2} \tag{2.15}$$

It remains to determine λ; this may be accomplished by relating λ to $\overline{v^2}$. To do this, Eqs. (2.14) must be extended to the three-dimensional case by adding an equation expressing the distribution for $\phi(v_z)$. Upon carrying out once again the steps that led to Eq. (2.13), it is found that

$$\phi(v)\phi^2(0) = \phi(v_x)\phi(v_y)\phi(v_z)$$

where $v^2 = v_x^2 + v_y^2 + v_z^2$. The corresponding number of molecules in a volume increment in three-dimensional momentum space is

$$d^3n = n\phi(v_x)\phi(v_y)\phi(v_z) \, dv_x \, dv_y \, dv_z$$
$$= n\phi(v)\phi^2(0) \, dv_x \, dv_y \, dv_z$$

or, upon transforming to spherical coordinates and integrating over a spherical shell, the number of molecules in this shell is

$$dn = 4\pi n v^2 \phi^2(0)\phi(v) \, dv \tag{2.16}$$

From definition (2.5) the average of the square of the velocities is given by

$$\overline{v^2} = 4\pi\phi^2(0) \int_0^{\infty} v^4 \phi(v) \, dv$$

and upon substituting Eqs. (2.14) and (2.15) into this equation,

$$\overline{v^2} = 4\pi\left(\frac{\lambda}{2\pi}\right)^{3/2} \int_0^{\infty} v^4 \exp\left(-\frac{\lambda}{2}v^2\right) dv = \frac{3}{\lambda}$$

This result may be combined with Eq. (2.8) applied to molecules of identical mass with the result

$$\lambda = m/kT$$

If this is substituted back into the distribution function (2.14), the one-

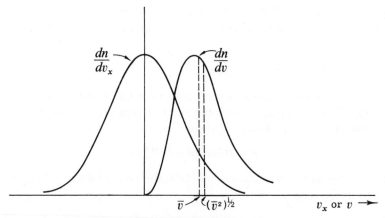

FIG. 2.4. The one-dimensional velocity distribution (v_x) and the three-dimensional velocity distribution (v) represented by Eqs. (2.17) and (2.18), respectively.

dimensional distribution is found to be

$$\phi(v_x) = \left(\frac{m}{2\pi kT}\right)^{1/2} \exp\left(-\frac{mv_x^2}{2kT}\right)$$

or, from Eq. (2.11)

$$\frac{dn}{dv_x} = n\left(\frac{m}{2\pi kT}\right)^{1/2} \exp\left(-\frac{mv_x^2}{2kT}\right) \tag{2.17}$$

The three-dimensional velocity distribution is obtained by substituting Eq. (2.17) into (2.16) with the result

$$\frac{dn}{dv} = 4\pi nv^2\left(\frac{m}{2\pi kT}\right)^{3/2} \exp\left(-\frac{mv^2}{2kT}\right) \tag{2.18}$$

Equations (2.17) and (2.18) are represented graphically in Fig. 2.4. An application of the velocity distribution is given in Problem 4. Figure 2.4 shows that substantial proportions of the molecules have kinetic energies very different from the mean value and that the mean energy is greater than the most probable energy.

2.6 The Atmosphere in Equilibrium

We are now in a position to investigate the effect of the earth's field of gravity on the energy or velocity distribution of the molecules and on the height dependence of density or pressure. For this purpose substitute the equation of state (2.10) into the hydrostatic equation (1.19), yielding

$$\frac{\partial p}{\partial \Phi} = -\frac{p}{R_m T} \tag{2.19}$$

For the case of temperature independent of height (isothermal case), Eq. (2.19) can be integrated yielding the vertical pressure distribution in the form known as the *barometric equation*

$$p = p(0) \exp(-\Phi/R_m T)$$

and, by using again the equation of state, the vertical density distribution is found in the form

$$\rho = \rho(0) \exp(-\Phi/R_m T) \tag{2.20}$$

The molecules traveling upward in the field of gravity lose kinetic energy and gain potential energy and vice versa; it might be inferred that the velocity distribution should be a function of height, or geopotential, and that the atmosphere in equilibrium should not be isothermal. However, we shall show below that an atmosphere in equilibrium is also isothermal.

Because the density is proportional to the number of molecules per unit volume, Eq. (2.20) requires that

$$n = n(0) \exp(-\Phi/R_m T) = n(0) \exp(-m\Phi/kT) \tag{2.21}$$

The form of the exponent is similar to the exponent in the velocity distribution function [Eq. (2.17)]. The difference is that in Eq. (2.21) the potential energy takes the place of the kinetic energy, $\frac{1}{2}mv^2$. Consider now two levels of constant geopotential Φ_1 and Φ_2 so close together that collisions occurring between the planes may be neglected. The number of molecules passing through level Φ_1 per unit area per unit time with velocities between v_z and $v_z + dv_z$ is expressed by Eq. (2.17) as

$$v_z(\Phi_1)\, dn_1 = n_1 v_z(\Phi_1)\left(\frac{m}{2\pi k T_1}\right)^{1/2} \exp\left(-\frac{m v_z^2(\Phi_1)}{2k T_1}\right) dv_{z1} \tag{2.22}$$

The same number of molecules pass through level Φ_2. Because the molecules form part of the distribution at the new level, the number is expressed by

$$v_z(\Phi_1)\, dn_1 = v_z(\Phi_2)\, dn_2 = n_2 v_z(\Phi_2)\left(\frac{m}{2\pi k T_2}\right)^{1/2} \exp\left(-\frac{m v_z^2(\Phi_2)}{2k T_2}\right) dv_{z2} \tag{2.23}$$

where the index 2 refers to the level Φ_2. The z component of the velocity has been affected by the field of gravity in going from level Φ_1 to Φ_2. There-fore,

$$\tfrac{1}{2}v_z^2(\Phi_2) = \tfrac{1}{2}v_z^2(\Phi_1) - \Delta\Phi \tag{2.24}$$

where $\Delta\Phi = \Phi_2 - \Phi_1$. Differentiation yields

$$v_z(\Phi_2)\,dv_{z2} = v_z(\Phi_1)\,dv_{z1} \tag{2.25}$$

If Eqs. (2.24) and (2.25) are substituted into Eq. (2.23)

$$v_z(\Phi_1)\,dn_1 = n_2 \exp\!\left(\frac{m\Delta\Phi}{kT_2}\right) v_z(\Phi_1)\left(\frac{m}{2\pi kT_2}\right)^{1/2} \exp\!\left(-\frac{mv_z^2(\Phi_1)}{2kT_2}\right) dv_{z1}$$

Now, provided $T_2 = T_1$ and $n_2 = n_1 \exp(-m\Delta\Phi/kT_1)$, this equation re-duces to Eq. (2.22). Therefore, compression of the gas with decreasing height occurs at the expense of potential energy only, and an atmosphere in equi-librium is also an atmosphere with uniform temperature. This conclusion holds for a gas in equilibrium in any potential field.

As a consequence of Dalton's law, which states that under equilibrium conditions the partial pressure of each component of a mixture of gases is the same as the pressure exerted by that gas if it occupied the volume alone, the hydrostatic equation may be integrated independently for each con-stituent of a mixture of gases. The partial densities of two constituents are then represented by

$$\rho_1 = \rho_1(0) \exp(-\Phi/R_{m1}T) \qquad \rho_2 = \rho_2(0) \exp(-\Phi/R_{m2}T)$$

where the indices 1, 2 refer to the two constituents. Additional similar equations may be written for other constituents. From a series of these equations the height distribution of each constituent may be calculated easily. The heavy gases have highest concentrations near the ground, and the light gases are most abundant at higher levels. An exercise concerning *diffusive equilibrium* is given in Problem 5.

Diffusive equilibrium is not found in the lowest 90 km of the atmosphere because turbulent mixing maintains constant proportions of the permanent gases. However, the tendency toward diffusive equilibrium becomes in-creasingly important with increasing height above 90 km.

2.7   Conservation of Mass

The principle of conservation of mass simply states that *the mass of a closed system remains constant*. This applies to all closed systems for which mass–energy transformations are unimportant, i.e., systems in which no mass is created or destroyed. The same principle applied to an open system may be recognized as stating that *the net mass flux into the system equals the rate of increase of mass of the system.* This is expressed by

$$\oiint \rho \mathbf{v} \cdot \mathbf{n} \, dA = -\iiint \frac{\partial \rho}{\partial t} \, dx \, dy \, dz \qquad (2.26)$$

where $\mathbf{n}$ represents the normal unit vector outwardly directed from the surface. In Appendix I.G the surface and volume integrals are related by Gauss's theorem in the form

$$\oiint \rho \mathbf{v} \cdot \mathbf{n} \, dA = \iiint \mathbf{V} \cdot (\rho \mathbf{v}) \, dx \, dy \, dz \qquad (2.27)$$

Equations (2.26) and (2.27) may be combined for any volume; therefore,

$$\boxed{\mathbf{V} \cdot (\rho \mathbf{v}) = -\frac{\partial \rho}{\partial t}} \qquad (2.28a)$$

Equation (2.28a) is known as the *equation of continuity.* It may be applied to any *conservative* scalar quantity. The left-hand side may be expanded to $\rho \mathbf{V} \cdot \mathbf{v} + \mathbf{v} \cdot \mathbf{V} \rho$; and, upon recognizing the expansion of the total derivative, Eq. (2.28a) may also be written

$$\mathbf{V} \cdot \mathbf{v} = -\frac{1}{\rho} \frac{d\rho}{dt} \qquad (2.28b)$$

2.8   Conservation of Energy

The principle of conservation of energy states that *it is not possible to create or destroy energy.* This is an independent fundamental principle which, like the other fundamental principles already introduced, is based upon intuitive insight into nature. It cannot be derived from other principles or laws.

The principle of conservation of energy requires, however, that we think in particular ways about transformations of energy from one form to another. We must recognize that two systems can exchange energy through work done by one on the other and by heat transfer and that the energy transferred from one by these transfer processes (work and heat) must appear as energy in the other. This insight developed gradually during the first half of the nineteenth century as the result of observations from widely different fields. Rumford in 1798 drew attention to the concepts later to be known as work, heat, and internal energy as a result of observations of cannon boring. R. J. Mayer made the first clear statement of the principle in 1842 from physiological observations, and Joule determined accurately from laboratory measurements the mechanical equivalent of heat in 1847, thereby verifying the principle.

Energy may take many different forms: kinetic, internal, potential (of various kinds), chemical, nuclear, etc. We shall consider only kinetic energy of the mean motion, internal energy (which represents the kinetic energy of the molecular motions), and the potential energy associated with the field of gravity; and we shall assume that other forms of energy do not change during atmospheric processes. The total energy is the sum of these three energies, and change in the specific total energy is expressed by

$$de_t = du + d\Phi + \tfrac{1}{2} dv^2 \tag{2.29}$$

where u represents internal energy per unit mass.

The principle of conservation of energy requires that the change in total energy equals the sum of heat transferred to the system and work done on the system, and it may be written in the form

$$\boxed{de_t = dh + dw} \tag{2.30}$$

where h represents the heat added per unit mass and w work done on the system per unit mass. One should be alert to the fact that in some texts w is defined as positive for work done *by* the system.

To evaluate the work done *on* the system (dW) it is most convenient to consider the rate at which work is done on a parcel enclosed at a particular instant within the boundaries shown in Fig. 2.5. The rate of work done at each face may be represented by the product of the normal velocity and the pressure at the face. Therefore, upon expanding the pressure and velocity on the right y, z face in a Taylor series, the rate of work done at the two y, z faces is

$$\frac{dW_x}{dt} = p(x, t)v_x(x, t)\, dy\, dz - \left[\left(p + \frac{\partial p}{\partial x} dx\right)\left(v_x + \frac{\partial v_x}{\partial x} dx\right)\right] dy\, dz$$

or

$$\frac{dW_x}{dt} = -\left(p \frac{\partial v_x}{\partial x} + v_x \frac{\partial p}{\partial x}\right) dx\, dy\, dz$$

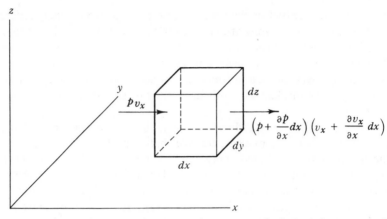

FIG. 2.5.   Product of pressure (p) and normal component of velocity (v_x) on opposite faces of a differential fluid element.

This equation may be extended to three dimensions by writing similar expressions for the other faces and adding. This yields for the total rate of work done on the air parcel

$$\frac{dW}{dt} = -(p\mathbf{\nabla} \cdot \mathbf{v} + \mathbf{v} \cdot \mathbf{\nabla}p)\, dx\, dy\, dz$$

and upon dividing by the mass of the parcel, the work rate per unit mass or the *specific work* rate is

$$\frac{dw}{dt} = -\frac{1}{\rho}(p\mathbf{\nabla} \cdot \mathbf{v} + \mathbf{v} \cdot \mathbf{\nabla}p) \qquad (2.31)$$

The first term on the right represents rate of work done by compression (external work), and the second term represents rate of work done by the pressure field (internal work). Equation (2.31) is a complete statement of the rate of work done on the parcel. This combined with Eq. (2.28b) leads to a formulation of the first law of thermodynamics.

2.9   First Law of Thermodynamics

The rate of work done on the parcel may now be expressed by introducing Eq. (2.28b) into (2.31) in the form

$$\frac{dw}{dt} = \frac{p}{\rho^2}\frac{d\rho}{dt} - \frac{1}{\rho}\mathbf{v} \cdot \mathbf{\nabla}p$$

or

$$\frac{dw}{dt} = -p\frac{d\alpha}{dt} - \alpha\mathbf{v} \cdot \nabla p \qquad (2.32)$$

which is, like Eq. (2.31), a general expression for the rate of work done. The second term is negligible when the pressure is uniform (in most laboratory work, for example) but, in general, work done by the atmospheric pressure field is not negligible.

If Eqs. (2.29) and (2.32) are introduced into (2.30), the statement of conservation of energy becomes

$$dh = du + d\Phi + \tfrac{1}{2}\,dv^2 + p\,d\alpha + \alpha\mathbf{v} \cdot \nabla p\,dt \qquad (2.33)$$

This equation suggests that there are many possible energy transformations. However, the number of possible transformations is limited by Newton's second law and the second law of thermodynamics. If Newton's second law in the form of Eq. (1.20) is multiplied by v_z and if friction is neglected, the rates of change of kinetic and potential energy are related to the rate at which work is done by the pressure field according to

$$\frac{1}{2}\frac{dv_z^2}{dt} + \frac{d\Phi}{dt} = -\alpha v_z\frac{\partial p}{\partial z}$$

This equation may be generalized for the three-dimensional case to the form

$$\tfrac{1}{2}\,dv^2 + d\Phi = -\alpha\mathbf{v} \cdot \nabla p\,dt$$

This equation states that the sum of kinetic and potential energies is changed only by virtue of the work done by the pressure field. Each term in this equation appears in Eq. (2.33), and combination therefore yields

$$dh = du + p\,d\alpha \qquad (2.34)$$

This result is called the *first law of thermodynamics*. It may be used as the definition of heat added to the system per unit mass.

The internal energy is by definition the kinetic energy of molecular motions of the molecules, and therefore is proportional to the temperature. If the volume is held constant, $d\alpha = 0$, and all the heat added to the system is used for the increase of internal energy. For the ideal gas discussed earlier, Eq. (2.8) shows that

$$dh = du = \tfrac{1}{2}mn^*\,d(\overline{v^2}) = \tfrac{3}{2}n^*k\,dT \qquad (2.35)$$

where n^* represents the number of molecules per unit mass. The relation between internal energy and temperature will be further discussed in Sections 2.10 and 2.11.

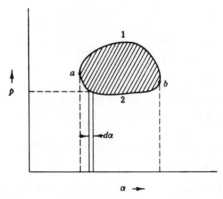

FIG. 2.6.   Work represented on the pressure–specific volume (p–α) diagram.

The last term in Eq. (2.34), the expansion work, can be represented in a graph of p versus α as illustrated in Fig. 2.6. If the state is changed from a to b along curve 1, then the area under the curve represents the expansion work done by the system

$$\int_a^b p \, d\alpha$$

If the state now is changed from b to a along curve 2, then an amount of work is done on the system which corresponds to the area under this new curve. The area enclosed by the two paths, 1 from a to b and 2 from b to a, represents the work per unit mass performed by the system in a complete cycle. The complete integration described is called the *line integral* around a closed path and is written

$$\oint p \, d\alpha$$

An example is provided by Problem 6.

2.10   Equipartition of Energy

The idealized point masses utilized in Sections 2.3 and 2.4 exhibit energy which may be specified by the three degrees of freedom corresponding to the three velocity components. If Eq. (2.8) is written in the form

$$\tfrac{1}{2}m(v_x^2 + v_y^2 + v_z^2) = \tfrac{3}{2}kT$$

it is clear that the average energy per degree of freedom (df) is $\tfrac{1}{2}kT$. From the discussion of distribution of molecular velocities in Section 2.5, it is also clear that the total energy must be equally divided among the 3 df. This is a special case of *equipartition of energy*.

In the case of a more complex molecule whose energy is shared by rotational as well as translational motion, the molecule is said to possess more than 3 df. The same intuitive statement made in Section 2.5 in establishing the uniform distributions of velocity components may be extended to include the rotational degrees of freedom. It follows that the *total energy is equally divided among each degree of freedom, and each component of energy is equal to* $\frac{1}{2}kT$. This is a statement of the law of equipartition of energy. The law holds accurately for molecules which obey classical mechanics; it *does not* apply to degrees of freedom corresponding to internal vibrations, in which case quantum mechanical interactions are important. It now may be recognized that the general form of the dependence of internal energy change on temperature change is, in place of Eq. (2.35),

$$du = \tfrac{1}{2}jn^*k \, dT \qquad (2.36)$$

where j represents the number of degrees of freedom. Whereas a monatomic molecule has 3 df (translational), a diatomic molecule has, in addition, 2 df corresponding to rotation about the two axes perpendicular to the line connecting the atoms. Similarly, a triatomic molecule possesses 3 df (rotational). The vibrational energies of air molecules except H_2O and CO_2 are usually negligible.

2.11 Specific Heat

The heat required to raise the temperature of a unit mass of a substance by one degree is called the *specific heat* of that substance; it is defined by

$$c \equiv \frac{dh}{dT}$$

This definition is not sufficient to determine a unique value for c because the circumstances under which the heat is supplied to the system are not yet specified. It may be that the system expands, contracts, or maintains a constant volume. In each of an infinite number of possible cases a different value for the specific heat would be found. In the case of an ideal gas Eqs. (2.34) and (2.36) lead to the general form

$$c = \frac{\tfrac{1}{2}jn^*k \, dT + p \, d\alpha}{dT} \qquad (2.37)$$

It is convenient to distinguish between the specific heat at constant volume (c_v) and the specific heat at constant pressure (c_p). It follows from Eq. (2.37) that the *specific heat at constant volume* is defined by

$$c_v \equiv \tfrac{1}{2}jn^*k$$

Upon combining this with Eq. (2.36) the differential of internal energy is

$$du = c_v \, dT \tag{2.38}$$

and because $n^*k = R_m$

$$c_v = \tfrac{1}{2}jR_m \tag{2.39}$$

The *specific heat at constant pressure* can be determined if $p \, d\alpha$ is eliminated from Eq. (2.37) by use of the equation of state and if pressure is then held constant. This yields

$$c_p = (\tfrac{1}{2}j + 1)R_m \tag{2.40}$$

Two useful relations can be derived from Eqs. (2.39) and (2.40). Subtraction gives

$$c_p - c_v = R_m \tag{2.41}$$

TABLE 2.3

PHYSICAL PROPERTIES OF ATMOSPHERIC GASES AT $0°$ C[a]

Gas	Degrees of freedom			10^3 J kg^{-1} K^{-1}			
	trans.	rot.	vib.	R_m	c_p	c_v	c_p/c_v
Monatomic gases							
Helium (He)	3			2.076	5.37	3.29	1.63
Argon (Ar)	3			0.2081	0.520	0.312	1.667
Diatomic gases							
Hydrogen (H$_2$)	3	2		4.124	14.25	10.13	1.407
Nitrogen (N$_2$)	3	2		0.2967	1.037	0.740	1.401
Oxygen (O$_2$)	3	2		0.2598	0.909	0.649	1.400
Triatomic gases							
Water vapor (H$_2$O)	3	3	0.3	0.4615	1.847	1.386	1.333
Carbon							
dioxide (CO$_2$)	3	3	0.3	0.1889	0.820	0.630	1.300
Ozone (O$_3$)	3	3		0.1732	0.770	0.597	1.290
Dry air				0.2871	1.004	0.717	1.400

[a] Adapted from G. W. C. Kaye and T. H. Laby, *Tables of Physical and Chemical Constants*, 14th ed., John Wiley and Sons, New York, 1973, and W. E. Forsythe, *Smithsonian Physical Tables*, 9th revised ed., Smithsonian Institution, Washington, D.C., 1959.

and division gives

$$\frac{c_p}{c_v} = \frac{j + 2}{j}$$

The physical properties of molecules discussed so far are summarized for the major constituents of the atmosphere in Table 2.3. From this table it is clear that the specific heat as computed from Eqs. (2.39) and (2.40) corresponds closely to the observed values.

2.12 Entropy

From the discussion of Section 2.9 it follows that work done by a system depends on the process it goes through and not only on the initial and final states of the system; therefore dw is not expressible as an exact differential by expansion in terms of the state variables. The same consideration holds for heat. In Appendix I.E properties of the exact differential are discussed.

In order to transform the first law of thermodynamics [Eq. (2.34)] to a form containing only exact differentials, divide this equation by the temperature and eliminate the pressure by use of the equation of state. This yields

$$\frac{dh}{T} = c_v \frac{dT}{T} + R_m \frac{d\alpha}{\alpha} \tag{2.42}$$

The right-hand side of Eq. (2.42) is the sum of two exact differentials, so it follows that the left-hand side is also an exact differential. It is convenient, therefore, to define the differential of the *specific entropy* by

$$\frac{dh}{T} \equiv ds \tag{2.43}$$

For any cyclic process

$$\oint ds = 0$$

Entropy will prove to be very useful in discussing the second law of thermodynamics in Section 2.15 and in deriving the Clausius–Clapeyron equation in Section 2.16.

2.13   Isentropic Processes and Potential Temperature

A process in which no heat is added to the system is called an *adiabatic* process; if the adiabatic process is also reversible the entropy remains constant and the process is called *isentropic*. In this case

$$dh = ds = 0$$

Atmospheric processes are seldom strictly adiabatic or strictly reversible, but many cases occur in which the isentropic approximation is useful and sufficiently accurate. For this reason the terms adiabatic and isentropic are often used interchangeably.

If Eq. (2.38) is introduced into Eq. (2.34), the isentropic form of the first law is written

$$c_v \, dT + p \, d\alpha = 0 \tag{2.44}$$

Upon combining the equation of state with Eq. (2.42), the temperature, the pressure, or the specific volume can be eliminated. After some simple algebraic manipulations the following isentropic equations are obtained:

$$\frac{c_v}{R_m} \frac{dT}{T} + \frac{d\alpha}{\alpha} = 0 \tag{2.45a}$$

$$\frac{c_p}{R_m} \frac{dT}{T} - \frac{dp}{p} = 0 \tag{2.45b}$$

and

$$\frac{c_p}{c_v} \frac{d\alpha}{\alpha} + \frac{dp}{p} = 0 \tag{2.45c}$$

These equations can be integrated giving the following results, first obtained by Poisson in 1823:

$$T\alpha^{R_m/c_v} = \text{const} \tag{2.46a}$$

$$Tp^{-R_m/c_p} = \text{const} \tag{2.46b}$$

$$p\alpha^{c_p/c_v} = \text{const} \tag{2.46c}$$

These equations are widely used in thermodynamics. Equations (2.46b) and (2.46c) are the forms most often used in atmospheric applications. An example is provided by Problem 7.

Potential Temperature

Because the pressure varies with height according to the hydrostatic equation, an air parcel moving vertically with constant entropy experiences a change in temperature specified by Eq. (2.46b). The constant of integration can be evaluated from a particular state of the system. If a pressure surface is chosen as a reference, then the characteristic constant is determined by the temperature at that level. The *potential temperature* (θ) is defined as that temperature which would result if the air were brought isentropically to a standard pressure p_0. Equation (2.46b) then can be written

$$T_\mathrm{p}^{-\kappa} = \theta p_0^{-\kappa} = \text{const}$$

where κ represents $R_\mathrm{m}/c_\mathrm{p}$ or

$$\theta = T(p_0/p)^\kappa \tag{2.47}$$

Poisson's equation is most frequently used in this form. The standard pressure p_0 is usually taken as 100 kPa (1000 mb). The close connection between potential temperature and entropy is illustrated by Problem 8.

The "Adiabatic" Chart

Since θ and P_0 are constants, Eq. (2.47) may be treated as a linear relation between T and p^κ and expressed by

$$T = \theta(p/p_0)^\kappa$$

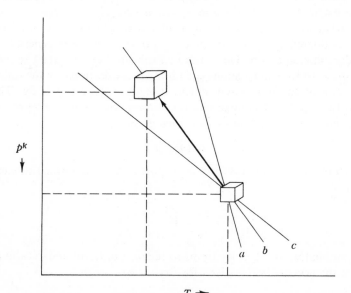

$$T \longrightarrow$$

FIG. 2.7.   Path followed on the *adiabatic chart* by a parcel displaced isentropically upward. Lines a, b, and c represent three possible vertical distributions (the stable, the adiabatic, and the unstable, respectively). The coordinates are T (temperature) and p^κ (pressure raised to the $R_\mathrm{m}/c_\mathrm{p}$ power).

Now, if T and p^κ are used as coordinates, lines of constant potential temperature may be drawn having a slope given by

$$\theta p_0^{-\kappa}$$

To determine the potential temperature corresponding to any combination of p and T, it is necessary only to find the intersection of the appropriate p and T lines and to read the value of the potential temperature line passing through that point. In this way potential temperature may be determined graphically on the *"adiabatic" chart* as illustrated by Fig. 2.7. Graphical determination of the change in temperature experienced by a parcel in moving isentropically through a certain pressure interval also is illustrated.

2.14   Static Stability

We propose to examine the problem of vertical displacement in an atmosphere in which vertical accelerations are negligible, that is, in an atmosphere in which the hydrostatic equation holds. If the force acting on the displaced parcel in Fig. 2.7 tends to bring it back to its original position, the atmosphere is called statically *stable*; if the parcel remains in balance with its surroundings, the atmosphere is called *adiabatic*, and if the parcel tends to move away from its original position, the atmosphere is statically *unstable*.

The net force on a parcel of air displaced from its equilibrium position may be evaluated by use of Archimedes' principle or, more generally, from the hydrostatic equation. The result of Problem 9 of Chapter I shows that the net upward force on a submerged body is proportional to the difference in densities of the fluid environment and the submerged body. The net upward force per unit volume (*buoyancy force*) on an air parcel of density ρ_1 in an environment of density ρ is then given by

$$F = (\rho - \rho_1)g$$

By Newton's second law this force is proportional to an upward acceleration given by

$$a = \frac{F}{\rho_1} = \left(\frac{\rho - \rho_1}{\rho_1}\right)g$$

Upon eliminating the density by introducing Eq. (2.10) and recalling that $\alpha = 1/\rho$, the acceleration is given for dry air by

$$a = g\frac{(p/T) - (p_1/T_1)}{p_1/T_1}$$

But it has been pointed out earlier that fluid in equilibrium must assume the pressure of its surroundings. Although acceleration is involved in this problem, in many cases this departure from equilibrium conditions does not affect significantly the pressure of the moving fluid. Therefore, $p \approx p_1$, and

$$a = g\left(\frac{T_1 - T}{T}\right) \tag{2.48}$$

If a parcel having the same initial temperature (T_0) as its surroundings is displaced vertically from height z_0, and if heat transfer from the parcel is negligible, the temperature changes according to Eq. (2.45b). If the pressure is replaced as the vertical coordinate by use of the hydrostatic equation, the *adiabatic lapse rate* is given by

$$\Gamma \equiv -\left(\frac{dT}{dz}\right)_{ad} = \frac{g}{c_p} \tag{2.49}$$

Integration of Eq. (2.49) gives for the temperature of the displaced parcel

$$T_1 = T_0 - \Gamma \Delta z$$

where Δz represents the vertical displacement. The surrounding temperature changes with height at a rate defined by $\gamma \equiv -\partial T/\partial z$. In this case integration gives for the temperature of the environment at the height of the displaced parcel

$$T = T_0 - \gamma \Delta z$$

If these two temperatures are introduced into Eq. (2.48), the acceleration of the displaced parcel is expressed by

$$a = -\frac{g\Delta z}{T}(\Gamma - \gamma)$$

or upon defining the *static stability* by

$$s_z \equiv -\frac{a}{g\Delta z} = -\frac{a}{\Delta \Phi}$$

it follows that

$$s_z = (1/T)(\Gamma - \gamma) \tag{2.50}$$

The static stability is the downward acceleration experienced by the displaced parcel per unit of geopotential. A refinement of Eq. (2.50) is discussed in

Problem 9. Although the static stability has been derived using the adiabatic assumption, Eq. (2.50) holds also in the more general case if Γ is interpreted as the rate of change of temperature with height following the parcel as it is displaced. The effects of humidity and condensation are discussed in Sections 2.19 and 2.20.

It is often useful to express the stability in terms of the potential temperature. Differentiation of Eq. (2.47) with respect to z gives

$$\frac{1}{T}\frac{\partial T}{\partial z} = \frac{1}{\theta}\frac{\partial \theta}{\partial z} + \frac{R_m}{c_p}\frac{\partial p}{p\,\partial z}$$

Upon introducing the hydrostatic equation, the static stability is

$$S_z = \frac{1}{T}\left(\frac{g}{c_p} + \frac{\partial T}{\partial z}\right) = \frac{1}{\theta}\frac{\partial \theta}{\partial z} = \frac{\partial(\ln \theta)}{\partial z} \tag{2.51}$$

Equation (2.51) shows that where the potential temperature increases with height the atmosphere is statically stable, and vice versa.

2.15   Thermodynamic Probability and Entropy†

Although the development so far has concerned systems in equilibrium, processes in which the state of the system changes must introduce departures from the equilibrium condition. Atmospheric processes seldom involve such great departures from equilibrium that the equations developed in preceding sections become invalid, but there is something to be learned from a detailed look at molecular behavior and its relation to equilibrium. Only the case of a monatomic gas will be considered, and it will be assumed that no heat is added or work done on the gas.

The molecules which constitute the system move with various velocities in a random fashion and make elastic collisions with other molecules. In order to completely describe the state, the positions and the momenta of all the molecules must be specified. The state of each molecule can be specified by six numbers indicating the three coordinates in space and the three momentum components. We introduce now a six-dimensional space, the *phase space*, whose coordinates are those of ordinary space and those of *momentum space*. The state of each molecule is now specified by a point in phase space

---

† This section contains advanced material which is used in Section 5.3 but is otherwise not essential for understanding the following sections and chapters.

(x, y, z, p_x, p_y, p_z) where p_x, p_y, and p_z are the three momentum components, mv_x, mv_y, mv_z, respectively. Now divide the phase space into many small cells

$$\Delta x \Delta y \Delta z \Delta p_x \Delta p_y \Delta p_z \equiv H$$

and count the number of molecules belonging to each cell. This number is called the *occupation number*.

The state of the system could be described by giving the complete set of occupation numbers corresponding to all the cells in the system. This describes a *microstate* of the system. However, observational techniques are insufficiently precise to give a complete account of the microstate; only the number of particles in large groups of cells can be estimated. This gross specification is called the *macrostate*.

Obviously there are many possible microstates corresponding to a particular macrostate. In order to assign a probability to a macrostate, it is necessary to know the relative probabilities of individual microstates. It is assumed that each microstate is equally probable. The probability of a macrostate is then proportional to the number of microstates that belong to it. The *thermodynamic probability* (W) of the macrostate is now defined as the number of microstates which belong to it. Because a system in equilibrium will most of the time be in or close to the macrostate with the highest probability, the immediate problem is to find an expression for the number of microstates belonging to a macrostate, and then to find the macrostate which makes this number a maximum. If two systems A and B are combined into a single system $A + B$ each microstate of component A of the system can combine with all the microstates of component B to give a new microstate for the new system. The number of microstates of the components therefore multiply together to give that of the combined system. In other words the thermodynamic probability W_{A+B} of a macrostate of system $A + B$ is equal to the product of the thermodynamic probabilities of system A and system B, so

$$W_{A+B} = W_A \cdot W_B \tag{2.52}$$

To find the number of microstates contained by a macrostate it is necessary to divide the large number of cells into groups in such a way that the macrostates of all the groups together define the macrostate of the system. Suppose that each group contains g cells and the total number of cells is G, then the number of groups is G/g. Consider now the ith group; all the cells of this group are numbered from i_1 to i_g. The cells are arranged in a series in such a way that after each cell symbol the number of identical particles contained by that cell is indicated by a number of zeros. For example, this series might look as follows:

$$i_1 i_2 i_3 0 i_4 0 i_5 0 0 i_6 i_7 \ldots i_g 0 \tag{2.53}$$

We see, for instance, that cell i_5 contains two particles, i_1 zero, etc. The order of the symbols and zeros may be changed in all possible ways without change in the macrostate, except that the series cannot begin with a zero. Therefore, there are $g(g + n_i - 1)!$ sequences of series (2.53) belonging to the same macrostate, where n_i represents the total number of particles in group i. Each sequence represents a microstate, but there are many repetitions. Permuting the is is not necessary, as all the microstates can be represented by sequences in which the cells are in the order of their numbers. Furthermore, permuting the zeros makes no difference, because the particles are identical. It follows that the thermodynamic probability of the ith group is represented by

$$W_i = \frac{g(g + n_i - 1)!}{g! \, n_i!} = \frac{(g + n_i - 1)!}{(g - 1)! \, n_i!} \tag{2.54}$$

When the ith group is combined with the other groups forming the system, their probabilities multiply and the probability of the system is given by

$$W = \prod_{i=1}^{G/g} W_i = \prod_{i=1}^{G/g} \frac{(g + n_i - 1)!}{(g - 1)! \, n_i!} \tag{2.55}$$

where the symbol $\prod_{i=1}^{G/g}$ means that the product of all W_is must be taken between the limit 1 and G/g.

Now recall Stirling's approximation in the form

$$\ln(P!) = \sum_{j=1}^{P} \ln j \approx \int_1^P \ln j \, dj = P \ln P - P$$

and apply to Eq. (2.54) with the result

$$\ln W_i = (g + n_i - 1) \ln(g + n_i - 1) - (g - 1) \ln(g - 1) - n_i \ln n_i$$

By assuming $g \gg 1$ this may be written

$$\ln W_i = g[(1 + N_i) \ln(1 + N_i) - N_i \ln N_i] \tag{2.56}$$

where $N_i = n_i/g$ represents the average occupation number. Substituting Eq. (2.56) into (2.55) after taking the logarithm, the logarithm of the probability becomes

$$\ln W = \sum_{i=1}^{G/g} g[(1 + N_i)\ln(1 + N_i) - N_i \ln N_i]$$

$$= \sum_{1}^{G} [(1 + N_i)\ln(1 + N_i - N_i \ln N_i] \qquad (2.57)$$

because the quantity summed is identical over each cell group.

In order that Eq. (2.57) express a maximum

$$\frac{\delta \ln W}{\delta N_i} = 0$$

This condition applied to the right-hand side of Eq. (2.57) yields

$$\delta \ln W = \sum [\delta(1 + N_i)\ln(1 + N_i) + (1 + N_i)\delta \ln(1 + N_i)$$

$$- \delta N_i \ln N_i - N_i \delta \ln N_i] = \sum \ln\left(1 + \frac{1}{N_i}\right)\delta N_i = 0 \quad (2.58)$$

Because the total number of molecules (N) does not change

$$\sum \delta N_i = 0 \qquad (2.59)$$

Also, the internal energy of the system does not change because it is assumed that no heat is supplied to and no work is done on the system. The internal energy can be given by adding the energies of the individual molecules (u_i) associated with each cell and then adding all the energies of the cells. Therefore

$$U = \sum u_i N_i$$

and because $\delta(U) = 0$

$$\sum u_i \delta N_i = 0 \qquad (2.60)$$

In order to determine the distribution N_i, Lagrange's method of undetermined multipliers may be used. Equations (2.59) and (2.60) are multiplied, respectively, by the undetermined constants α' and β. Addition of these equations to (2.58) then gives

$$\sum \left\{\ln\left(1 + \frac{1}{N_i}\right) - \alpha' - \beta u_i\right\}\delta N_i = 0$$

The choice of the δN_i is independent of the expression in brackets, so for any value of i the bracket must vanish, and

$$N_i = \frac{1}{\exp(\alpha' + \beta u_i) - 1} \tag{2.61}$$

For an ideal gas the cells in phase space may be made so small that $N_i \ll 1$, which means that $1/N_i \gg 1$, so that unity in Eq. (2.58) can be neglected. Therefore, Eq. (2.61) may be written

$$N_i = \exp(-\alpha' - \beta u_i)$$

And because $u_i = \frac{1}{2}mv_i^2$ for a monatomic gas

$$N_i = \exp[-(\alpha' + \frac{1}{2}\beta mv_i^2)] \tag{2.62}$$

It remains to evaluate α' and β.

If all cells occupying the spherical shell in phase space defined by the velocity interval between v and $v + dv$ are considered, the number of cells may be expressed by

$$dG = \frac{4\pi V m^3 v^2}{H} \, dv$$

The number of molecules in the velocity interval is therefore

$$N_i \, dG = \frac{4\pi V m^3 N_i v^2}{H} dv = \frac{4\pi V m^3 v^2}{H} \exp\left[-\left(\alpha' + \frac{\beta mv_i^2}{2}\right)\right] dv$$

This number is easily obtained from Eq. (2.18) if it is assumed that the velocity distribution of the gas in equilibrium corresponds to the most probable distribution. Therefore

$$\frac{4\pi m^3 N_i v^2}{H} = 4\pi n v^2 \left(\frac{m}{2\pi kT}\right)^{3/2} \exp\left(-\frac{mv^2}{2kT}\right)$$

or, because v and v_i are identical

$$N_i = nH(2\pi mkT)^{-3/2} \exp(-mv_i^2/2kT) \tag{2.63}$$

Equation (2.63) is identical to Eq. (2.62) if α' is chosen to be given by

$$\alpha' = -\ln\{nH(2\pi mkT)^{-3/2}\}$$

and

$$\beta = 1/kT \tag{2.64}$$

The thermodynamic probability now may be computed from Eqs. (2.57) and (2.63). If it is again assumed that the cells are so small that $N_i \ll 1$, Eq. (2.57) may be simplified to

$$\ln W = N - \sum N_i \ln N_i$$

Upon substituting Eq. (2.63) into the logarithm on the right-hand side and recognizing that $\sum N_i(mv_i^2/2kT) = \frac{3}{2}N$, the summation yields

$$\ln W = N\{\tfrac{5}{2} - \ln(nH) + \tfrac{3}{2}\ln(2\pi mk) + \tfrac{3}{2}\ln T\} \qquad (2.65)$$

Now if Eq. (2.65) is differentiated and it is recognized that $dN = 0$ and that $dn/n = -d\alpha/\alpha$ where α represents specific volume, then

$$\frac{k}{Nm}d(\ln W) = kn^*\left(\frac{3}{2}\frac{dT}{T} + \frac{d\alpha}{\alpha}\right)$$

The right-hand side looks familiar, and upon referring to the definitions of Section 2.11 and Eqs. (2.42) and (2.43),

$$\frac{k}{Nm}d(\ln W) = ds \qquad (2.66)$$

Equation (2.66) shows that a system having a maximum probable state also must have maximum entropy. Consequently, a system in equilibrium has maximum entropy, and if the thermodynamic probability and the entropy have not reached their maxima, the system is not in equilibrium. The state of the system must be expected to change until this maximum is reached. This insight is the foundation for the second law of thermodynamics.

The theory developed here does not give a clue to how fast equilibrium will be reached from a specified nonequilibrium state. In Section 2.16 a simple model of transport mechanism is introduced which permits quantitative prediction of the rate at which equilibrium is approached.

2.16 Second Law of Thermodynamics and Transfer Processes

It is common experience that heat flows from a warm to a cold system. We wait before drinking a hot cup of coffee because we know that the hot coffee will lose heat to its surroundings and therefore will become cooler. The change takes place only in one direction and the change continues until equilibrium is reached; that is, until the coffee has the same temperature as its surroundings.

Observations like this form the basis for the *second law of thermodynamics*. In the middle of the nineteenth century this law was formulated in a number of equivalent statements by Carnot, Clausius, and Lord Kelvin. One form of the second law states that *it is impossible to construct a device whose sole effect is the transfer of heat from a cooler to a hotter body.*

No exceptions to the second law are known. However, there is a subtle difference between the status of the second law and the status of the other fundamental principles stated earlier. For example, conservation of energy is presumed to hold for all systems no matter how small and for all time increments. On the other hand, the second law of thermodynamics (and also the zeroth law) is a statement of probability which has meaning only if applied to a statistically significant number of molecules. The certainty and precision of the second law increase with the number of molecules contained in the system and with the time interval to which the law is applied.

If the second law is applied to the entire universe, it must be concluded that the universe is striving towards an equilibrium characterized by uniform temperature and maximum entropy. In such a state no further thermodynamic processes are possible, and the universe has reached a "heat death." The second law states only the direction in which a process takes place, but not how fast the changes occur or how much time must elapse before equilibrium is established. We are now in sight of an important generalization.

Changes in the state of a gas come about through transfer by random molecular motions. The general theory of the transfer process is very complex, but elementary kinetic theory again provides useful and instructive results.

Consider a system in which the temperature and the concentration of molecules varies linearly in the x direction as illustrated in Fig. 2.8. Each molecule carries with it a characteristic property (for example, its mass or

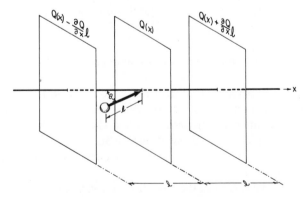

FIG. 2.8.   Distribution of the property Q and the contribution to transport of Q made by a molecule moving through the distance l at the angle θ to the x axis.

momentum or kinetic energy) which may be designated by Q. The rate of transfer of Q by a single molecule is expressed by vQ. These molecules travel with equal probability in all directions. The number per unit volume traveling in a specific direction confined by a solid angle $d\omega$ is given by $n\,d\omega/4\pi$. For simplicity we assume that all molecules travel with the "average" speed $\bar{v}$. The molecules that travel in the positive x direction and pass through a unit area at x may be visualized as having traveled in the average a distance l, called the *mean free path*, since their last collision. Therefore, the molecules that travel at an angle θ to the x axis pass the unit area with properties characteristic of the region a distance $l\cos\theta$ to the left of the unit area. The first two terms of a Taylor series expansion for the rate of transport of the property Q from the left through the unit area can be found with the aid of Appendix I.I to be

$$\frac{1}{2}\int_0^{\pi/2}\left(n\bar{v}Q - \frac{\partial(n\bar{v}Q)}{\partial x}l\cos\theta\right)\cos\theta\sin\theta\,d\theta = \tfrac{1}{4}n\bar{v}Q - \tfrac{1}{6}l\frac{\partial(n\bar{v}Q)}{\partial x}$$

The transport of Q by those molecules moving in the negative x direction is

$$-\tfrac{1}{4}n\bar{v}Q - \tfrac{1}{6}l\frac{\partial}{\partial x}(n\bar{v}Q)$$

There is no net mass flux in the x direction, which means that $\partial n\bar{v}/\partial x = 0$, so the net rate of transport per unit area (sometimes called the flux density) is

$$E_Q = -\tfrac{1}{3}n\bar{v}l\frac{\partial Q}{\partial x} \tag{2.67}$$

For three dimensions, the flux density of Q is given by

$$\mathbf{E}_Q = -\tfrac{1}{3}n\bar{v}l\,\nabla Q \tag{2.68}$$

Equation (2.68) shows that the transport of Q is represented by the product of the gradient of Q and a transfer coefficient which depends only on molecular properties ($\tfrac{1}{3}n\bar{v}l$). This *molecular transfer coefficient* can be further developed as follows.

The mean free path may be calculated by considering a molecule to move through a field of stationary identical molecules as illustrated in Fig. 2.9. Collisions occur with all molecules whose centers are within a distance $2r$ of the center line of the moving molecule. Therefore, in time t there occur $4\pi r^2 nvt$ collisions, and the average distance between successive collisions is given by

$$l = \frac{vt}{4\pi r^2 nvt} = \frac{1}{4\pi r^2 n}$$

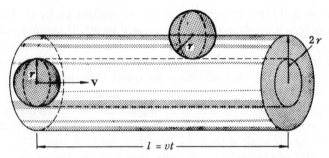

$$l = vt$$

FIG. 2.9.   Volume swept out by a molecule of radius r moving through a field of similar molecules.

This equation is correct only for an imaginary gas with all but one molecule at rest. In a real gas the molecules on the average move less than the "average" distance between collisions because, in effect, they are struck from the sides and rear by faster moving molecules. Computation for the Maxwellian distribution of velocities yields

$$l = \frac{1}{4\sqrt{2}\,\pi r^2 n} \tag{2.69}$$

From this we see that the molecular transfer coefficient is independent of n. This holds as long as l is small in comparison to the dimensions of the system.

Equation (2.68) may now be applied to obtain simplified expressions for thermal conductivity, viscosity, and diffusivity. The coefficients derived in this way from the elementary kinetic theory of gases can be compared with measurements. It turns out that values computed in this way are only approximately correct. The actual values may differ from the computed values by as much as a factor of two. However, more complete treatments of kinetic theory yield computed values of molecular transfer coefficients which are much more accurate.

Heat Conduction

The conduction of heat is related to the transfer of molecular kinetic energy, $\frac{1}{2}jmv^2$. Substituting this quantity for Q in Eq. (2.68) and using Eq. (2.8) we find

$$\mathbf{E}_h = -\tfrac{1}{6}l\bar{v}jnk\,\nabla T$$

Using the identity $R_m = k/m = kn^*$ and Eq. (2.39), this equation may be written

$$\mathbf{E}_h = -\tfrac{1}{3}l\bar{v}nmc_v\,\nabla T \tag{2.70}$$

Because the ratio of heat flux per unit area to temperature gradient defines the thermal conductivity (λ), Eq. (2.70) shows that

$$\lambda = \tfrac{1}{3}l\bar{v}\rho c_v$$

and $\mathbf{E}_h = -\lambda\,\nabla T$. After substituting Eq. (2.69)

$$\lambda = \frac{mc_v\bar{v}}{12\sqrt{2}\,\pi r^2} \tag{2.71}$$

Equation (2.71) relates the thermal conductivity to the molecular properties of the gas. Other things being equal, conductivity is inversely proportional to molecular size, and this is consistent with observations. In fact, a more exact form of Eq. (2.71) has been applied to measurements of heat conductivity in order to estimate molecular size.

Shear Stress

Consider a macroscopic velocity distribution in which the x component varies linearly with distance from the surface as shown in Fig. 2.10. The discussion is limited to a single plane because the physical principle can be demonstrated adequately in this way and because evaluation of the three-dimensional momentum flux requires use of second-order tensors which are not used in this book. For the important atmospheric case of vertical shear of the x component, Eq. (2.68) takes the form

$$E(mv_x)_z = -\tfrac{1}{3}nm\bar{v}l\frac{\partial v_x}{\partial z} = -\tfrac{1}{3}\rho\bar{v}l\frac{\partial v_x}{\partial z} \tag{2.72}$$

This flux of momentum per unit area may be visualized with the aid of Fig.

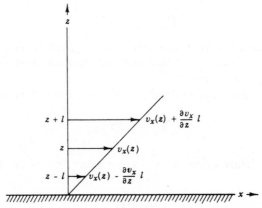

FIG. 2.10. Velocity profile $[v_x(z)]$ for the case of constant shear.

2.10 as follows. Downward moving molecules pass through a horizontal plane with more x momentum than those which move upward through the plane. Subsequent collisions then impart a positive x momentum to the region below the plane and a negative x momentum to the region above the plane. Newton's second law shows that the time rate of change of momentum must equal a "viscous" force, and Eq. (2.72) shows that this force must be proportional to the mean velocity shear. The viscous force per unit area, which is called the *shearing stress*, therefore may be expressed by

$$\tau'_x = \mu \frac{\partial v_x}{\partial z} \tag{2.73}$$

where μ is called the *viscosity* of the fluid and $\tau'_x = -E_{(mv_x)_z}$. Equations (2.72), (2.73), and (2.69) together show that the viscosity is given by

$$\mu = \tfrac{1}{3}\rho \bar{v} l = \frac{m\bar{v}}{4\sqrt{2}\,\pi r^2}$$

Diffusion

The previous examples of molecular transport may be applied to a gas consisting of a single type of molecule. Diffusion deals with the transport of at least two different types of molecules into each other. This process also can be treated by elementary kinetic theory if one of the components is present in relatively small concentrations such as is the case for water vapor in air. The molecular transport of water vapor in air depends on the gradient in concentration of water vapor molecules; hence Eq. (2.68) may be written in this case as

$$\mathbf{E}_w = -\tfrac{1}{3}l_w \bar{v}_w \, \nabla\rho_w \tag{2.74}$$

where subscript w refers to water vapor. There is a similar flux of air molecules in the opposite direction so that the total number of molecules per unit volume remains the same.

It is convenient to define the *diffusion coefficient* (D) by the equation

$$\mathbf{E}_w \equiv -D \, \nabla\rho_w \tag{2.75}$$

from which it follows that the molecular equivalent of the diffusion coefficient is

$$D = \tfrac{1}{3}l_w \bar{v}_w$$

The Heat Conduction and Diffusion Equations

A general differential equation governing the time and space variations of any conservative quantity may be derived by writing the continuity equation [Eq. (2.28a)] for the conservative molecular property Q in the form

$$\frac{\partial nQ}{\partial t} = -\nabla \cdot \mathbf{E}_Q \tag{2.76}$$

For the special case in which Q represents the mass of the water vapor molecule, nQ represents the water vapor density (ρ_w), and Eq. (2.75) may be substituted into Eq. (2.76) with the result

$$\frac{\partial \rho_w}{\partial t} = \nabla \cdot (D \nabla \rho_w) \tag{2.77}$$

For uniform diffusivity

$$\frac{\partial \rho_w}{\partial t} = D \nabla^2 \rho_w \tag{2.78}$$

which is the *diffusion equation* for water vapor in air.

For the important though special case of heat flux in air which is free to expand at constant pressure, the work done in expanding must be considered. The well-known box, this time closed, is considered to expand in response to heat flux through the walls as shown in Fig. 2.11. For constant

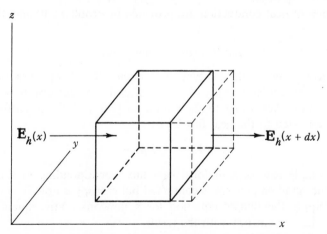

FIG. 2.11. The flux of molecular kinetic energy through opposite faces of a differential closed system. The dashed outline depicts expansion of the box at constant pressure as a result of increased temperature.

pressure the heat added to the system per unit mass may be expressed by

$$dh = d(u + p\alpha)$$

The term *specific enthalpy* is used to designate $u + p\alpha$. For an ideal gas

$$dh = c_p \, dT$$

so that $c_p T$ represents the specific enthalpy for an ideal gas. Now it is clear that in the example under consideration, although the heat flux density through the walls is expressed by Eq. (2.70), it is the enthalpy which is conserved within the box rather than the internal energy. When heat is supplied to the box because of thermal conduction, the conservation equation may be expressed by [using Eqs. (2.70) and (2.71)]

$$\rho c_p \frac{\partial T}{\partial t} = \nabla \cdot \lambda \nabla T$$

where the mass of the system $(\rho \, dV)$ has been kept constant. For uniform conductivity this takes the form

$$\frac{\partial T}{\partial t} = \frac{\lambda}{\rho c_p} \nabla^2 T \tag{2.79}$$

This is referred to as the *heat conduction equation*, although it is clear that it is a special case. The combination $\lambda/\rho c_p$ is called the *thermal diffusivity* for air. It plays the same role in the heat conduction equation as D plays in the diffusion equation.

Examples of heat conduction are provided in Problems 10 and 11.

Heat Conduction and Entropy

The second law of thermodynamics implies that heat flows from the warmer to the cooler regions of a system and that entropy flows in the same direction. According to the definition [Eq. (2.43)] the entropy flux $\mathbf{E}_s$ may be related to the heat flux $\mathbf{E}_h$ by

$$\mathbf{E}_s = (1/T)\mathbf{E}_h \tag{2.80}$$

For the simple case of a uniform heat flux, corresponding to a uniform temperature gradient, energy is conserved but entropy is not. To show this, we may apply the budget equation for a nonconservative quantity (see Appendix I.F.) to entropy with the result

$$\frac{ds}{dt} = \frac{\partial s}{\partial t} + \frac{1}{\rho} \nabla \cdot \mathbf{E}_s \tag{2.81}$$

Because $\mathbf{E}_h$ in our example is uniform, $\mathbf{V} \cdot \mathbf{E}_h = 0$ and $\partial T/\partial t = 0$. Consequently, $\partial s/\partial t = 0$ also. Thus by substituting Eq. (2.80) into (2.81) we obtain [also considering Eqs. (2.70) and (2.71)]

$$\rho \frac{ds}{dt} = \frac{\lambda}{T^2} (\nabla T)^2 \qquad (2.82)$$

In this case the rate of production of entropy is proportional to the square of the temperature gradient and to the conductivity. When the temperature gradient vanishes, the entropy production stops, and the entropy has reached its maximum.

2.17 Real Gases and Changes of Phase

The thermodynamic properties of an ideal gas apply with good accuracy to unsaturated air. However, when air is saturated and change of phase occurs, there are important deviations from the ideal gas behavior. In this case a more realistic model is needed than the elementary one presented in Sections 2.3 and 2.4. The elementary theory and the derived equation of state for an ideal gas imply that the volume may be made as small as desired by increasing the pressure or by decreasing the temperature. The gas may be imagined as composed of a group of mass points which individually and in the sum occupy no volume. These points can collide with the walls and exert forces on it; but they cannot collide with each other. The first modification to the molecular model, therefore, is to introduce a finite volume for the molecules.

The assumption that the molecules exert no forces upon each other except when they collide is another oversimplification. Experiment shows that attractive forces exist between molecules, and that the attractive force increases with decreasing separation. However, when the separation is reduced beyond approximately one "diameter," the molecules strongly repel each other. Figure 2.12 illustrates the intermolecular force as a function of the distance, and this subject is discussed again in Section 3.1. The separation (r_0) at which the force vanishes represents a position of minimum potential energy and therefore represents the mean distance between molecular centers in the liquid state. Molecules which are displaced from the equilibrium separate oscillate about r_0 just as two weights attached to a spring oscillate. The oscillations about r_0 are referred to as *thermal motion* of the molecules. If the temperature increases, the oscillations become more and more violent until the kinetic energy of the molecule at r_0 is larger than the potential energy required to overcome the attracting force of the other molecule, and the two molecules fly apart. Therefore, the second modification to the mo-

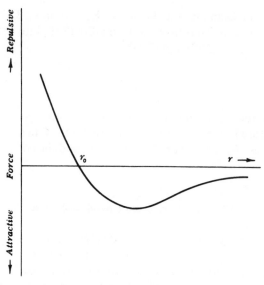

FIG. 2.12.   The force between two molecules as a function of the separation of their centers.

lecular model is to assume that the molecules exert an attractive force on each other.

Van der Waals' Equation

An equation of state taking into account the finite volume of the molecules and the attractive force between molecules was first deduced by van der Waals in 1873. He reasoned that as pressure increases or temperature decreases the volume must approach a small but finite value b, and that therefore α should be replaced in the equation of state by $\alpha - b$. Also, he represented the pressure at a point within the fluid as the sum of the pressure exerted by the walls (p) and the pressure exerted by the intermolecular attraction (p_i). To evaluate p_i consider the molecules near the walls. Each molecule is subject to an internal force which is proportional to the number of attracting molecules and therefore to the density. But the number of molecules near the boundary subject to the attractive force also is proportional to the density. It follows that the internal pressure due to the intermolecular force field is proportional to the square of the density. Consequently, van der Waals' equation of state may be written

$$\left(p + \frac{a}{\alpha^2}\right)(\alpha - b) = R_m T \qquad \text{or} \qquad p = \frac{R_m T}{\alpha - b} - \frac{a}{\alpha^2} \qquad (2.83)$$

where a and b are constants which depend upon the gas. It should be pointed out that van der Waals' equation represents a second approximation but not an exact statement of the equation of state for real gases, as indeed should be obvious from the largely intuitive derivation of the equation.

Van der Waals' equation is a cubic equation in α and consequently may have a maximum and a minimum. At high temperature the curves approach those computed from the equation of state for an ideal gas, whereas for low temperatures considerable difference exists. In Fig. 2.13 the $10°$ C isotherm for water vapor may be compared with the $10°$ C van der Waals' isotherm using values of a and b computed from Problem 12.

Experimental Behavior of Real Gases

If a real gas under low pressure is compressed isothermally at temperature T, the volume decreases to the point α_2 as illustrated in Fig. 2.13. At this time drops of liquid begin to appear, and further decrease in volume occurs with no increase in pressure. When the volume indicated by α_1 is reached, all the gas has condensed into liquid; a *change of phase* is said to have occurred between α_2 and α_1. Appreciable reduction in volume beyond α_1 is possible only under great pressure. Conversion from the liquid phase to the gaseous phase occurs if the pressure is released isothermally. Van der Waals' equation is obeyed closely in the liquid and gaseous phases; however, between α_2 and α_1 the pressure is constant, so the van der Waals' equation is not valid during change of phase.

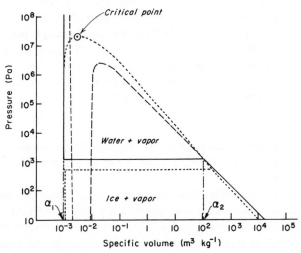

FIG. 2.13. The $10°$ C isotherm on the pressure–specific volume (p–α) diagram observed for water (solid line) and computed from van der Waals' equation (dashed line). The regions of mixed phases are bounded by dotted lines.

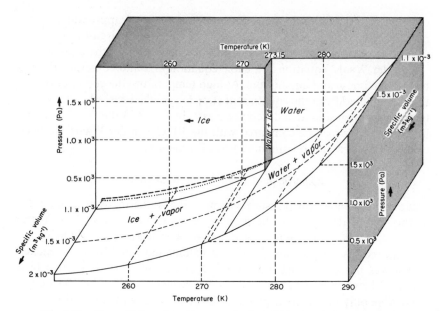

Fig. 2.14.   Phase diagram for water vapor in the neighborhood of the triple point.

If the change of phase occurs at a temperature T' which is higher than T, it is found that the change in volume $(\alpha_2' - \alpha_1')$ is less than $\alpha_2 - \alpha_1$. It follows that the work expended or the heat absorbed in vaporization (latent heat) varies inversely with the temperature. At a certain critical temperature T_c, $(\partial p/\partial \alpha)_T$ and $(\partial^2 p/\partial \alpha^2)_T$ both vanish at a single point called the *critical point*. At this point it is not possible to distinguish sharply between the liquid and gaseous phases. At higher temperatures the gas cannot be liquefied by increase of pressure. The critical point is a characteristic property of all real gases. For water this point is specified by

$$T_c = 647 \text{ K} \qquad p_c = 2.26 \times 10^4 \text{ kPa} \qquad \alpha_c = 3.1 \times 10^{-3} \text{ m}^3 \text{ kg}^{-1}[\dagger]$$

Problem 12 requires calculation of the van der Waals' constants for water vapor from these values. Condensation in the atmosphere is discussed further in Sections 3.2–3.4.

So far only the transition from gaseous to liquid phase or vice versa has been considered, but other phase changes are also important. In general, the transition between liquid and solid occurs at a lower temperature than does the transition from vapor to liquid. However, at the *triple point* all

† Osborne, N. S., Stimson, H. F., Ginnings, D. C., *J. Res. Nat. Bur. Stand.* **18**, 389–448 (RP 983) 1937.

three phases may exist in equilibrium. For water, the triple point occurs at a pressure of 0.611 kPa and a temperature of 0.0098° C. At pressures below the triple point water vapor may be converted directly into ice without passing through the liquid phase. The boundaries between the three phases are shown in Fig. 2.14. The dashed line represents the equilibrium vapor pressure over liquid water at temperatures below 273 K (*supercooled water*). The fact that the equilibrium vapor pressure over supercooled water is higher than the equilibrium vapor pressure over ice at the same temperature has an important consequence which is discussed in Section 3.10.

Latent Heat of Vaporization

Suppose a system is in the liquid state indicated by the point α_1 in Fig. 2.13. In order to change the phase from liquid to vapor at constant temperature and constant pressure, a characteristic quantity of heat, called the *latent heat of vaporization* (L), must be added. The first law of thermodynamics gives

$$L = \int_{h_1}^{h_2} dh = \int_{u_1}^{u_2} du + p_s \int_{\alpha_1}^{\alpha_2} d\alpha = u_2 - u_1 + p_s(\alpha_2 - \alpha_1) \quad (2.84)$$

where index 1 refers to the liquid phase and index 2 to the vapor phase. The latent heat of vaporization consists of two terms: a change in internal energy of the system and expansion work done by the system.

2.18   Clausius–Clapeyron Equation

The vapor pressure at which change of phase occurs at constant temperature is called the *saturation* or *equilibrium vapor pressure*. In order to determine the dependence of saturation vapor pressure on temperature, Eq. (2.84) may be combined with definition (2.43) with the result

$$L = T(s_2 - s_1) = u_2 - u_1 + p_s(\alpha_2 - \alpha_1) \quad (2.85)$$

Therefore,

$$u_1 + p_s\alpha_1 - Ts_1 = u_2 + p_s\alpha_2 - Ts_2 \quad (2.86)$$

where p_s represents saturation vapor pressure. The combination $u + p_s\alpha - Ts$ is called the *Gibbs function* and is represented by G. The Gibbs function is a function of the state only, and Eq. (2.86) shows that it is constant during an isothermal change of phase.

Now consider the isothermal change of phase at the temperature, $T + dT$, and the corresponding pressure, $p_s + dp_s$. The Gibbs function now has another constant value, $G + dG$. From the definition of G

$$dG = du + p_s \, d\alpha - T \, ds + \alpha \, dp_s - s \, dT$$

Upon combining this with Eqs. (2.34) and (2.43)

$$dG = \alpha \, dp_s - s \, dT$$

But $G_1 + dG_1 = G_2 + dG_2$, so also $dG_1 = dG_2$ and

$$\alpha_1 \, dp_s - s_1 \, dT = \alpha_2 \, dp_s - s_2 \, dT$$

This combined with Eq. (2.85) yields the Clausius–Clapeyron equation in the form

$$\frac{dp_s}{dT} = \frac{L}{T(\alpha_2 - \alpha_1)} \tag{2.87}$$

This equation can be applied successfully to water vapor. Under normal atmospheric conditions $\alpha_2 \gg \alpha_1$, so α_1 may be neglected. If it is assumed that water vapor behaves closely as an ideal gas, then Eq. (2.87) can be combined with (2.10) yielding

$$\frac{de_s}{e_s} = \frac{L}{R_{mw}} \frac{dT}{T^2} \tag{2.88}$$

where p_s is replaced by e_s, the saturation water vapor pressure, and R_{mw} is the specific gas constant for water vapor. A convenient approximation may be developed by integrating Eq. (2.88) assuming L constant. Using $L = 2.500 \times 10^6 \text{ J kg}^{-1}$, $R_{mw} = 461.7 \text{ J kg}^{-1} \text{ K}^{-1}$ and saturation vapor pressure of 0.611 kPa at 273.155 K, Eq. (2.88) yields

$$\log_{10} e_s = 11.40 - 2353/T \tag{2.89}$$

This equation can be used in graphical determination of e_s as a function of T. The accuracy of Eq. (2.89) is examined in Problem 13.

An equation similar to Eq. (2.89) can be derived for the saturation vapor pressure over ice; this is examined in Problem 14.

2.19 The Moist Atmosphere

The proportions of dry air, water vapor, and liquid or solid water vary within rather wide limits in the atmosphere as results of evaporation, condensation, and precipitation. The amount of moisture contained by a certain parcel of air may be considered one of the specific properties of the parcel which, like temperature and pressure, specify its state.

Humidity Variables

The amount of water vapor the air contains can be expressed in the following ways:

The *specific humidity* (q) is the mass of water vapor per unit mass of air.

The *mixing ratio* (w) is the mass of water vapor contained by a unit mass of dry air.

The mixing ratio and specific humidity are related by

$$w = \frac{q}{1 - q} \quad \text{and} \quad q = \frac{w}{1 + w}$$

which is easily verified. Both quantities are usually expressed in the units grams per kilogram (g/kg).

The *vapor pressure* (e) is the partial pressure exerted by water vapor. Assuming that the equation of state for an ideal gas can be applied to water vapor, it is possible to express the vapor pressure in terms of the mixing ratio or the specific humidity. The error introduced in this way is examined in Problem 15. The equation of state for an ideal gas applied to water vapor is

$$e\alpha_{\mathrm{w}} = R_{\mathrm{mw}}T \tag{2.90}$$

where α_{w} is the specific volume of water vapor. If Eq. (2.10) is divided by Eq. (2.90)

$$w \equiv \frac{\rho_{\mathrm{w}}}{\rho_{\mathrm{a}}} = \frac{R_{\mathrm{ma}}}{R_{\mathrm{mw}}} \frac{e}{p - e} \equiv \varepsilon \frac{e}{p - e} \tag{2.91}$$

The molecular masses of water vapor and dry air are, respectively, 18.015 and 28.97, from which it follows that $\varepsilon = 0.622$. Because $e \ll p$, $w \ll 1$, and $q \ll 1$

$$w \approx q \approx \varepsilon \frac{e}{p} \quad \text{and} \quad \frac{dw}{w} \approx \frac{de}{e} \tag{2.92}$$

The *relative humidity* (r) is defined in many classical and modern texts as the ratio of the actual vapor pressure to the saturation vapor pressure, but the International Meteorological Organization in 1947 adopted the ratio of mixing ratio to saturation ratio as the definition of relative humidity. The difference is very slight, so that we may write

$$r = \frac{e}{e_{\mathrm{s}}} \approx \frac{w}{w_{\mathrm{s}}} \approx \frac{q}{q_{\mathrm{s}}} \tag{2.93}$$

Among the several humidity variables, most frequent use is made of vapor pressure and mixing ratio, but the other variables are useful in certain problems. For example, relative humidity is an important parameter in the study of the environment of plants and animals because it is a rough indicator of the rate of evaporation from animal or vegetable tissue. For this reason, we are more immediately aware of changes in relative humidity than of changes in specific humidity or vapor pressure.

The Gas Constant and Specific Heats of Unsaturated Air

Because the proportion of water vapor in the atmosphere is variable, the gas constant for moist (that is, natural) air is variable. This is evident from the equation of state for moist unsaturated air written in the form

$$p\alpha_{wa} = R_{mwa}T \tag{2.94}$$

where the symbol R_{mwa} is slightly shorter than the five words it represents and is equivalent to

$$R_{ma}(1 - q) + R_{mw}q = [1 + (1/\varepsilon - 1)q]R_{ma}$$
$$= [1 + 0.61q]R_{ma} \approx (1 + 0.61w)R_{ma}$$

In order that the equation of state contain only three variables rather than the four in Eq. (2.94), the variation of temperature and of humidity may be combined by defining the *virtual temperature* according to

$$T_v \equiv (1 + 0.61q)T \tag{2.95}$$

The equation of state may then be written

$$p\alpha_{wa} = R_{ma}T_v$$

The virtual temperature is the temperature of dry air having the same pressure and specific volume as the moist air. For a specific humidity of 10 g/kg the virtual temperature exceeds the temperature by about 2° C. The difference between temperature and virtual temperature is important in calculation of the vertical pressure distribution by integration of the hydrostatic equation. Because the virtual temperature accounts for the fluctuations in density due to humidity, it must be used instead of the temperature to calculate static stability. Thus Eq. (2.48) must be generalized to read

$$a = g\left(\frac{T_{v1} - T_v}{T_v}\right) \tag{2.96}$$

and Eq. (2.51), which expresses the static stability for dry air, must be replaced by

$$s_z = \frac{1}{T_v}\left(\frac{\partial T_v}{\partial z} + \Gamma\right) = \frac{1}{\theta_v}\frac{\partial \theta_v}{\partial z} \tag{2.97}$$

where θ_v is the *virtual potential temperature* defined by $\theta_v \equiv \theta(1 + 0.61q)$.

The influence of humidity on the specific heats is expressed by the equations

and
$$\left.\begin{array}{l} c_p = (1 + 0.84q)c_{pa} \approx (1 + 0.84w)c_{pa} \\ c_v = (1 + 0.93q)c_{va} \approx (1 + 0.93w)c_{va} \end{array}\right\} \tag{2.98}$$

where the subscript a refers to dry air.

2.20   Saturation Adiabatic Processes

In a *saturation adiabatic process* condensation (or evaporation) and consequent release of latent heat occurs within the system, but no heat is added to or taken from the system. For example, an air parcel lifted adiabatically experiences decrease in temperature at the adiabatic rate until saturated. Further lifting is accompanied by release of latent heat within the parcel; consequently, the rate of decrease in temperature is less than the adiabatic rate. The first law of thermodynamics for this process may be written

$$-L\,dw_s = c_v\,dT + p\,d\alpha \tag{2.99a}$$

where dw_s represents the mass of water condensed per unit mass of air. This expression combined with the equation of state in differential form and Eq. (2.41) yields

$$-L\,dw_s = c_p\,dT - \alpha\,dp \tag{2.99b}$$

and upon introducing the hydrostatic equation,

$$\frac{dT}{dz} = -\frac{g}{c_p} - \frac{L}{c_p}\frac{dw_s}{dz} \tag{2.100}$$

But

$$\frac{dw_s}{dz} = \frac{\partial w_s}{\partial T}\frac{dT}{dz} + \frac{\partial w_s}{\partial p}\frac{dp}{dz}$$

which with the aid of Eqs. (2.92) and (2.88) may be written

$$\frac{dw_s}{dz} \simeq \frac{Lw_s}{R_{mw}T^2}\frac{dT}{dz} + \frac{w_s g}{RT}$$

Substituting this expression into Eq. (2.100) and solving for $-(dT/dz)$ yields

$$\Gamma_s = -\frac{dT}{dz_s} \simeq \frac{g}{c_p}\left(1 + \frac{Lw_s}{R_{ma}T}\right)\left(1 + \frac{L^2w_s}{c_pR_{mw}T^2}\right)^{-1} \tag{2.101}$$

which is a good approximation of the *saturation adiabatic lapse rate*. This quantity can be calculated for any point on a thermodynamic diagram which displays T and w_s. Since supersaturation rarely exceeds 1%, saturated air which is rising or sinking follows rather closely the rate of temperature change given in Eq. (2.101). When water in liquid or solid form is removed from the system by precipitation, the change of state is referred to as *pseudo-adiabatic*. The equations for the pseudo-adiabatic process differ only slightly from the corresponding equations for the saturation adiabatic process, and this difference usually is ignored.

Stability of Saturated Air

Equation (2.50) expresses the stability of a layer of air as a function of the difference between the vertical rate of change of temperature experienced by rising air and the vertical rate of temperature change within the layer. If the rising air is saturated, the stability may be written

$$s_z = (1/T)(\Gamma_s - \gamma)$$

A lapse rate greater than the pseudo-adiabatic but less than the adiabatic is statically stable for unsaturated air but is statically unstable for saturated air and is referred to as *conditionally unstable*.

The Pseudo-Adiabatic Chart

Lines which describe the temperature change experienced by saturated air in rising or sinking in the atmosphere may be added to the adiabatic chart by introducing appropriate numerical values in (2.101) or other equations derived from it. The resulting *pseudo-adiabatic chart* may be used to determine temperature change experienced by either saturated or unsaturated air. The chart usually includes lines which indicate w_s as a function of temperature and pressure (height). The w_s lines may be constructed by using Eq. (2.90) combined with the Clausius–Clapeyron equation in the form of Eq. (2.89). Upon eliminating e_s, w_s is expressed as a function of p and T, and then may be plotted on the pseudo-adiabatic chart as illustrated in Fig. 2.15. By using the dry adiabats and the lines of constant saturation mixing ratio it is possible to determine graphically the point at which unsaturated air becomes saturated during ascent. The pseudo-adiabat determines the temperature change during ascent of a saturated parcel of air which approaches asymptotically a dry adiabat when all the moisture has condensed and precipitated out of the air, i.e., when $w_s = 0$. The potential temperature of this asymptotic dry adiabat is called the *equivalent potential temperature*

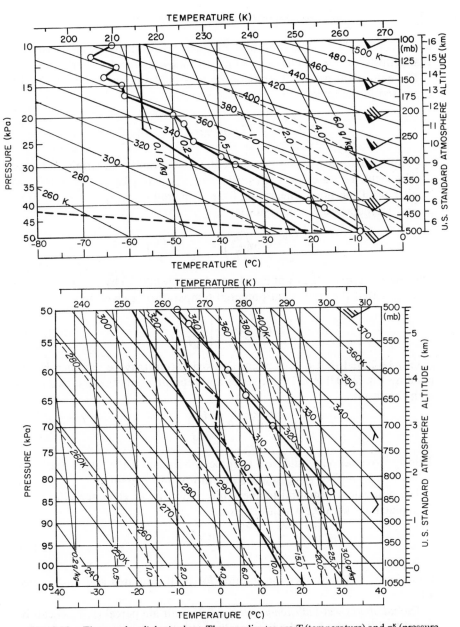

FIG. 2.15. The *pseudo-adiabatic chart*. The coordinates are T (temperature) and p^κ (pressure raised to the R_m/c_p power). Solid sloping lines labeled in degrees Kelvin denote *dry adiabats* or lines of constant potential temperature (θ). Solid lines labeled in grams of water vapor per kilogram of dry air refer to saturation mixing ratios (w_s). Dashed lines denote *pseudo-adiabats* or lines of constant equivalent potential temperature (θ_e) labeled in degrees Kelvin. The single heavy solid line represents the U. S. Standard Atmosphere. The solid and dashed lines connecting data points represent actual soundings of temperature and dewpoint, the flags represent wind velocity. These soundings were taken at 00 Greenwich Mean Time, 5 August 1976 at Stapleton Airport, Denver, Colorado. They were taken about 4 hr prior to the photograph shown in Fig. 3.17.

and is denoted by θ_e. The value of θ_e uniquely labels the pseudo-adiabat, but is also invariant for an unsaturated parcel following the dry adiabat.

The equivalent potential temperature may be developed from Eq. (2.99b) expressed in the form

$$-L\,d\left(\frac{w_s}{T}\right) = \left(c_p + \frac{Lw_s}{T}\right)\frac{dT}{T} - R_m\frac{dp}{p}$$

An approximation is made by neglecting the second term within brackets on the right, because $Lw_s/c_pT \sim 10^{-1}$ for w_s of $10\,\mathrm{g\,kg^{-1}}$ and is less for smaller w_s. Then upon integrating from T, p, and w_s to $w_s = 0$ at pressure p_0 and temperature θ_e, the result is

$$\frac{Lw_s}{c_pT} = \ln\left(\frac{\theta_e}{T}\right) - \frac{R_m}{c_p}\ln\left(\frac{p_0}{p}\right)$$

The second term on the right may be replaced by $\ln(\theta/T)$ by using Eq. (2.47). It follows that the equivalent potential temperature is given by

$$\theta_e = \theta\exp\left[\frac{Lw_s}{c_pT}\right] \qquad (2.102)$$

It may be noted that a convenient approximation may be obtained by expanding Eq. (2.102) in a Taylor series and retaining only the first two terms with the result

$$\theta_e = \theta + \frac{Lw_s}{c_pT}$$

This approximation would also have been obtained if the latent heat from w_s had been released at constant temperature and pressure.

The equivalent potential temperature allows us to generalize Eq. (2.51) by replacing θ with θ_e, which results in

$$s_z = \frac{\partial \ln \theta_e}{\partial z} \qquad (2.103)$$

Equation (2.103) describes the stability for both saturated and unsaturated air.

The *adiabatic wet-bulb temperature* (T_w) of an air parcel (p, T, w) can be found by lifting the parcel adiabatically until it reaches saturation, then returning it pseudo-adiabatically to pressure p. The resulting temperature (T_w) is very close to the temperature of a well-ventilated wet bulb placed in air at p, T, w (the *isobaric wet-bulb temperature*). The wet-bulb depression for an isobaric process can be developed from Eq. (2.99b) together with the diffusion equation (2.78) and the heat conduction equation (2.79). A similar development is carried out in Section 3.6. As Fig. 2.15 indicates, T_w and T together specify the mixing ratio of the original air parcel.

The heat released in condensation within clouds represents an important source of energy for the atmosphere. In the following simple example the pseudo-adiabatic chart provides an estimate of the heating brought about by condensation. Air which is forced upward by mountains cools adiabatically until the vapor pressure equals the equilibrium vapor pressure. This point is determined by the intersection of the appropriate adiabatic line with the w_s line equal to the mixing ratio of the rising air. From this point to the top of the mountain the rising air cools pseudo-adiabatically, and a cloud is formed. If precipitation falls from the cloud, the total water per unit mass is reduced; therefore, if the air descends on the leeward side of the mountain, the air warms pseudo-adiabatically only until the new value of w_s is reached. From this point the air warms adiabatically. Consequently, the temperature on the leeward side of a mountain may be considerably higher than on the windward side. An example of such a sequence is examined in Problem 16. Although the strong dry wind in the lee of a mountain, called *foehn* or *chinook*, is often attributed to this mechanism, in the most marked foehns air descends from far above the top of the mountain and is warmed adiabatically on the entire descent.

2.21 Distribution of Temperature and Water Vapor

Equation (2.19) and the pressure at the surface together specify the vertical pressure distribution as a function of the temperature distribution. Because, as is illustrated in Fig. 2.16, the temperature distribution is usually a complicated function of height, Eq. (2.19) must be integrated numerically. An example of such an integration is given in Problem 17.

The Average Vertical Temperature Distribution

The atmosphere consists of a series of nearly spherical layers each characterized by a distinctive vertical temperature distribution. These layers usually can be identified in soundings of the temperature, like the one labeled *a* in Fig. 2.16. The characteristic features are even more clearly defined in curve *b* of Fig. 2.16, which represents the U.S. Standard Atmosphere. Although many observed temperature profiles resemble closely the U.S. Standard Atmosphere, deviations of 20 K or more are observed frequently. In Appendix II.E the U.S. Standard Atmosphere and the International Reference Atmosphere have been summarized.

Figure 2.17 indicates the gross features of the temperature distribution to 100 km. The lowest layer, characterized for the most part by decreasing temperature with height, is called the *troposphere*, which means the turning or changing sphere. This layer contains about 80% of the total atmospheric mass. The troposphere is most closely influenced by the energy transfer that takes place at the earth's surface through evaporation and heat conduction.

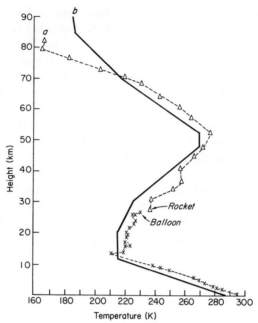

FIG. 2.16.   (a) Temperature as a function of height observed at Fort Churchill, Canada (59°N) at 2330 CST on July 23, 1957 (after W. G. Stroud, W. Nordberg, W. R. Bandeen, F. L. Batman, and P. Titus, *J. Geophys. Res.*, **65**, 2307, 1960; copyrighted by American Geophysical Union). (b) Temperature of the U.S. Standard Atmosphere (see Appendix II.E).

These processes often create horizontal and vertical temperature gradients which may lead to the development of atmospheric motions and the upward transport of heat and water vapor. But rising air tends to cool adiabatically or pseudo-adiabatically; as a result temperature decreases with height and clouds form in the troposphere. A warning is in order; the motions which are referred to here are extremely complex and should not be imagined simply as closed circuits in a vertical plane. These topics are developed more fully in later chapters.

The vertical extent of the troposphere varies with season and latitude. In tropical regions it is usually 16–18 km. Over the poles the extent in summer is about 8–10 km, but in the winter the troposphere may be entirely absent.

The troposphere is usually bounded at the top by a remarkably abrupt increase of static stability with height. The surface formed by this virtual discontinuity of lapse rate is called the *tropopause*; it is particularly clearly defined in tropical regions. In middle and high latitudes above large scale storms the tropopause often is not clearly defined.

The statically stable layer above the troposphere is called the *stratosphere*

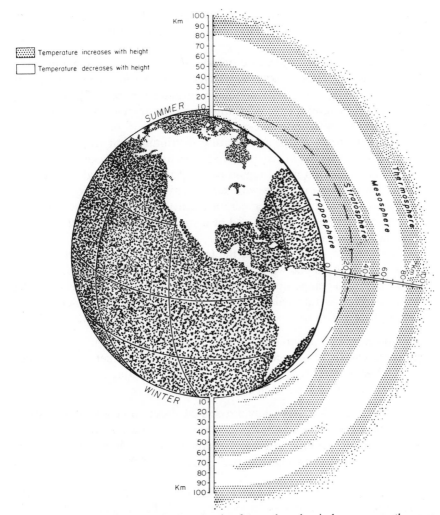

FIG. 2.17. Meridional cross section showing for northern hemisphere summer the gross features of the temperature distribution.

(stratified sphere). It extends upward to a height of about 50 km where the temperature is comparable to the earth's surface temperature. Above the tropopause the temperature increases first slowly with height to about 20 km and then much more rapidly. This temperature distribution is associated with absorption of ultraviolet solar radiation by the ozone which is present between heights of 20 and 50 km. The top of the stratosphere is a surface of maximum temperature, which is called the *stratopause*. The state and processes of the stratosphere are observed only with considerable difficulty, and important aspects of stratospheric phenomena remain incompletely understood.

Above 50 km the temperature decreases with height to a minimum of about 180 K at about 85 km. This layer of decreasing temperature with height is called the *mesosphere* (middle spere) and has been less fully observed than the regions above or below it. Information gathered from meteor trails and rockets indicates that wind speeds up to 150 m s^{-1} occur in this region. The similarity between the temperature distributions in troposphere and mesosphere suggests the existence of somewhat similar processes. Absorption of solar radiation in the region of the stratopause provides the energy source, and resulting motions with accompanying adiabatic expansion and cooling may account for the temperature distribution in this layer. Noctilucent clouds are sometimes observed near the region of lowest temperature, called the *mesopause*.

Above a height of 80 or 90 km the temperature increases with height to a temperature varying between 600 and 2000 K at a height of about 500 km. The variation in temperature is related to the activity on the sun. This extensive region is referred to as the *thermosphere*. The character of the thermosphere differs from the lower layers because

(a) ionization of air molecules and atoms occurs,

(b) dissociation of molecular oxygen and other constituents takes place,

(c) diffusion is more important than mixing, resulting in relatively high concentrations of light gases in the upper layers.

Molecular oxygen is increasingly dissociated above about 80 km, and above 130 km most oxygen is in atomic form. At these heights, because of the large mean free path, diffusion becomes more important than mixing, and in accordance with Dalton's law the heavier gases tend to be less concentrated with increasing height.

Above about 110 kms atomic oxygen increases in relative abundance with height while molecular nitrogen decreases. Between 500 and 1000 km helium replaces atomic oxygen as the dominant constituent, and above about 1500 km atomic hydrogen replaces helium. These estimates are based on diffusive equilibrium in an isothermal temperature distribution of about 1500 K above 500 km. These conditions permit the existence of a population of neutral hydrogen atoms which are formed by dissociation of water vapor and methane in the lower thermosphere. The neutral atoms diffuse upward through the *thermopause* to the base of the *exosphere* (perhaps 300–500 km), and the more energetic may then escape from the earth's gravitational field. Helium escapes more slowly, and the rate of escape of the heavier gases is probably negligible.

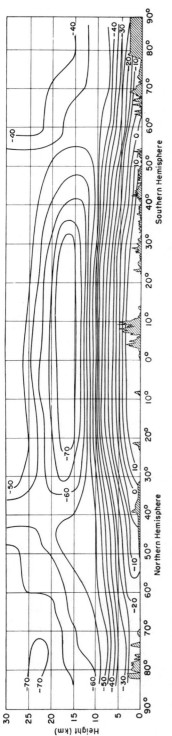

Fig. 2.18. Meridional cross section showing the average temperature distribution for January 1958 between the earth and a height of 30 km at 75°W longitude (after U.S. Weather Bureau, *Monthly Mean Aerological Cross Sections*. U.S. Govt. Printing Office, Washington, D.C., 1961).

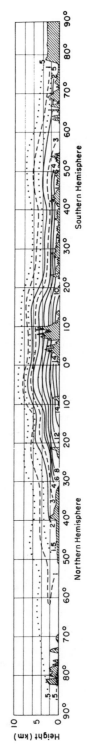

Fig. 2.19. Meridional cross section showing the average mixing ratio distribution for January 1958 between the earth and a height of 10 km at 75°W longitude (after U.S. Weather Bureau, *Monthly Mean Aerological Cross Sections*. U.S. Govt. Printing Office, Washington, D.C., 1961).

Average Latitudinal Temperature and Water Vapor Distribution

The average temperature distribution plotted on an atmospheric cross section on a plane through the axis of the earth is illustrated in Fig. 2.18. The troposphere extends to a maximum height near the equator and to minimum height over the poles. We see also that in summer the troposphere is deeper than in winter.

The average latitudinal water vapor distribution is illustrated in Fig. 2.19. Highest concentration of water vapor is found at low latitudes near the earth's surface where mixing ratio as high as 40 g kg^{-1} or more may be observed. Water vapor concentration normally decreases upward and toward the poles. The distribution is, like the temperature, highly variable in time and space. Variation by one or two orders of magnitude occurs frequently over a vertical distance of 10 km or a horizontal distance of 10,000 km.

List of Symbols

		First used in Section
a	Acceleration	2.14
A	Area	2.3
a, b	Constants in Van der Waals' equation	2.17
c	Specific heat	2.11
c_p	Specific heat at constant pressure	2.11
c_v	Specific heat at constant volume	2.11
D	Diffusivity	2.16
e	Vapor pressure	2.18
e_t	Total energy per unit mass	2.8
$\mathbf{E}_Q$	Flux density of property Q	2.16
F	Force	2.4
g	Force of gravity per unit mass (magnitude)	2.14
G	Number of cells, Gibbs function	2.15, 2.18
h	Heat added per unit mass	2.8
H	Volume of cell in phase space	2.15
j	Number of degrees of freedom	2.10
k	Boltzmann's constant	2.4
l	Mean free path length	2.16
L	Latent heat of vaporization	2.17
m	Mass of a molecule	2.4
M	Molecular mass, based on the C_{12} nucleus	2.4
n	Number of molecules per unit volume	2.3
n^*	Number of molecules per unit mass	2.9
N	Number of molecules in a system	2.3
N_0	Avogadro's number	2.1
p	Pressure	2.4

		First used in Section
p_x, p_y, p_z	Components of momentum in momentum space	2.15
q	Specific humidity	2.19
Q	Conservative molecular property	2.16
r	Radius of molecule, relative humidity	2.16, 2.19
R	Universal gas constant	2.4
R_m	Specific gas constant for a gas with molecular mass M	2.4
s	Specific entropy	2.12
s_z	Static stability	2.14
t	Time	2.3
T	Temperature	2.4
T_v	Virtual temperature	2.19
T_w	Adiabatic wet-bulb temperature	2.18
u	Specific internal energy	2.7
v	Speed of a molecule	2.3
$\bar{v}$	Mean speed of molecules in a system	2.3
$\mathbf{v}$	Velocity vector	2.7
V	Volume of a system	2.3
w	Work per unit mass, mixing ratio	2.8, 2.19
W	Work, thermodynamic probability of a system	2.8, 2.15
x, y, z	Cartesian coordinates	2.5
α	Specific volume	2.4
α_m	Molar specific volume	2.4
γ	Lapse rate	2.14
Γ	Adiabatic lapse rate	2.14
Γ_s	Pseudo-adiabatic lapse rate	2.20
ε	Ratio of specific gas constants of air and water vapor	2.19
θ	Angle between normal to area and direction in which molecule moves, potential temperature	2.3, 2.13
θ_e	Equivalent potential temperature	2.20
θ_v	Virtual potential temperature	2.19
κ	R_m/c_p	2.13
λ	Thermal conductivity	2.16
μ	Coefficient of viscosity	2.16
ρ	Density	2.14
ϕ	Azimuth angle, distribution function	2.3, 2.5
Φ	Geopotential	2.8

Subscripts

a	Dry air
c	Critical
s	Saturation
w	Water vapor
wa	Moist air
x, y, z	Components corresponding to axes
1	Liquid phase
2	Vapor phase

Problems

1. If 10^6 molecules are required in order to ensure a statistically uniform distribution of velocities in all directions, what is the minimum volume in which the state can be defined at standard atmospheric conditions (101.3 kPa = 1013 mb and 0° C)?

2. Show that the number of collisions per unit wall area per unit time and per unit solid angle can be expressed by

$$\frac{1}{4\pi}\, n\bar{v}\cos\theta$$

3. At what height would a hydrogen molecule reach the escape speed in the vertical direction if the temperature of the molecule is 5000 K?

4. If the temperature is 2000 K at a height of 500 km, what proportion of the molecules with molecular mass 2 will have at least the escape velocity?

5. Suppose that the atmosphere is completely at rest and in diffusive equilibrium. Compute at what height 10% O_2 and 90% N_2 is found when at the surface the temperature is 300 K, the pressure is 100 kPa (1000 mb), and the composition is 20% O_2 and 80% N_2.

6. What is the expansion work done by a system in the cyclic process shown in the adjoining figure? Expansion occurs at constant pressure p_1 from α_1 to α_2, compression occurs at

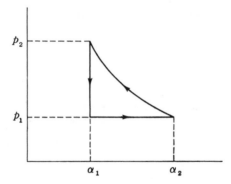

constant temperature from α_2 to α_1, and cooling occurs at constant volume from p_2 to p_1. Assume that $\alpha_1 = 1\ \mathrm{m}^3\ \mathrm{kg}^{-1}$, $\alpha_2 = 2\ \mathrm{m}^3\ \mathrm{kg}^{-1}$, and T at α_2 and p_1 is 500 K. The system contains air with molecular mass 29.

7. A cyclic reversible process following two isotherms and two adiabats is called a Carnot cycle. Compute the work done per unit mass by a system containing dry air in completing a Carnot cycle defined by (a) an isothermal expansion at 300 K from 100 to 50 kPa followed by (b) an adiabatic expansion until the temperature is 200 K and (c) an isothermal compression and adiabatic compression in such a manner that the original point of 300 K and 100 kPa is reached. How much heat has been supplied to the system in completing the cycle?

8. Show that a change in potential temperature is related to a change in specific entropy by

$$ds = c_{\mathrm{p}}\, d\ln\theta$$

9. Show that when a parcel with temperature T' moves isentropically in air of temperature T' the lapse rate following the parcel is given by

$$-\frac{dT'}{dz} = \frac{T'}{T}\Gamma$$

10. Solve Eq. (2.79) for steady state conduction of heat through a wall of thickness D and conductivity k if the temperature is maintained at one side at T_0 and at the other side at T_1. Choose the x axis perpendicular to the planes of the wall.

11. Assume that the earth is a homogeneous sphere with thermal conductivity $= 8 \times 10^{-5}$ W m^{-1} K^{-1} and constant surface temperature $T = 285$ K. If the temperature near the surface increases downward by 1° C (100 m)$^{-1}$, what is the temperature 1000 km from the center? Assume steady-state conditions and also that all the heat that flows out is generated within the core of 1000 km radius.

12. Show that the Van der Waals' constants may be expressed by

$$a = \frac{27}{64} \frac{(R_m T_c)^2}{p_c} \qquad b = \frac{R_m T_c}{8 p_c}$$

and evaluate a and b for water vapor. Rembember that at the critical point $\partial^2 p / \partial \alpha^2 = 0$ and $\partial p / \partial \alpha = 0$.

13. The observed saturation vapor pressure for water vapor at 20° and at 30° C is 2338 and 4243 Pa, respectively. Compute the error made by using Eq. (2.89) at these temperatures.

14. Derive an equation similar to Eq. (2.89) for the saturation vapor pressure over ice. Use for the latent heat of vaporization and the specific gas constant for water vapor

$$L = 2.834 \times 10^6 \text{ J kg}^{-1} \qquad \text{and} \qquad R_{mw} = 461.7 \text{ J kg}^{-1} \text{ K}^{-1}$$

15. Compute the error that is made in the specific volume of water vapor when the equation of state for an ideal gas is used instead of Van der Waals' equation for temperature of 300 K and vapor pressure of 1.5 kPa (15 mb).

$$T_c = 647 \text{ K} \qquad\qquad p_c = 2.26 \times 10^4 \text{ kPa}$$

$$\alpha_c = 3.1 \times 10^{-3} \text{ m}^3 \text{ kg}^{-1} \qquad R_{mw} = 461.7 \text{ J kg}^{-1} \text{ K}^{-1}$$

16. Air at a temperature of 20° C and mixing ratio $w = 10$ g kg^{-1} is lifted from 100 to 70 kPa by moving over a mountain barrier. Determine from a pseudo-adiabatic chart its temperature after descending to 90 kPa on the other side of the mountain. Assume that 10% of the water vapor is removed by precipitation during the ascent. Describe the thermodynamic state of the air at each significant point.

17. Compute the height of the 70, 50, 30, 10, 5, and 2.5 kPa surfaces from the following radiosonde sounding, taken at Ft. Churchill, 00 GMT, 24 July, 1957. The height of the ground surface is 30 m above sea level and the surface pressure is 100.7 kPa. Use $R_m = 287$ K kg^{-1} K^{-1} and $g = 9.81$ m s^{-2}.

Pressure (kPa)	Temperature (°C)	Pressure (kPa)	Temperature (°C)
100.7	27.5	21.4	−58.2
100.0	27.8	20.0	−57.5
98.8	28.1	17.6	−59.6
85.0	15.9	16.4	−52.1
75.5	8.0	15.0	−54.1
70.0	5.1	14.1	−55.6
64.8	2.5	10.0	−50.4
50.0	−11.5	5.0	−46.9
46.8	−14.2	4.0	−47.0
40.0	−23.9	2.5	−43.4
30.0	−40.0	1.2	−33.7
25.0	−50.1		

Solutions

1. The volume occupied by 10^6 molecules at standard pressure and temperature calculated from Eq. (2.1) is 37.21×10^{-21} m^3. This is represented by a cube 3.34×10^{-7} m on a side.

2. The number of molecules with speed v in the direction ϕ, θ as expressed in Section 2.3 can be divided by the differentials of solid angle ($\sin \theta \, d\theta \, d\phi$), of wall area ($dA$), and time ($dt$). This gives

$$\frac{v \, dn_v \cos \theta}{4\pi}$$

Integrating over the number of molecules and substituting from Eq. (2.3) gives

$$\frac{1}{4\pi} n \bar{v} \cos \theta$$

3. From Eq. (1.6) and (1.10), the energy per unit mass required to escape from height h is

$$\frac{g_0^* R^2}{R + h}$$

Equating this to the kinetic energy of the molecules given by Eq. (2.8) leads to

$$h = \frac{2 g_0^* R_m^2}{3kT} - R$$

The numerical result is 6.43×10^6 m or a little more than one earth radius.

4. From Eq. (2.18) the number of molecules having velocities greater than v_{es} is

$$n_{es} = 4\pi n \left(\frac{m}{2\pi kT} \right)^{3/2} \int_{v_{es}}^{\infty} v^2 \exp\left(-\frac{mv^2}{2kT} \right) dv$$

Let

$$x \equiv \left(\frac{m}{2kT} \right)^{1/2} v$$

Then

$$\frac{n_{es}}{n} = \frac{4}{\sqrt{\pi}} \int_{x_{es}}^{\infty} x^2 e^{-x^2} \, dx$$

The integral can be expanded to the following terms

$$-\tfrac{1}{2} \int_{x_{es}}^{\infty} x \, d(e^{-x^2}) = -\tfrac{1}{2} \int_{x_{es}}^{\infty} d(x e^{-x^2}) + \tfrac{1}{2} \int_{x_{es}}^{\infty} e^{-x^2} \, dx$$

The value of x_{es} is found by equating the kinetic energy of escape to the geopotential energy at height h. This gives

$$v_{es} = \left(\frac{2 g_0^* R^2}{R + h} \right)^{1/2} \quad \text{and} \quad x_{es} = \left[\frac{m g_0^* R^2}{kT(R + h)} \right]^{1/2}$$

The numerical value of x_{es} is 2.64. The first integral on the right then is $\tfrac{1}{2} \times 2.64 \exp(-6.99) =$

0.0012. The second integral can be evaluated by substituing $x = y/\sqrt{2}$. This results in the error integral

$$\tfrac{1}{2}\sqrt{2} \int_{3.73}^{\infty} e^{-1/2y^2}\, dy$$

which from tables of the error function is about 0.0001, or an order of magnitude smaller than the first integral. Therefore,

$$n_{es}/n \approx 2.26 \times 0.0013 = 0.0029$$

5. From Eq. (2.21) the ratio of O_2 to N_2 molecules can be expressed by

$$\frac{n_O}{n_N} = \frac{n_O(0)}{n_N(0)} \exp\left[\frac{\Phi}{kT}(m_N - m_O)\right]$$

Therefore, upon substituting $\Phi = gz$,

$$z = \frac{kT}{g(m_N - m_O)} \ln\left[\frac{n_O\, n_N(0)}{n_N\, n_O(0)}\right]$$

The numerical values result in $z = 52$ km.

6. The line integral around the closed figure (W) can be expressed by

$$W = \int_{\alpha_1}^{\alpha_2} p_1\, d\alpha - \frac{RT}{M}\int_{p_1}^{p_2}\frac{dp}{p} + 0 = p_1(\alpha_2 - \alpha_1) + \frac{RT}{M}\ln\frac{p_1}{p_2}$$

For the stated conditions

$$p_1 = \frac{RT}{M\alpha_2} \quad \text{and} \quad p_2 = \frac{RT}{M\alpha_1}$$

Therefore,

$$W = \frac{RT}{M}\left[1 - \frac{\alpha_1}{\alpha_2} + \ln\frac{\alpha_1}{\alpha_2}\right]$$

Substitution of values yields $W = -28 \times 10^3$ J kg^{-1}. The negative sign indicates that work is done on the system.

7. The work during the isothermal processes is

$$R_m(T_1 - T_2)\ln\frac{p_1}{p_2} = 20.0 \times 10^3 \text{ J kg}^{-1}$$

This is equivalent to the heat supplied to the system. During the adiabatic expansion the work done is equal and opposite to the work done during the adiabatic compression so there is no contribution from these processes.

8. From Eqs. (2.41)–(2.43)

$$ds = c_p\frac{dT}{T} - R_m\frac{dp}{p}$$

Upon introducing Eq. (2.47)

$$ds = c_p\frac{d\theta}{\theta} = c_p d\ln\theta$$

9. The potential temperature of the parcel of temperature T' is

$$\theta' = T'(100 \text{ kPa}/p)^{\kappa}$$

The bracket can be eliminated by using Eq. (2.47) with the result

$$\theta' = (T'/T)\theta$$

Therefore,

$$\frac{d\theta'}{dz} = 0 = \frac{\theta}{T^2}\left(T\frac{dT'}{dz} - T'\frac{dT}{dz}\right)$$

and

$$\frac{dT'}{dz} = \frac{T'}{T}\frac{dT}{dz} = -\frac{T'}{T}\Gamma$$

10. Equation (2.79) for this case reduces to

$$\frac{\partial^2 T}{\partial x^2} = 0$$

Integrating gives the general solution

$$T = c_1 x + c_2$$

The boundary conditions are $T = T_0$ at $x = 0$ and $T = T_1$ at $x = D$. Therefore,

$$c_1 = \frac{T_1 - T_0}{D} \qquad c_2 = T_0 \qquad T = T_0 + \frac{T_1 - T_0}{D}x$$

11. The radial heat flux is expressed from Eq. (2.70) by $E_h = -\lambda(dT/dr)$. The total heat transferred radially from $r = 100$ km (r_1) to the earth's surface $(r_0 = 6.4 \times 10^6$ m$)$ is independent of radius. Therefore

$$4\pi r^2 \lambda\left(\frac{dT}{dr}\right) = C \qquad \text{(const)}$$

Upon integrating from r_1 to r_0

$$T_1 = T_0 + \frac{C}{4\pi\lambda}\left(\frac{1}{r_1} - \frac{1}{r_0}\right)$$

The data given yield $C = 4.12 \times 10^8$ W and $T_1 = 344 \times 10^3$ K.

12. Differentiate Eq. (2.83) with respect to α twice, set the two resulting equations equal to zero, eliminate α, and solve for a and b. Using the critical values of temperature and pressure given in Section 2.17

$$a = 1665.7 \text{ m}^5 \text{ kg}^{-1} \text{ s}^{-2}$$

$$b = 1.652 \times 10^{-2} \text{ m}^3 \text{ kg}^{-1}$$

13. Error at 20° C: 25 Pa   (1.05% too large)
           30° C: 102 Pa   (2.41% too large)

14. $\log_{10} e_s = 12.545 - 2665.8/T$

15. Differentiate Eq. (2.83) treating $d\alpha$ as the error resulting from change from $a = b = 0$ to the values derived in Problem 12. The proportional error then can be expressed by

$$\frac{d\alpha}{\alpha} = \frac{bR_{mw}T - a[1 - (be/R_{mw}T)]}{(R_{mw}^2 T^2/e) - 2bR_{mw}T + a} \frac{bR_{mw}T - a}{R_{mw}^2 T^2/e}$$

For the specified numerical values

$$\frac{d\alpha}{\alpha} = -1.09 \times 10^{-3} = -0.1\%$$

16. On ascent condensation is reached at 91.5 kPa, 12.5° C (cloud base). At 70 kPa; $T = 1°$ C; $w = 5.9\%_{oo}$; liquid water $= 4.1\%_{oo}$. Descent occurs to cloud base where $w = 9.6\%_{oo}$; $p = 88.5$ kPa; $T = 11.4°$ C. Further descent occurs to $p = 90.0$ kPa; $T = 12.5°$ C; $w = 9.6\%_{oo}$.

17.

Pressure (kPa)	Height above sea level (m)
100.7	30.0
70.0	3104.8
50.0	5768.6
30.0	9470.2
10.0	16515.6
5.0	21069.2
2.5	25680.6

General References

Allis and Herlin, *Thermodynamic and Statistical Mechanics*, gives an excellent introduction to statistical mechanics and thermodynamic probability.

Banks and Kockarts, *Aeronomy*, gives a detailed description of the upper atmosphere.

Holmboe, Forsythe, and Gustin, *Dynamic Meteorology*, gives a useful account of the thermodynamic effects of water vapor.

Kennard, *Kinetic Theory of Gases*, is a solid book on a more advanced level. It contains a wealth of information and is clearly written.

Morse, *Thermal Physics*, gives a slightly more advanced treatment of thermodynamics and includes a good account of nonequilibrium thermodynamics.

Sears, *An Introduction to Thermodynamics, the Kinetic Theory of Gases and Statistical Mechanics*, covers about the same material as Allis and Herlin and gives a clear account of the kinetic theory of gases.

Sommerfeld, *Thermodynamics and Statistical Mechanics*, is a beautiful book giving a more advanced account of the principles used in this chapter.

CHAPTER III

Properties and Behavior of Cloud Particles

PART I: GROWTH

3.1 Intermolecular Force and Surface Tension

"The parts of all homogeneal hard Bodies which fully touch one another, stick together very strongly . . . their Particles attract one another by some Force, which in immediate Contact is exceeding strong, at small distances performs the chymical Operations above mentioned, and reaches not far from the Particles with any sensible Effect."† In these words Sir Isaac Newton described the intermolecular force which he recognized as underlying chemical reactions, wetting or resistance to wetting, capillary phenomena, solubility, crystallization, and evaporation. It has required the full development of electrodynamics and quantum mechanics to elucidate these forces to the level that predictions are possible for a few relatively simple systems. Here we shall be satisfied to give a partial, intuitive, and brief account of these forces. Figure 2.12 indicates the dependence of the total intermolecular force on separation.

At very small separations, so small that the electron clouds of the molecules overlap, strong repulsive *valence* forces dominate. In Fig. 2.12 the force shown at separation less than r_0 is dominated by the valence repulsive forces. The theory of the valence force, which is developed from quantum mechanics, predicts that this force is approximately an exponential function of distance. Beyond the effective range of the valence force several forces combine to give a complex dependence on separation. *Electrostatic* contributions may be calculated from Coulomb's law if the geometrical arrangement and electrical nature of the molecules are known. *Induction* contributions may arise through dipoles induced in nonpolar molecules by nearby charges or polar molecules. As illustrated in Fig. 3.1a, the induction force is attractive regardless of the orientation of the dipole. If the dipole in Fig. 3.1a were replaced by a nonpolar molecule, as illustrated in Fig. 3.1b, there would still arise an attractive force between the molecules. This force may be understood from the fact that the electrons in a nonpolar molecule are constantly in oscillation, so that at a certain instant the molecule assumes a dipole configuration. An induced dipole is created from the adjacent

† Sir Isaac Newton, *Opticks* (4th ed., London, 1730), Dover, New York, 1952.

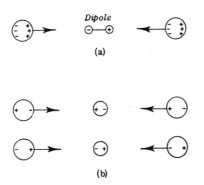

FIG. 3.1. (a) Induced charges and induced forces in two nonpolar molecules in the neighborhood of a neutral dipole. (b) Induced charges and induced forces produced by the instantaneous dipole configurations of a nonpolar molecule.

molecule, and an attractive force arises between them. In both examples of induction the force is inversely proportional to the seventh power of the distance, so at small separations large attractive forces exist. There are other contributions to the intermolecular force, but the induction forces are probably of greatest importance in understanding surface properties of liquids.

The water molecules which at a certain instant form the free surface of a body of water are subjected to intermolecular attractive forces exerted by the nearby molecules just below the surface. The opposing forces exerted on the surface molecules by the air molecules outside are very much less. If the surface area is increased by changing the shape of the container or by other means, molecules must be moved from the interior into the surface layer, and work must be done against the intermolecular force. The work required to increase the surface area by one unit stores potential energy in the surface; it is called *surface energy* or *surface tension* and has units of energy per unit area or force per unit length.

Because of surface tension a volume of liquid tends to assume a shape with minimum area-to-volume ratio. Therefore, small masses tend strongly to assume spherical shapes; in the case of larger masses, forces which are mass dependent may distort or destroy the spherical shape. Spherical drops experience an internal pressure due to surface tension which may be calculated in the following way.

Imagine a small spherical drop of radius r which is divided in half by a hypothetical plane as shown in Fig. 3.2. Surface tension acting across the plane holds the edges of the sphere together; the total force may be expressed by $2\pi r\sigma$ where σ represents surface tension or surface energy per unit area. The two halves of the sphere are held apart by the equal pressure forces

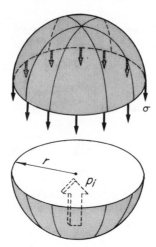

FIG. 3.2.   The surface tension force per unit length (σ) and the internal pressure force per unit area (p_i) for a dissected drop of radius r.

exerted normal to the plane surface and given by $\pi r^2 p_i$ where p_i represents the internal pressure due to surface tension. For equilibrium between the two forces, the internal pressure due to surface tension is given by

$$p_i = 2\sigma/r$$

The surface tension of pure water at $0°$ C is about 0.075 N m^{-1} and is very slightly dependent on radius. This equation shows that a droplet of 1 μm radius experiences an internal pressure of about 1.5 atm.

3.2   Equilibrium Vapor Pressure over a Curved Surface

The surface energy associated with a curved surface has an important effect on the equilibrium vapor pressure and the rate of evaporation from droplets. The difference between the equilibrium vapor pressure over a flat surface (e_s) and over a spherical (or curved) surface (e_c) may be calculated from consideration of the change in surface energy which accompanies decrease of surface area.

The energy released by a droplet during a change in radius (dr) is given by the surface energy multiplied by the change in area and therefore is equal to

$$8\pi\sigma r\, dr$$

And the energy released per unit mass of evaporated water is consequently given by $(2\sigma/r)\alpha_1$, where α_1 represents the specific volume of the liquid water in the drop. The heat required to evaporate 1 kg liquid water is now, instead of Eq. (2.85), expressed by

$$L = T(s_2 - s_1) = u_2 - u_1 + e_c(\alpha_2 - \alpha_1) - (2\sigma/r)\alpha_1$$

or in a form analogous to Eq. (2.86),

$$u_1 + (e_c + 2\sigma/r)\alpha_1 - Ts_1 = u_2 + e_c\alpha_2 - Ts_1 = G$$

The term $e_c + 2\sigma/r$ represents the pressure inside a droplet of radius r. Evidently, the Gibbs function is constant during an isothermal change of phase.

We now consider an isothermal change of phase at the same temperature, but for a droplet of radius $r + dr$. The Gibbs function now has another constant value $G + dG$. Following a similar argument to that given in Section 2.18, the difference in Gibbs function between the two cases may be expressed by

$$dG = \alpha_1 \, de_c - 2\sigma\alpha_1 \frac{dr}{r^2} = \alpha_2 \, de_c$$

and because $\alpha_2 - \alpha_1 \approx \alpha_2 \approx R_w T / e_c$,

$$R_w T \frac{de_c}{e_c} = -2\sigma\alpha_1 \frac{dr}{r^2}$$

where R_w represents the specific gas constant for water vapor. The equilibrium vapor pressure over the droplet can be determined by integrating from a flat surface $(r = \infty)$ to droplet radius r because over a flat surface the equilibrium vapor pressure is identical to the saturation vapor pressure. This yields

$$R_w T \ln \frac{e_c}{e_s} = \frac{2\sigma\alpha_1}{r_0}$$

or

$$e_c = e_s \exp\left(\frac{2\sigma}{\rho_w R_w T r}\right) \tag{3.1}$$

where ρ_w represents the water density. Equation (3.1) shows that the vapor pressure required for condensation on very small droplets may be very large. Surface tension itself is somewhat dependent on curvature, but this does not invalidate the conclusion that very large *supersaturation* (excess of vapor pressure over equilibrium vapor pressure over a flat surface) is required to make minute droplets of pure water grow. Problem 1 requires a calculation of the equilibrium vapor pressure as a function of radius.

The original formulation of Eq. (3.1) is due to Lord Kelvin, who deduced it from observations of the capillary rise of water in small tubes. Problem 2 describes the experiment and requires the reader to reproduce Lord Kelvin's deduction.

3.3   Homogeneous Condensation

When saturated air is subjected to rapid expansion in the laboratory, the temperature of the air drops and large supersaturation can be achieved. Experiments show that using air which is free of foreign particles and ions, the saturation ratio (the ratio of vapor pressure to the saturation vapor pressure over a plane surface of water) must reach about 5 or perhaps more in order for droplets to form from the vapor. This result is interpreted by imagining that in the highly supersaturated air random collisions of water molecules result in the occasional formation of clusters of molecules or droplet *embryos*. If the saturation ratio exceeds the critical value (~ 5), the embryos continue to grow by condensation, whereas for a smaller ambient vapor pressure, they evaporate. This process is called *homogeneous* or *spontaneous* condensation.

Condensation occurs in clean air on negative ions at a saturation ratio of about 4 and on positive ions at a saturation ratio of about 6. These ions are formed by cosmic rays or radioactive decay products striking neutral air molecules. In the Wilson cloud chamber the paths of ionizing particles are clearly defined by minute water droplets formed when the chamber is suddenly expanded and cooled adiabatically, in this way producing a large supersaturation.

Supersaturation in the atmosphere is seldom observed to exceed about one percent, so that it is evident that homogeneous condensation does not occur in the atmosphere. We must look for another effect to explain the formation of cloud droplets.

3.4   Condensation Nuclei and the Equilibrium Vapor Pressure
over Solutions

There are in the atmosphere innumerable small solid and liquid particles on some of which condensation can proceed without the powerful inhibiting effect of curved surface and surface tension. Some of these particles are soluble in water (*hygroscopic*), others are wettable but insoluble, and others are water resistant (*hydrophobic*). The hygroscopic particles or nuclei are the most favorable for condensation; for this reason their effect will be discussed.

Imagine a substance having essentially zero vapor pressure to be dissolved in water. The molecules of the solute are distributed uniformly through the water and some of them occupy positions in the surface layer. Now compare a unit surface before and after solute is added to the water. The proportion of the surface area occupied by water molecules will be reduced in the approximate ratio $n/(n + n')$ where n represents the number of molecules of solute and n' the number of molecules of water. This assumes that the molecules retain their identity in solution (are undissociated), that the cross-sectional areas of individual particles of solute and water are the same, and that both types of particles are distributed uniformly over the surface. It follows that the number of escaping molecules and the equilibrium vapor pressure should be reduced in the same ratio. These considerations are expressed as Raoult's law for ideal solutions in the form

$$\frac{e_s - e_h}{e_s} = \frac{n}{n + n'}$$

where e_h represents the equilibrium vapor pressure over the solution. For dilute solutions Raoult's law may be written

$$\frac{e_h}{e_s} = 1 - \frac{n}{n'} \tag{3.2}$$

For dilute solutions in which the dissolved molecules are dissociated Eq. (3.2) must be modified by multiplying n by i, the number of dissociated ions per molecule. In dilute solution the NaCl molecule dissociates into two ions and the $(NH_4)_2SO_4$ molecule dissociates into three ions. The number of dissociated ions of solute of mass M may now be expressed by $iN_0 M/m_s$ where N_0 represents Avogadro's number and m_s the molecular mass of solute. The number of water molecules (n') in mass m may be expressed by $N_0 m/m_w$ where m_w represents the molecular mass of water. For a spherical water drop, Eq. (3.2) may now be written

$$\frac{e_h}{e_s} = 1 - \frac{3im_w M}{4\pi r^3 \rho_w m_w} \tag{3.3}$$

Equation (3.3) shows that for a specific mass of dissolved material the equilibrium vapor pressure over a solution decreases rapidly with decreasing radius of the drop.

The effects of surface tension and of hygroscopic substance are of opposite sign and both increase in importance with decreasing size of the droplet.

A general equation which represents both effects may be written by combining Eqs. (3.1) and (3.3) and expanding the exponential function in a Taylor series. Where e_{hc} represents the equilibrium vapor pressure over a droplet of solution, the leading terms of this expansion are

$$\frac{e_{hc}}{e_s} = 1 + \frac{2\sigma}{r\rho_w R_w T} - \frac{3im_w M}{4\pi r^3 \rho_w m_s} \tag{3.4}$$

Equation (3.4) is illustrated in Fig. 3.3. The maximum of each curve represents a critical radius which can be calculated from Eq. (3.4). Problem 3 provides an exercise using Eq. (3.4).

Several conclusions may be drawn from Fig. 3.3. (a) In an isolated population of droplets containing equal masses of solute, the very small droplets must grow at the expense of the larger droplets. (b) Beyond a critical size the larger droplets may grow at the expense of the smaller. (c) If the supersaturation of the vapor is limited, say to about 0.1%, droplets must contain more than 10^{-15} g of NaCl (or other similar solute) in order to reach critical size. (d) Small droplets may exist at relative humidities far below 100% if the droplets contain solute in sufficient concentration.

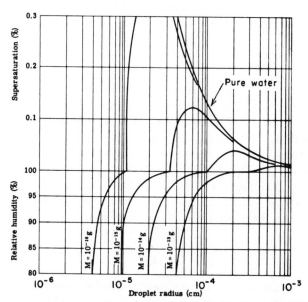

FIG. 3.3. Ratio of equilibrium vapor pressure over droplet containing fully dissociated NaCl of mass M to that over flat water surface (expressed in per cent) plotted as a function of radius as computed from Eq. (3.4). Surface tension of 0.075 N m^{-1} and temperature of 273 K have been used.

3.5 Distribution and Properties of Aerosols

Suspended solid and liquid particles (aerosols) are present in the atmosphere in enormous numbers, and their concentration varies by several orders of magnitude with time and in space. They vary from about 5×10^{-3} to 20 μm in "effective" radius. These particles play crucial roles both in condensation and in the formation of ice crystals. In addition, the aerosols participate in chemical processes, they influence the electrical properties of the atmosphere, and in large concentrations they may be annoying, dangerous, and even lethal. Radioactive aerosols may be used as tracers of the air motion but also have their uniquely hazardous aspects.

Representative distributions by number and volume are shown in Figs. 3.4 and 3.5, respectively. Except for areas near industries or other sources, the distributions exhibit remarkable similarity, which may be explained by the physical processes of *coagulation* and *fallout*. If one imagines an initial distribution with equal numbers of particles of every size, the particles of radius less than 10^{-2} μm quickly become attached to larger particles due to Brownian motion, while the particles of radius greater than 20 μm are sufficiently heavy to precipitate out. Both processes are rather sharply dependent on size, so that they produce fixed limits to the distribution as shown in Fig. 3.4. Between the upper and lower limits the distribution of particles by mass is roughly uniform as shown in Fig. 3.5. Division into the three categories: Aitken, large, and giant, is arbitrary but corresponds roughly to differences in technique of observation. *Aitken* nuclei are named after the physicist who first studied the small particles using a counter based on the rapid expansion of a chamber containing saturated air. Expansion produces adiabatic cooling until the Aitken particles act as condensation nuclei; the minute droplets so formed grow almost instantaneously to visible size. The large nuclei, which are far fewer than the number of Aitken nuclei, can be removed by allowing the larger droplets to settle out. As the curves of Fig. 3.3 suggest, the critical supersaturation above which hygroscopic nuclei can grow to become visible droplets is less than about 0.1% for a nucleus of 0.1 μm radius. The Aitken nucleus counter is the standard instrument for detection of particles less than 0.2 μm in radius. Particles in the range from 0.2 to 1 μm, called *large* nuclei, have been detected by optical means and have been collected by thermal and electrostatic precipitation, by filters on coated slides, and on spider webs. Particles larger than 1 μm, called *giant* nuclei, have been collected on coated slides by impactor (air jets which project the particles at high speed onto a slide), and certain types such as NaCl have been detected by burning in a hydrogen flame. After collection, the large and giant nuclei may be analyzed using an optical or electron microscope.

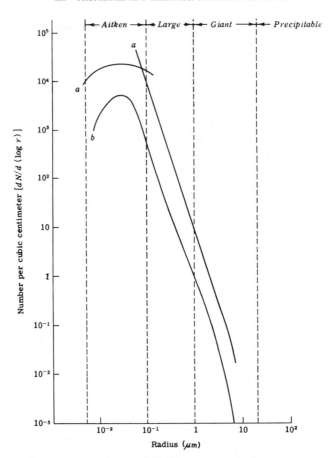

FIG. 3.4.   Size distribution of aerosol particles for (a) Frankfurt, Germany and (b) Zugspitze, Germany (after C. E. Junge, *Advances in Geophys.* **4,** 8, 1958).

Evidence concerning the composition and origin of Aitken nuclei is largely indirect. It is presumed that they are formed for the most part from gases produced by combustion, either through gas-to-particle conversion of supersaturated vapors, or through chemical reactions in cloud droplets involving trace gases such as SO_2, NH_3, and Cl_2. Natural processes, including volcanic emissions and decomposition of organic matter, also are believed to contribute. Concentrations in air near the earth's surface vary from the order of 10^5 cm^{-3} or more in heavily industrial areas to 10^3 cm^{-3} or less over the oceans. Concentrations decrease with height to the order of 10^2 cm^{-3} in the midtroposphere and perhaps 10 cm^{-3} in the lower stratosphere. Occasionally counts near the surface as low as 2 cm^{-3} have

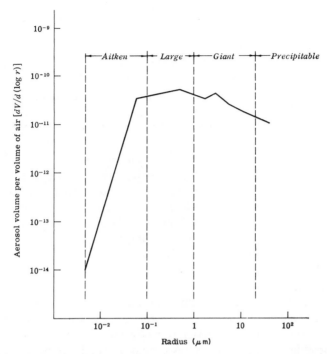

FIG. 3.5. Volume distribution of aerosol particles for Frankfurt, Germany (after C. E. Junge, *Advances in Geophys.* **4**, 8, 1958).

been reported, possibly as a result of descent of stratospheric air which occurs in middle latitude cyclones. Due to their very small size, Aitken nuclei do not compete effectively for water vapor with larger nuclei, and therefore they usually do not constitute the dominant source of condensation nuclei for cloud droplets. They may coagulate to form larger particles, and they may adhere to larger particles.

Large and giant nuclei vary in concentration from about 10^2 cm^{-3} in industrial areas to 1–10 cm^{-3} over the sea. Concentrations typically vary with size according to the empirical and approximate relation

$$\frac{dN}{d(\log r)} = \frac{A}{r^3}$$

where A is a constant. This implies that volume of aerosol per unit volume of air is independent of particle size, as is indicated approximately in Fig. 3.5. In industrial cities large nuclei appear to be mostly ammonium sulfate, whereas giant nuclei contain either considerable sulfuric acid (of industrial origin) or chlorides (from the ocean). Over continents, windblown dust from the land and pollens and spores probably are significant sources. Over oceans, evaporation of droplets originating in breaking waves accounts for the formation of airborne sodium chloride particles in concentrations of $1-10$ cm^{-3}. As a result of these differences in nucleus concentrations, cloud droplet concentrations vary widely; in air of marine origin concentrations less than 10^2 cm^{-3} are normal, whereas in continental clouds concentrations of 10^2-10^3 cm^{-3} occur frequently. Droplet sizes tend to be greater in marine clouds than in continental clouds.

When major volcanic eruptions occur, nucleus concentrations at high levels in the atmosphere may be greatly increased for several years after the event with consequent influence on cloud processes and climate.

When cooling of air brings about supersaturation, the largest hygroscopic nuclei are, of course, the first to act as cloud condensation nuclei, and they are responsible for development of the largest droplets. Although few in number, the giant nuclei play an important role in creation of the broad drop size spectrum which, as is shown later in Section 3.13, is necessary for efficient coalescence of droplets and for the formation of precipitation from water clouds.

Measurements of cloud condensation nuclei are, as yet, too limited in number and spatial distribution to permit either a detailed chronology of changes in average concentrations over large areas or a synoptic mapping of the global distribution of nuclei. A great deal remains to be learned about possible relationships between cloud condensation nucleus concentrations and types and changes in clouds and precipitation.

3.6   Growth of Droplets by Condensation

A droplet will grow by condensation if the vapor pressure surrounding it exceeds the equilibrium vapor pressure at the surface of the droplet. As it grows, the latent heat of condensation tends to raise the temperature of the drop, and consequently also the equilibrium vapor pressure. If the droplet temperature is higher than the air temperature surrounding the droplet, heat is conducted away from the droplet. These processes are represented by diffusion equations developed in the following discussion.

The diffusive flux of vapor of mass (m) toward the droplet may be expressed from Eq. (2.75)

$$\frac{dm}{dt} = 4\pi \frac{r^2 D}{R_w T} \frac{de}{dr}$$

upon neglecting the small temperature term. Here D represents the diffusion coefficient of water vapor in air. Under steady state conditions the flux will be the same across any sphere concentric with the center of the droplet, that is, dm/dt is independent of r. Therefore, upon integrating from the droplet radius r_0 to the radius r_∞ where the vapor pressure is unaffected by diffusion to the droplet, we get

$$\frac{dm}{dt} = \frac{4\pi D r_0}{R_w T}(e_\infty - e_0)$$

where $1/r_\infty$ has been neglected compared to $1/r_0$. This is equivalent to

$$r_0 \frac{dr_0}{dt} = \frac{D(e_\infty - e_0)}{\rho_w R_w T} \tag{3.5}$$

where ρ_w represents the density of liquid water.

Latent heat liberated by condensation at the droplet surface is diffused away from the drop according to Eq. (2.70) in the form

$$L\frac{dm}{dt} = -4\pi r_0^2 \lambda \frac{dT}{dr}$$

Upon combining this equation with Eq. (3.5) and integrating

$$T_0 - T_\infty = \frac{DL(e_\infty - e_0)}{\lambda R_w T} \tag{3.6}$$

where $T \approx T_0 \approx T_\infty$.

Equations (3.4)–(3.6) and the Clausius–Clapeyron equation (2.88) are a set of four simultaneous equations in the variables e_0, e_s, T_0, and r_0. If the mass of the solute and the vapor pressure and temperature of the environment are specified, the three unknowns can be calculated for any value of r_0. Then r_0 can be calculated as a function of time by numerical integration. Sample calculations illustrated in Fig. 3.6 show that an NaCl nucleus of about 0.1 μm radius (6×10^{-15} g) in an environment of 0.05% supersaturation grows to a radius of 1 μm in about 1 s, to 8 μm in 10 min, and to 20 μm in 1 h. Subsequent growth by condensation is quite slow.

For droplets larger than about 3 μm the effects of curvature and of dissolved salt are not important, and Eqs. (3.5), (3.6), and (2.88) yield

$$\int_{3\,\mu m}^{r_0} r_0 \, dr_0 = At \tag{3.7}$$

where

$$A \equiv \frac{\lambda D R_w T_\infty^2 (e_\infty - e_{\infty s})}{\rho_w D L^2 e_{\infty s} + k R_w^2 T_\infty^3}$$

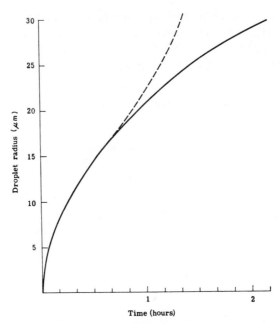

FIG. 3.6.   *Solid:* Radius of droplet as a function of time for a NaCl nucleus of 6×10^{-15} g, supersaturation of 0.05% temperature of 273 K, and pressure of 80 kPa as calculated from Eqs. (2.88), (3.4), (3.5), and (3.6). *Dashed:* Estimated radius of droplet as a function of time for growth by coalescence and condensation.

and A has been taken as constant. Upon integrating Eq. (3.7) the radius is seen to increase as the square root of time.

When many droplets are present, growth occurs at different rates for the different cloud condensation nuclei and different droplet sizes. Droplets compete for water vapor, so that the supersaturation becomes a dependent variable to be evaluated along with the individual droplet sizes. If the cloud is "static," a population of droplets may quickly reach equilibrium. However, if there is an updraft, water vapor is supplied to the cloud, and the droplets will grow and the size distribution will evolve as they are carried in the updraft. The problem requires the simultaneous solution of equations representing the vapor pressure of the droplets (3.4), the rate of growth of droplets by condensation (3.5), the conduction of heat from the droplets (3.6), the rate of change of temperature following the rising air [derived from Eq. (2.99)], the conservation of total water substance, and an equation for liquid water in terms of the numbers of droplets of each initial category.

The result of a calculation is illustrated in Fig. 3.7. In this example ten groups of salt nuclei were assumed to range from 5×10^{-16} g to 10^{-11} g

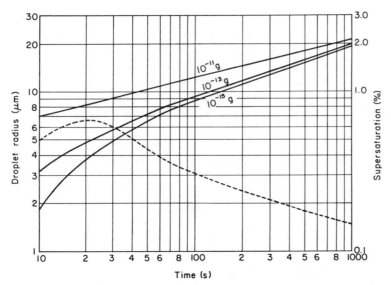

FIG. 3.7. Growth curves for three classes of nuclei and supersaturation for an arbitrary set of conditions as described in the text. The dashed curve describes supersaturation as a function of time (after B. J. Mason, *Clouds, Rain and Rainmaking*, 2nd ed., p. 55. Cambridge Univ. Press, 1975).

with the concentrations inversely proportional to their masses and ranging from 3.8×10^4 g^{-1} to 1.9 g^{-1}. A steady updraft of 1 m s^{-1} was assumed, and growth curves for only three of the ten classes of nuclei are shown. The horizontal scale represents both the time from introduction of the nuclei at the bottom of the cloud and the height of the droplets above the base of the cloud. Figure 3.7 shows that as the air rises and cools pseudo-adiabatically the supersaturation increases to a maximum of about 0.66% after 20 s. Subsequently, water vapor condenses on the droplets faster than it becomes available through cooling; this is indicated by the decrease in supersaturation with time and height. For the first few seconds of ascent, before supersaturation reaches the critical value, the rate of growth of droplets is slow. This is not shown in Fig. 3.7. Subsequently, those nuclei which reach critical size grow rapidly while supersaturation is increasing, then more slowly as supersaturation falls. In consequence, the droplet size distribution tends to become more uniform with time and height. Calculations of this kind using realistic cloud condensation nucleus concentrations and vertical velocities show that supersaturation normally reaches a maximum of only a few tenths of one percent. The theory of growth by condensation as outlined here describes growth of cloud droplets reasonably well for the first several hundred seconds or to droplet radii of about 10 μm.

However, this model predicts too narrow a drop-size spectrum (distribution of number of droplets as a function of size). More realistic results have been achieved by considering the effects of exchange of heat, water vapor, and cloud condensation nuclei between the cloud and the drier environment by turbulent air motions. Models of cumulus convection cells which rise through the residues of previously evaporated cells have been shown to produce bimodal spectra with significant numbers of droplets larger than 20 μm.

3.7 Growth of Droplets by Collision and Coalescence

Growth of droplets by condensation is not likely to account for the development of full grown raindrops because, as indicated by Figs. 3.6 and 3.7, the time required to reach precipitable size ($\sim 100 \mu$m) would exceed the normal lifetime of most clouds. In this section the process of growth by collision and coalescence with other droplets of different sizes and different fall velocities will be examined.

A falling droplet is acted on by the force of gravity and by the friction or drag force exerted by the air. The downward force due to gravity is expressed by

$$\tfrac{4}{3}\pi r^3 (\rho_w - \rho)g$$

where ρ represents air density and r represents the radius of the droplet. Intuition suggests that the drag force for small droplets must increase with viscosity, with radius of the droplet, and with fall speed. This is borne out by Stokes's law which states that the drag force is represented by

$$6\pi\eta r w_t$$

where η represents the dynamic viscosity of air and w_t the fall velocity of the droplet. Stokes's law holds accurately for droplets of radii less than 50 μm. Stokes's law is derived in Joos, "Theoretical Physics," p. 206 (see Bibliography), and elsewhere. When the droplet is falling at its terminal (constant) velocity, the drag force and the force of gravity are equal; and upon neglecting ρ compared to ρ_w, the terminal velocity is given by

$$w_t = \tfrac{2}{9}\frac{r^2\rho_w g}{\eta} \tag{3.8}$$

For droplets with radius larger than 50 μm, Eq. (3.8) overestimates the fall velocity. The computation then becomes considerably more complicated. Theoretical and experimental values are given in Fig. 3.8.

As a relatively large falling drop approaches a droplet, the air flows

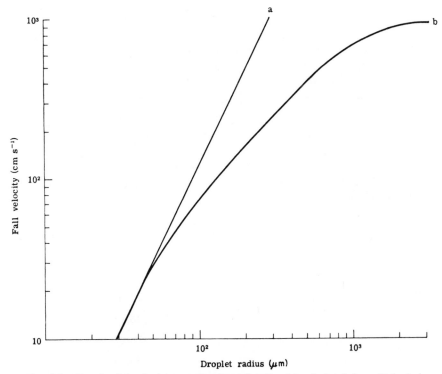

FIG. 3.8.   Droplet fall velocity as a function of radius (a) calculated from Stokes's law (Eq. 3.8), (b) observed by R. Gunn and G. D. Kinzer, *J. Meteorol.* **6**, 243, 1949.

around the large drop as indicated in Fig. 3.9, tending to carry the smaller droplet with it. However, the inertia of the droplet causes it to cross the streamlines in regions where they are curved, and consequently the trajectory of the droplet lies between a straight line and a streamline. An exercise is provided in Problem 4. We define as the *collision efficiency* (ε_c) the proportion of all the small droplets in the volume swept out by the falling drop which collide with the drop. It is easily shown that the radius of the circle which encloses all droplets which collide with the larger drop can be expressed by $\sqrt{\varepsilon_c}\, r_0$, where the radius of the droplet is neglected compared to r_0. Collision efficiencies can be calculated from the equations of viscous flow under certain simplifying assumptions. Results of such calculations are shown in Fig. 3.10 as a function of ratio of droplet to drop sizes. The following properties are evident: (a) efficiencies are very small for very small ratios, (b) the collecting drop must be at least about 30 μm in radius in order that collision efficiencies can approach unity, (c) as drop size in-

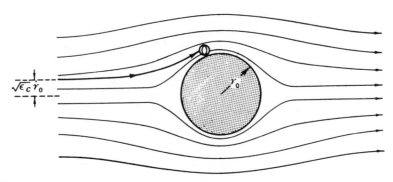

FIG. 3.9.   Streamlines for air flowing slowly around a sphere and the limiting trajectory of the droplets which collide with the sphere.

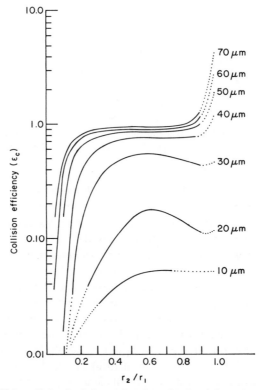

FIG. 3.10.   Collision efficiencies (ε_c) as a function of ratio of sizes of droplets to collecting drop (r_2/r_1) for various size of collecting drop (after J. D. Klett, and W. H. Davis, *J. Atmos. Sci.* **30**, 112, 1973).

creases collision efficiency approaches unity for a wide range of droplet sizes, (d) collision efficiencies for drops of the same size falling at about the same speed may exceed unity (they attract each other due to acceleration of air between them).

Not all droplets that collide with the larger drop adhere to it. The ratio of the number of droplets that adhere to the number of collisions, called the *coalescence efficiency*, can be determined by comparing laboratory measurements of *collection efficiency* with calculated collision efficiencies. These measurements show that coalescence efficiencies are near unity for drop size ratios less than about 0.1, but they fall rapidly as the ratio approaches unity. Coalescence can be enhanced in a strong electric field. Model calculations which treat growth of cloud particles and cloud electrification simultaneously indicate that electrical processes may influence precipitation strongly.† This is further discussed in Section 3.18.

Since collection efficiency (product of collision and coalescence efficiencies) increases with radius of the collecting drop, and relative velocity increases with radius, rate of growth by collection proceeds more and more rapidly as size increases. Growth by collection is illustrated by the dashed curve in Fig. 3.6 which suggests that at radii greater than about 25 μm rate of growth by collection exceeds rate of growth by condensation.

Growth of droplets by coalescence has been calculated for an initial cloud droplet population by introducing more or less simplified models of the growth process. In the *continuous* models a few uniformly large collector drops are considered to fall relative to a cloud of fine droplets and to sweep up a certain proportion of all droplets in their paths. The rate of growth of mass (m) for this process can be represented by

$$\frac{dm}{dt} = \pi r_0^2 \varepsilon w_d' \rho_d \tag{3.9}$$

where r_0 represents radius of the collector drop, ε the collection efficiency, w_d' the difference in fall velocity of the collecting drop and the small droplets, and ρ_d the mass of liquid water contained in the small droplets per unit volume of air. However, the continuous models are unrealistic in starting with just two sizes of droplets and in ignoring coalescences between small droplets.

More realistic models consider that droplet sizes are initially distributed statistically in a known manner. Collisions and coalescences can occur between droplets of various sizes, and these are discrete events. Because by chance some droplets will experience more than the average number of coalescences, these statistically fortunate droplets will grow faster than the

† W. D. Scott, and Z. Levin, *J. Atmos. Sci.* **32**, 1814, 1975.

average. As a result, the drop-size spectrum will be broadened and the largest droplets will be likely to grow still more rapidly. Models which account for these probabilistic aspects of cloud development are called *stochastic* models; they require numerical calculations using large capacity computers. The results depend strongly on the initial drop-size spectrum; a broad spectrum of fairly large droplets, typical of marine clouds, may require only 10 min or so to produce drops of 1 mm radius, whereas a cloud of equal water content having a narrow spectrum of smaller droplets, typical of continental clouds, may require much longer to produce 1 mm drops by coalescence. Thus the initial drop-size spectrum is an important factor in determining rate of growth by the coalescence process. Observations of drop-size spectra are made rarely, so that growth rates cannot be reliably calculated for most clouds.

In a cloud initially containing a narrow drop-size spectrum condensation

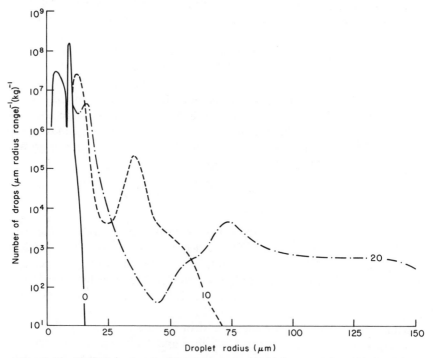

FIG. 3.11.   Evolution of a model cumulus cloud initially containing 530 droplets/mg. Supersaturation of 0.22% and an entrainment parameter of $9 \times 10^{-4}\,\mathrm{s}^{-1}$ have been used. The three curves describe the drop-size spectra at the initial time (0) and after 10 and 20 min (after P. R. Jonas and B. J. Mason, *Quart. J. Roy. Met. Soc.* **100,** 286, 1974).

growth beyond a size of about 20 μm is extremely slow, while growth by coalescence for droplets smaller than about 40 μm is ineffective. In order to bridge this gap and to account for the observed growth of cloud droplets, models have been developed which treat condensation and coalescence simultaneously. An example for a typical continental cumulus cloud is shown in Fig. 3.11. The results show that an initial cloud containing only droplets less than about 20 μm in radius can evolve through condensation and coalescence into a cloud containing precipitation-size drops (> 100 μm) in about 20 min. The results are not sensitive to the assumed entrainment parameter (proportional rate of change of total mass of cloud).

3.8 Supercooling of Droplets

Cloud droplets which are cooled to temperatures below 0° C commonly remain as liquid droplets and often do not freeze until temperatures of −20° C or lower are reached. Observations show that at temperatures of −14° C, about 20% of clouds contain only liquid droplets and at −8° C about 50% contain only liquid droplets.

In the laboratory very pure water droplets can be cooled to about −40° C before freezing occurs. The formation of ice crystals from supercooled droplets at this temperature is called *homogeneous* or *spontaneous* nucleation. At temperatures above −40° C a droplet freezes only if it contains a foreign particle, called an *ice nucleus*. If a cloud of droplets is cooled gradually below 0° C, ice crystals may appear at a variety of temperatures, indicating the presence of ice nuclei effective at different temperatures. Initiation of freezing by an ice nucleus is called *heterogeneous* nucleation.

Freezing of droplets may be explained on the basis of statistical considerations. Water molecules in thermal agitation may come into temporary alignment similar to that of an ice crystal. Such molecular aggregates may grow but also may be destroyed by random molecular motions. If an aggregate happens to grow to a size such that it is stable in the presence of thermal agitation, the whole droplet quickly freezes. The probability of growth of an aggregate to this critical size increases as temperature decreases. The presence of a foreign particle makes the initial growth of the aggregate more probable by attracting a surface layer of water molecules on which the ice crystal lattice can form more readily than in the interior of the liquid. Freezing of a droplet requires that only one aggregate reach critical size, and therefore probability of freezing increases with volume. The probability that at least one foreign particle is present also increases with volume, so freezing temperature depends markedly on volume.

3.9   Ice Nuclei

Ice nuclei make up a very small part of the total atmospheric aerosol. Their concentration can be determined by observing ice crystal scintillations in a light beam in a refrigerated box, or expansion chamber, or on a chilled metal plate. By these methods measurements have been made which show that ice nuclei become effective at different temperatures depending on their chemical composition. For example, dust particles of montmorillonite are effective at about $-25°$ C, gypsum at $-16°$ C, volcanic ash at $-13°$ C, kaolinite at $-9°$ C, and covellite at $-5°$ C. Observations in natural air near the ground show that ice nuclei effective at temperatures as high as $-5°$ C are only rarely present and in concentrations of only 1 m^{-3} or so. The number of effective ice nuclei increases with lowered temperature, and at $-20°$ C concentrations of 10^2–10^3 m^{-3} are often present. At temperatures below $-25°$ C more than 10^6 ice nuclei m^{-3} have been observed. Concentrations at one temperature vary greatly in space and time. This natural variation in the concentration of ice nuclei is the basis for most efforts which have been made to increase precipitation by "cloud seeding." This subject is discussed in Section 3.14.

Ice nuclei may be effective in three different processes. An ice nucleus embedded within a droplet may activate freezing of the droplet, or an ice nucleus outside the droplet may activate freezing by contact with the droplet surface. Ice crystals also may grow by direct sublimation or deposition of water vapor onto the surface of an ice nucleus.

Ice nuclei having a hexagonal crystalline structure and molecular structure similar to that of ice tend to be effective in initiating ice formation. Silver iodide is effective at $-4°$ C, while lead iodide and cupric sulfide are effective at $-6°$ C. In addition, many organic substances are effective as ice nuclei at temperatures only a few degrees below $0°$ C. Experiments have shown that various clay particles can act as natural nuclei and that kaolinite (aluminum silicate), which has a hexagonal crystalline structure, is probably the most important inorganic source of natural ice nuclei at temperatures above $-15°$ C.

When an ice crystal is evaporated from a nucleus and then the same nucleus is tested in a cold chamber, crystallization sometimes occurs at temperatures just below $0°$ C, that is at a temperature higher than the original activation temperature for the nucleus. This may result from the crystal retaining a surface film of ice or from small particles of ice which, remaining in crevices in the nucleus during evaporation, serve as ice nuclei during the subsequent cold cycle. Such nuclei are called "trained" ice nuclei, and they are likely to be present among the nuclei which are left when cirrus clouds evaporate.

It has been suggested that trained ice nuclei formed in this way may be very important in initiating freezing in supercooled clouds.

The concentrations of ice nuclei observed in some supercooled layer clouds are two to three orders of magnitude smaller than would be required to account for the fall of snow from these systems if one ice nucleus were required for each crystal. Also, the concentrations of ice crystals observed in older supercooled cumulus clouds are often two or more orders of magnitude larger than the measured concentrations of ice nuclei. These discrepancies between the concentrations of ice nuclei and ice crystals indicate that an effective process of ice multiplication must be active in clouds. Laboratory experiments help to elucidate how the number of ice crystals may multiply. When supercooled droplets collide with an ice surface, the droplets may adhere to the ice and freeze. In this process, which is referred to as *riming*, ice splinters are formed and are carried away in the air stream. Experiments show that a pronounced maximum of splinter formation occurs at a temperature of $-5°$ C and impact velocity of 2.5 m s^{-1} for droplets larger than 12 μm in radius. At $-5°$ C ice crystals grow as fragile ice needles or hollow columns which may be readily broken by an impacting droplet. These experiments suggest that ice multiplication in clouds may be critically dependent on occurrence of a favorable combination of temperature and particle size distribution.

3.10 Equilibrium Vapor Pressures over Ice and Water

From the molecular point of view, the fundamental distinction between the liquid and solid phases is a difference in energy level, that is, a molecule which makes up part of an ice crystal lattice is tightly held by intermolecular forces and is therefore in a state of low potential energy. A certain quantity of energy, called the latent heat of melting, is required to free the molecule from this low-energy state or to bring about the phase change from ice to liquid water. Still greater energy, the latent heat of vaporization, is required to bring about the phase change from liquid to water vapor. The latent heats have the following values near $0°$ C:

Melting: $\qquad L_{iw} = 3.34 \times 10^5$ J kg^{-1}
Vaporization: $\qquad L = (25.00 - 0.02274t°C) \times 10^5$ J kg^{-1}
Sublimation: $\qquad L_{iv} = (28.34 - 0.00149t°C) \times 10^5$ J kg^{-1}

It follows from the conservation of energy that at the triple point $L_{iv} = L + L_{iw}$.

The Clausius–Clapeyron equation shows that the equilibrium vapor pressure depends on the latent heat; it follows that the equilibrium saturation vapor pressure over ice differs from that over water at the same temperature. To calculate this difference, the Clausius–Clapeyron equation [Eq. (2.88)] may be integrated from the triple point, where the equilibrium saturation vapor pressures are equal, to the temperature T. The result is

$$e_s = e_0 \exp\left(\frac{T - T_0}{R_w T T_0} L\right) \tag{3.10}$$

where e_s represents the equilibrium saturation vapor pressure at temperature T, e_0 and T_0 the vapor pressure and temperature at the triple point, and L the latent heat. If a similar integral is formed for the ice–vapor and liquid–water transition, it follows that

$$e_{sl} - e_{si} = e_0\left[\exp\left(\frac{T - T_0}{R_w T T_0} L\right) - \exp\left(\frac{T - T_0}{R_w T T_0} L_{iv}\right)\right] \tag{3.11}$$

Equation (3.11) is illustrated in Fig. 3.12. Maximum difference in vapor

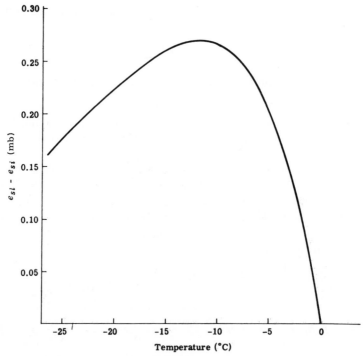

FIG. 3.12.   Difference between saturation vapor pressures over water and over ice as a function of temperature calculated from Eq. (3.11).

pressure occurs at about $-12°$ C. Here the difference amounts to 27 Pa (0.27 mb); this represents a supersaturation of 12.5% with respect to ice.

Now consider what happens in a cloud composed of supercooled water droplets when a few ice crystals are formed in it. The vapor pressure of the air in the cloud is likely to be within 0.1% of the equilibrium vapor pressure of the droplets, but, as has just been calculated, it may be higher than the equilibrium vapor pressure of the ice crystals by as much as 27 Pa. Consequently, the ice crystals grow rapidly, depleting the water vapor in the cloud, thereby causing the supercooled droplets to evaporate. In this way water substance is rapidly transferred from supercooled droplets to ice crystals.

3.11 Growth of Ice Crystals

The growth of ice crystals depends on diffusion of water vapor and conduction of heat just as does the growth of liquid droplets. However, ice crystal growth is simpler in that it does not directly involve change of equilibrium vapor pressure with size of the crystal [as occurs through Eq. (3.1) for droplets]; on the other hand, the complex geometry of the crystal introduces some difficulty into the calculation.

The surface of the crystal may be considered to have a uniform temperature and therefore a uniform equilibrium vapor pressure, and the vapor pressure at infinite distance (several crystal diameters) from the crystal is assumed uniform. The vapor pressure or vapor density in the neighborhood of the crystal may be represented by surfaces which tend to follow the contours of the crystal, but beyond the neighborhood of the crystal these surfaces approach the spherical shape as illustrated in Fig. 3.13.

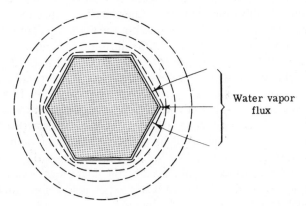

Water vapor flux

FIG. 3.13. Diffusion of water vapor toward one corner of a hexagonal ice crystal. Dashed lines represent surfaces of constant vapor density and arrows represent direction of flux.

Flux of water vapor by diffusion occurs in the direction normal to the surfaces of constant vapor density and with the magnitude given by $-D\,\partial p_v/\partial n$ where n represents the coordinate measured normal to lines of constant ρ_v. Consequently, in the neighborhood of a sharp point, vapor diffuses toward the point from many directions with the result that ice may accumulate more rapidly here than on flat surfaces. In spite of complexities of this sort, the diffusion occurring through a spherical surface of constant vapor density shown in Fig. 3.13 may be calculated using Eq. (2.78). The radius of the sphere is found to be represented by

$$\frac{R_w T\; dm/dt}{4\pi D(e_\infty - e_0)} = r \tag{3.12}$$

Evidently, the ratio of flux of vapor to vapor density increment is independent of the vapor density and is a characteristic of the sphere. This ratio may be called the *diffusion capacitance* of the sphere. Now consider the situation just outside the crystal. The diffusion capacitance of the crystal depends in a complicated way on the size and shape of the crystal, but in spite of this, the water vapor flux is the same as that through the spherical surface, and the capacitance may be defined by the ratio of rate of water vapor flux to $4\pi D$ times the difference in vapor density between environment and the crystal surface. Now if the capacitance of the crystal is represented by C, Eq. (3.12) may be written

$$\frac{dm}{dt} = 4\pi \frac{DC}{R_w T}(e_\infty - e_0) \tag{3.13}$$

The problem of crystal growth by diffusion is an analogue of the problem of the charging of an electrical capacitor in which electrical potential replaces the vapor density increment and electrical charge replaces the ratio of vapor flux to diffusion coefficient. The electrical capacitance of a complex geometrical shape can be measured in the laboratory from the charge-to-potential ratio just as, in principle, the diffusion capacitance can be measured from the flux to vapor pressure ratio given by Eq. (3.12). Therefore, existence of the electrical analogue to the diffusion problem makes it possible to determine the diffusion capacitance of any crystal by measuring the electrical capacitance of a conductor of the same shape and size.

Equation (3.13) permits calculation of the rate of growth of a crystal for any ambient vapor density if its shape and size and vapor density at the crystal surface are known. The latter, of course, depends on the temperature of the surface which is influenced by the latent heat of fusion released at the surface. Inward diffusion of water vapor is therefore limited by the rate at which latent heat can be dissipated.

If it is assumed that dissipation of the latent heat occurs solely by conduction into the air surrounding the crystal and that the crystal surface is at uniform temperature, the heat conduction problem becomes an analogue of the diffusion problem. Thus

$$L_{iv} \frac{dm}{dt} = 4\pi C\lambda(T_0 - T_\infty) \tag{3.14}$$

Using Eqs. (2.88), (3.13), and (3.14) the rate of crystal growth may be derived by eliminating T_0 and e_0 in the same way that they were eliminated in the case of droplet growth. The result may be written in the form

$$\frac{dm}{dt} = \frac{4\pi CD\lambda R_w T_\infty^2 (e_\infty - e_{si\infty})}{L_{iv}^2 e_{si\infty} D + R_w^2 T_\infty^3 \lambda} \tag{3.15}$$

where e_∞ represents ambient vapor pressure and $e_{si\infty}$ saturation vapor pressure over ice at the ambient temperature. Equation (3.15) shows that the rate of growth is directly proportional to the supersaturation of water vapor with respect to ice. It follows that Eq. (3.15) also describes the evaporation of ice crystals in unsaturated air. An example is given in Problem 5. The rate of growth calculated from Eq. (3.15) is very much greater than the rate of growth of water droplets, primarily due to the large supersaturation with respect to ice. A sample comparison shows that a $20\,\mu m$ water drop in air supersaturated by 0.05% requires about 500 times as long to grow to a droplet of $50\text{-}\mu m$ radius as does an ice crystal of equivalent mass at $-12°$ C. However, the rate of deposition of vapor on ice crystals is sufficient only to produce quite small rain drops in the time interval usually available to a falling ice particle.

3.12 Structure of Ice Crystals

The molecular structure of ice crystals has been determined by X-ray and neutron diffraction experiments. These results show that at atmospheric temperatures each oxygen atom is surrounded by four oxygen atoms forming a tetrahedron. These tetrahedrons are joined to form a hexagonal lattice with oxygen atoms alternatively above and below the base plane. Each oxygen is bonded to two hydrogens, and each hydrogen is bonded to two oxygens. Crystals may grow in various ways, for example, the hexagonal base may grow in two dimensions to form a hexagonal plate, or it may grow in the direction normal to the base to form a hexagonal prism. Mode of development is influenced by the external diffusion field, but it also probably depends on accidental arrangements of molecules and on impurities in the crystal lattice. The chief distinctive forms of ice crystals found abundantly in clouds are illustrated in Fig. 3.14.

Experiment sheds valuable light on the phenomenon. If a thread is suspended in a chamber in which the air temperature increases upward from, say −30° to 0° C, and water vapor is introduced into the chamber, ice crystals form in a very curious fashion on the thread. Near the top of the chamber hexagonal plates form, further down there are prisms and needles, still further down plates and dendrites, and finally prisms again near the bottom. The separation between the various crystal forms is quite sharp and corresponds closely to the following temperature ranges:

0 to −3° C	thin hexagonal plates
−3 to −5° C	needles
−5 to −8° C	hollow prismatic columns
−8 to −12° C	hexagonal sector plates (symmetrical assembly of six hexagonal plates)
−12 to −16° C	dendrites (crystals with fern-shaped arms)
−16 to −25° C	hexagonal sector plates
< −25° C	hollow prisms

Additional experiments show that this full array of crystal forms occurs only

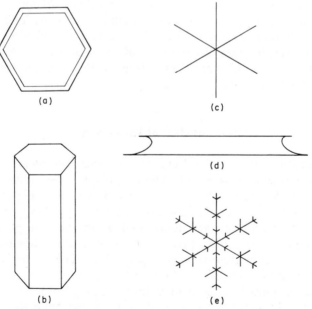

FIG. 3.14.   Typical forms of ice crystals. (a) Hexagonal plate, (b) hexagonal column, (c) stellar crystals, (d) needle, (e) dendrite.

when the humidity corresponds at least roughly to saturation with respect to water. At humidities between saturation with respect to ice and saturation with respect to water the array of crystal forms is more limited. In this case needles, sector plates, and dendrites do not appear; however, the transitions indicated at about $-3°$, $-8°$, and $-25°$ C are observed.

The various forms of ice crystals produced in the laboratory are found in the atmosphere, and it is possible to reconstruct from the observed structure of an ice particle something of its history as it grew within the cloud.

3.13 Precipitation

In the preceding sections various microphysical processes responsible for the growth of cloud particles have been discussed. In trying to understand the relation of these processes to precipitation, it is important to realize that the microphysical processes which govern the behavior of cloud particles are dependent on the large scale or macrophysical processes, therefore an account of factors which govern precipitation is likely to be a much more complex affair than is an account of the microphysical factors. However, it is possible to give a reasonable explanation of the formation of precipitation with the aid of calculations on simple models of the air motion.

The liquid or solid water contained in a cloud at a particular time is rarely sufficient to provide more than a trace of precipitation, as may be verified by solving Problem 6. Obviously, in order to produce a significant amount of precipitation an updraft of air must provide a continuous and plentiful supply of moisture to the cloud. Clouds which are supplied with adequate water vapor must generate precipitation-size particles by the processes described in earlier sections. In this connection it is worth noting that the volume of a 10-μm cloud droplet is 10^6 times that of a 0.1-μm cloud droplet. The cloud processes discussed in this chapter are responsible for growth by these large factors in time intervals in many cases ranging from a few tens of minutes to a few hours.

Condensation on small liquid droplets is important in the early stages of droplet growth, but acting alone it cannot produce raindrops. Ice crystals surrounded by supercooled water droplets may grow rapidly by vapor deposition, and under favorable conditions they may accumulate sufficient mass to become small raindrops after melting. Wegener in 1911 had already recognized the rapid growth of ice particles, but Bergeron in 1935 was the first to hold this mechanism responsible for the formation of rain. For some time after Bergeron's theory was published it was believed that precipitation could only be initiated in clouds which extend considerably above the height of melting. However, studies have shown that precipitation often falls from clouds whose tops are warmer than 0° C. In these cases precipitation can form only through the coalescence process.

In clouds which extend well above the height of melting, both processes are present. In the supercooled region of the cloud collisions between ice crystals and droplets may result in the collection and freezing of the droplets. This process of *riming* is responsible for the growth of *graupel* or *soft hail*. Individual ice crystals also may aggregate into snowflakes which are often observed to be formed of 10–100 or more crystals. Below the layer of melting growth may continue by coalescence of drops.

The Wegener–Bergeron mechanism is likely to be most effective in clouds with relatively weak updrafts (stratiform clouds) and in these cases may produce light precipitation of small snowflakes or drizzle. The coalescence process is likely to be most important in clouds with relatively large updrafts (cumiliform clouds), in these cases large raindrops, or possibly hail, may occur. And in many cases both processes may be significant. In the following parts of this section for simplicity methods for calculating rate of precipitation by the two mechanisms will be discussed separately.

The Wegener–Bergeron Mechanism

Consider a stratiform cloud in which a continuous supply of water vapor is provided by an updraft of the order of 0.1 m s^{-1}, and imagine that the updraft causes the temperature in the cloud to decrease at a uniform rate. When the temperature is slightly below 0° C condensation takes place on supercooled droplets but at a rather slow rate. At a temperature between $-10°$ and $-20°$ C, ice nuclei present in the cloud may initiate ice crystal formation. Because of their rapid growth at the expense of the supercooled water droplets, the ice particles quickly attain sufficient size to fall with respect to the surrounding air. The size attained during their fall through the cloud depends on the number of ice particles formed. If many particles compete, the supercooled droplets evaporate altogether, and the vapor pressure attains a value somewhere between saturation pressure over ice and saturation vapor pressure over water. This reduces the rate of growth of the ice crystals. If only a few ice crystals are formed, they grow at a rapid rate.

As the crystal falls, it usually enters air which is increasingly warm and moist. Therefore, a crystal at a temperature lower than $-12°$ C usually experiences an increased rate of growth by sublimation as it falls toward warmer air, whereas for a crystal which is warmer than $-12°$ C the reverse is likely to be true.

The rate of vertical displacement of the ice crystal may be expressed by

$$\frac{dz}{dt} = w - w_t \qquad (3.16)$$

where w represents the updraft velocity of the air in the cloud and w_t the terminal velocity of the ice crystal. Typical terminal velocities of ice crystals a few millimeters in diameter are about 0.5 m s^{-1}, while terminal velocities for rimed crystals and graupel are $1-2 \text{ m s}^{-1}$. If both w and w_t are considered constant, then Eq. (3.16) can be integrated to show, for example, that in a 0.3 m s^{-1} updraft an ice crystal with terminal velocity of 1 m s^{-1} would require about 23 min to fall 1 km. If the temperature and the vapor pressure are known through the entire cloud layer, then Eq. (3.15) may be integrated numerically to yield the mass of the crystal as a function of time. Before the crystal reaches the ground it may have melted into a water droplet of the same mass. During its fall some evaporation may take place because the air under the cloud may not be saturated. For this reason drops may decrease in size between the cloud base and the ground, and in many cases do not reach the ground at all. Neglecting evaporation between cloud and ground, the rate of precipitation can be determined if the updraft and the thickness of the cloud are known. The liquid water concentration (ρ_d) can be determined from the pseudo-adiabatic chart by assuming that it is equal to the decrease in saturation mixing ratio which occurs for ascent along the wet adiabatic lines from the cloud base. This tends to overestimate liquid water concentration because it does not take into account entrainment of unsaturated air from the sides of the clouds. Problem 7 provides an exercise.

Calculations using Eq. (3.15) indicate that crystals may attain a plate diameter or prism length of one to two millimeters in about one hour. If such a crystal falls below the melting level, it will produce a drop with radius of only about 0.25 mm. In most cases growth to raindrop size (1 to 2 mm) requires further growth by riming or by aggregation of crystals.

The Coalescence Mechanism

For simplicity the following discussion will be limited to the growth of liquid drops by coalescence in a cloud warmer than $0°$ C, though it should be recognized that frozen or partly frozen particles also grow by similar processes. It has been pointed out in Section 3.7 that droplets of $100 \mu m$ or larger in radius can be produced in cumulus clouds in about 20 min by the combined processes of condensation and coalescence. Here the emphasis is on growth to the size of raindrops occurring in shower clouds, typically 1 or 2 mm. Droplets may be visualized as introduced at the base of the cumulus cloud; as they are carried upward in the ascending air current, they grow by condensation and coalescence as described earlier. Ultimately, they reach a size such that their terminal velocity exceeds the updraft velocity of the air. They then begin to fall through the cloud and continue to grow until they reach the bottom of the cloud or until they reach a size such that they become mechanically unstable and break up into smaller drops. Drops larger than about 2.5 mm in radius are likely to break up.

In order to consider these processes in more detail, Eq. (3.9) may be written for the growing drop in the form

$$\frac{dr_0}{dt} = \frac{\varepsilon w'_d \rho_d}{4\rho_w} \tag{3.17}$$

And by combining Eqs. (3.16) and (3.17) the radius can be expressed as a function of height by

$$\frac{dr_0}{dz} = \frac{\varepsilon w'_d \rho_d}{4\rho_w(w - w_t)} \tag{3.18}$$

If we assume that the collecting drop interacts only with cloud droplets of negligible fall speed, then the relative velocity (w'_d) and the terminal velocity (w_t) are the same, and Eq. (3.18) may be written in the form

$$4\rho_w \int_{r_{01}}^{r_{02}} \frac{(w - w_t)}{\varepsilon w_t} dr_0 = \int_{z_1}^{z_2} \rho_d \, dz \tag{3.19}$$

If applied to a drop which enters the cloud at the base (z_1) and ultimately returns to this height, and if the model is further simplified by assuming ρ_d and w independent of height and time, the right side of Eq. (3.19) vanishes, with the result

$$r_{02} - r_{01} = w \int_{r_{01}}^{r_{02}} \frac{dr_0}{w_t} \tag{3.20}$$

Equation (3.20) shows that the final radius (r_{02}) of drops which reach a terminal velocity greater than the updraft velocity and therefore fall out of the cloud is a function only of the updraft velocity and the initial radius. Integration may be carried out numerically using the terminal velocities shown in Fig. 3.8. Results of calculations shown in Fig. 3.15 indicate that drops of 2.5 mm radius can develop in updrafts of a few meters per second. For a particular updraft the initially smaller droplets achieve larger size than the larger droplets; this surprising result occurs because the smaller droplets are carried higher into the cloud than the larger droplets, and therefor eventually grow to greater size. This of course requires a greater length of time, which may exceed the life of the cloud.

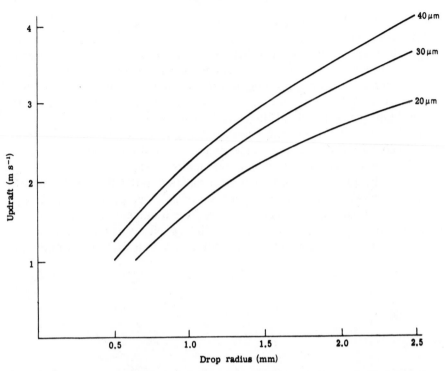

FIG. 3.15. Final raindrop radius as a function of updraft velocity as calculated from Eq. (3.20) for initial radii of 20, 30, and 40 μm (after F. H. Ludlam, *Quart. J. Roy. Meteorol. Soc.* **77**, 402, 1951).

In the preceeding discussion the time, liquid water content, and the depth of cloud required for growth have not been considered. If the liquid water content is too low or the depth of cloud too small, the drop may not grow large enough to fall downward against the updraft. An estimate of the minimum depth necessary for formation of showers can be based on the requirement that the upward moving drops must reach a radius of at least 150 μm by the time they reach the cloud tops. This size turns out to be sufficient for the drop to fall downward against an updraft near the top of the cloud which typically is about 1 m s^{-1}. The criterion for minimum cloud thickness h now can be developed from Eq. (3.19) in the form

$$\int_{r_0}^{150\,\mu m} \frac{w - w_t}{\varepsilon w_t}\, dr_0 = \frac{1}{4\rho_w} \int_0^h \rho_d\, dz \tag{3.21}$$

The liquid water concentration (ρ_d) can be determined from the pseudo-adiabatic chart by assuming that it is equal to the decrease in saturation mixing ratio which occurs for ascent along the wet adiabatic lines from the cloud base. This does not take account of entrainment of unsaturated air from the sides of the cloud and therefore tends to overestimate liquid water concentration. Results of these calculations, shown in Fig. 3.16, indicate that minimum thickness is roughly proportional to updraft velocity. For the warm cloud example the minimum depth for an updraft of 5 m s^{-1} is 2 or 3 km; the whole cloud therefore is well above freezing. On the other hand, for cold clouds the minimum depth is greater, so that the cloud tops are likely to be considerably colder than $0°$ C, and the drops may freeze. In this case the ice particles may grow rapidly by vapor deposition.

Use of the continuous model in these calculations rather than the stochastic model results in underestimating the rate of development of precipitation. On the other hand, omission of the process of entrainment results in an overestimate of rate of development. Although these effects tend to compensate, the resulting calculations clearly cannot be accurate in the full range of

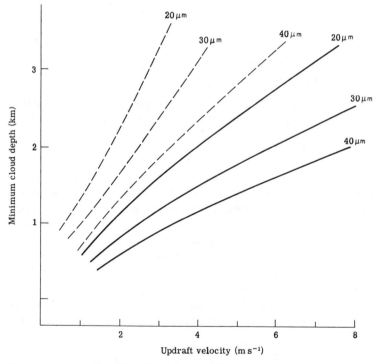

FIG. 3.16. Minimum cloud depth for drop formation by coalescence as a function of updraft velocity. Results are based on Eq. (3.21) using coalescence efficiency of 0.85 and cloud base temperatures of $-5°$ C (*dashed*) and $20°$ C (*solid*) (after F. H. Ludlam, *Quart. J. Roy. Meteorol.* **77**, 402, 1951).

cases occurring in nature. In more complete numerical models of cloud processes condensation nuclei concentrations are used as input data, and calculations are used to depict growth of droplets by condensation and by stochastic collisions. The results show that, starting with the nuclei concentrations typical of marine and continental clouds, under otherwise comparable conditions the marine clouds develop precipitable drops more quickly.

The Development of Hailstones

A falling raindrop cannot grow to a diameter greater than about 5 mm because at larger sizes surface tension is overcome by the drag force of the airstream, and the drop breaks into several fragments. However, if the drop freezes it may collect ice particles and liquid water which then may freeze, and growth may continue to a very large size. In this way hailstones grow to sizes ranging in diameter from 5 mm up to more than 10 cm. The largest hailstone recorded in the United States was about 15 cm in diameter (the size of a large grapefruit), and fell in Kansas in September 1970. Agricultural losses in the United States due to hail amount to about \$500 million per year on the average.

Hailstone structure reveals something of the history of their growth. At the center are found either graupel particles (rime ice) or frozen drops usually a few millimeters in diameter. Surrounding the centers are a number of concentric layers of ice of varying density and optical properties. The crystalline structure of the layers can be examined by viewing thin sections of a hailstone in transmitted polarized light. These observations can be interpreted to yield information (unfortunately, not entirely unambiguous) concerning the concentrations, size, temperature, and impact velocity of the supercooled droplets which produced the hailstone.

As supercooled droplets are collected by a hailstone, release of the latent heat of freezing raises the temperature of the surface above the surrounding air temperature; if the rate of collection and temperatures are suitable the surface temperature may reach 0° C, and some of the water may remain in liquid form. Some of this water may be shed in the wake of the hailstone, but much of it in some cases is retained in a meshlike structure of ice to form "spongy" hail.

Updrafts of 30 m s^{-1} or more occurring in thunderstorms are required to account for the final stage of growth of the very large hailstones. An example is shown in Fig. 3.17. Numerical models based on field observations have been used to describe plausible mechanisms of hailstone growth. However, the thunderstorms in which large hail occurs are so violent and their structure changes so rapidly that observational data is incomplete, and important questions remain unanswered.

Fig. 3.17. A very large thunderstorm as seen from Denver, Colorado at 7:30 p.m., 4 August 1976. Cloud base and top are probably at about 4.5 km and 12 km above sea level, respectively (photograph by Charles G. Summer, National Center for Atmospheric Research).

3.14   Artificial Cloud Modification

The mechanisms of cloud particle growth and precipitation can be influenced under certain conditions by *cloud seeding*, the deliberate addition of ice nuclei or condensation nuclei to existing clouds. Cloud seeding efforts depend for their effectiveness on instabilities within the precipitation mechanism, situations in which small efforts can exert large influences over subsequent developments. In Section 3.9 it was pointed out that naturally occurring ice nuclei vary widely in concentration and in effectiveness and that some cold clouds are deficient in ice nuclei effective at temperatures above about $-20°$ C. Several possibilities for modification of clouds and precipitation follow from this observation. Addition of artificial nuclei active at temperatures below $0°$ C may result in nucleation of supercooled droplets and thus may initiate the growth of ice particles and lead to precipitation. Or if an abundance of artificial ice nuclei is added to a supercooled cloud, so many crystals may form that a supercooled cloud is transformed to an ice crystal cloud in which further growth of crystals is inhibited by competition for water vapor between the myriad crystals. This is referred to as *overseeding* a cloud. The result may be to prevent precipitation from developing in a statically stable stratiform cloud. However, overseeding releases latent heat of freezing, and this adds buoyancy which, in the case of cumuliform clouds, may result in the cloud penetrating a stable layer and growing to greater heights in the unstable air above. In some cases the result may be to initiate precipitation where none would have occurred without seeding. Buoyancy is further discussed in Section 4.12. In addition to increasing buoyancy, the latent heat also tends to reduce the hydrostatic pressure in the underlying column of air. In this way the horizontal pressure distribution can be influenced, which in turn affects the wind distribution. The relation of pressure distribution to wind velocity is a major topic of Chapter IV.

In the case of warm clouds, growth of cloud droplets can be stimulated by deliberately broadening the drop-size spectrum by adding relatively large water droplets (~ 30–40-μm radius) to the base of a cumulus cloud or by adding giant hygroscopic condensation nuclei.

Results of Cloud Seeding

Most efforts to modify clouds by seeding have been directed at cold clouds. The first carefully observed experiments, which opened the way to cloud modification research, were carried out by Vincent J. Schaefer at the General Electric Laboratory in 1946. He observed the dramatic effect of dropping a bit of dry ice (solid CO_2) into a supercooled laboratory cloud, thereby producing some 10^8 ice crystals in a volume of about 0.13 m^3. Dry ice, whose temperature is $-78.5°$ C, cools the nearby air thus producing very high supersaturation and leading to production of many ice crystals by homogeneous nucleation. Subsequently, field observations demonstrated that dry ice dropped into supercooled layer clouds can convert large volumes of the clouds into glaciated clouds. In some cases precipitation fell, leaving clear air surrounded by cloud. Ice crystal clouds also have been formed by dropping dry ice into air which is saturated with respect to ice but not with respect to water.

Soon after Schaefer's demonstration of the effect of dry ice on supercooled cloud Bernard Vonnegut in the General Electric Laboratory discovered that silver iodide crystals are effective in producing ice crystals in water clouds at temperatures below about $-4°$ C. Subsequently, silver iodide was shown to be effective in modifying natural clouds. The importance of Vonnegut's discovery was that, although silver iodide is somewhat less effective than dry ice in nucleating ice crystals, it can be introduced into the atmosphere upwind of the target area and perhaps at the ground, where the temperature may be above $0°$ C, and can then be carried into supercooled clouds by natural air currents. Also, whereas dry ice creates ice crystals in large concentrations and therefore tends to overseed in the region of seeding, silver iodide concentrations can be to some extent controlled, so that optimal concentrations of ice nuclei (of the order of 10 per liter) may be possible.

a. Precipitation augumentation and distribution.   Field experiments in the early 1950's in the United States and Australia showed that seeding of winter supercooled orographic clouds (produced by airflow over mountains) seemed to result in average increases in snowfall of 10%–15%, somewhat less than the year to year variability of natural precipitation, but still an economically significant increment. Although a great deal has been learned about cloud physics in the years since these early experiments, evaluations of the most carefully conducted programs carried out in recent years in the United States and Israel indicate that seeding may result in increases or in decreases in winter orographic precipitation, but that the average increase over a season is still limited to about 10%–15%. Attempts also have been made to divert snowfall downwind by overseeding to reduce the number of supercooled droplets and thereby reduce the rate of growth and the fall speed of ice particles.

Results of seeding cumulus clouds have varied widely; some series of experiments have resulted in precipitation decreases while others have resulted in increases. The most encouraging results have been reported from a series of experiments carried out since 1968 in Florida by the U.S. National

Oceanic and Atmospheric Administration. In these experiments isolated cumulus clouds seeded to induce vertical growth (called "dynamic" seeding) were reported to produce about twice as much precipitation as the unseeded clouds used as a statistical control. However, it is still not known whether the net precipitation over large areas can be significantly increased in this way.

Attempts to increase precipitation from warm clouds by introducing water droplets or giant nuclei have been made in several countries, but the effectiveness has not been established. However, there is evidence that precipitation is greater downwind of industrial and urban areas than in the surrounding region. It is plausible that these increases may be due to increased concentrations of condensation nuclei, but other causes, such as the addition of heat and water vapor to the airstream, may be important also.

b. Reduction of hurricane winds. Numerical models of hurricanes suggest that it may be possible, through release of latent heat by overseeding just outside the cloud wall surrounding the hurricane "eye," to spread out the region of high wind velocity in the radial direction with resulting decrease of the highest speed. In the three field trials so far carried out by the U.S. National Oceanic and Atmospheric Administration, maximum wind speeds were observed to decrease following seeding, but the decreases were within the natural variability of hurricane winds. Consequently, this technique is considered promising but unproven, and further modeling and field research is necessary.

c. Hail suppression. Attempts have been made in many countries to reduce the average size of hailstones by deliberately increasing the number of small ice particles. Reports have ranged widely. In a carefully planned series of experiments carried out in northeast Colorado by the National Center for Atmospheric Research, no statistical evidence of suppression was found, and it appears that better understanding of the physics of thunderstorms and of hail growth is necessary. Since the economic value of rainfall from hailstorms is often greater than the losses due to hail, it is important that any hail suppression technique should not decrease precipitation.

d. Fog suppression. Supercooled fog, which may result in costly airport shut-downs, is routinely dissipated at the few U.S. airports where these fogs are common by introducing ice nuclei into the fog. However, only 5% of fogs occurring at U.S. airports are cold fogs. Warm fogs can be dissipated by introducing large quantities of heat produced by jet engines located along the runway, but this method is very expensive and it also is likely to produce dangerous turbulence unless carefully controlled. Other methods, including seeding with hygroscopic particles, have been tried, but a reliable economic method for dispersal of warm fog has not been developed.

   e. Lightning suppression.   Attempts have been made to reduce the intensity and frequency of lightning and the consequent damage. One method has been to introduce silver iodide particles into the base of the thunderstorm in order to produce additional ice crystals which could enhance coronal discharge and thereby reduce the electrical potential difference between the positive and negative charge centers (see Section 3.18). A few experiments by the U.S. Forest Service were reported to be encouraging, but they have not been carried to the point of yielding conclusive results. Efforts also have been made to discharge thunderstorms by releasing tiny metallic fibers in the bases of the clouds. Results in the cases reported were positive, but the technique has not yet been adequately tested.

PART II: ELECTRICAL CHARGE GENERATION AND ITS EFFECTS

3.15   Elementary Principles of Electricity

Coulomb's Law

   An electrical point charge (q) placed at a distance ($\mathbf{r}$) from a charge (Q) experiences a force given by Coulomb's law in the form

$$\mathbf{F}_c = \frac{qQ}{4\pi\varepsilon r^2}\frac{\mathbf{r}}{r} \tag{3.22}$$

where $\mathbf{r}$ is directed from Q to q, and ε, the *permittivity*, is a characteristic of the medium. Permittivity is defined for vacuum by $\varepsilon_0 = 10^7(4\pi c_0^2)^{-1}\,\mathrm{C}^2\,\mathrm{N}^{-1}\,\mathrm{m}^{-2}$, where c_0 represents the velocity of light in vacuum. For air $\varepsilon \approx \varepsilon_0$. When $\mathbf{F}_c = 1\,\mathrm{N}$, $r = 1\,\mathrm{m}$, $\varepsilon = \varepsilon_0$, and $q = Q$, then each charge is equal to 1.05×10^{-5} coulomb (C).

The Electric Field

   The force on a unit test charge due to another charge Q may be associated with an *electric field* which is proportional to Q. This bears close analogy to the gravitational field discussed in Chapter 1. The electric field is defined as the limit of the force per unit test charge as the test charge diminishes to zero, and is written

$$\mathbf{E} \equiv \lim_{q \to 0} \frac{\mathbf{F}}{q} \tag{3.23}$$

Equation (3.22) shows that the *electrostatic field* may be expressed by

$$\mathbf{E}_c \equiv \frac{Q}{4\pi\varepsilon r^2} \frac{\mathbf{r}}{r} \tag{3.24}$$

The electric field may arise from other sources than the electrostatic, and in Chapter VII it will be necessary to discuss this subject again and in more detail.

The potential energy of a unit charge in the electrostatic field is defined by the work necessary to bring the unit of charge from $r = \infty$ to $r = r_0$ against the electrostatic field. Thus the *electrostatic potential* or simply the *potential* at r_0 is expressed by

$$V_0 = \int_\infty^{r_0} \frac{Q}{4\pi\varepsilon r^2} \frac{\mathbf{r}}{r} \cdot d\mathbf{r} = \frac{Q}{4\pi\varepsilon r_0} \tag{3.25}$$

Equations (3.24) and (3.25) together show that the potential and the electrostatic field vector are related by

$$-\frac{\mathbf{r}}{r} \frac{\partial V}{\partial r} = \mathbf{E}_c$$

Surfaces of equal potential may be imagined which surround the charge Q; the test charge may be moved on a potential surface without work just as masses may be moved on a geopotential surface without work. It follows that the surface of a charged conductor must be an equipotential surface if the charges are at rest, and the potential must represent the work done in bringing a charge Q from infinite distance to the conductor. Therefore, the potential of a conductor is proportional to the charge, and the *capacitance* (C) may be defined by

$$V \equiv Q/C \quad \text{or} \quad Q = CV \tag{3.26}$$

Combining Eq. (3.25) and (3.26) yields for capacitance of a sphere in vacuum

$$C = 4\pi\varepsilon_0 r \tag{3.27}$$

Note that in the centimeter-gram-second system, $\varepsilon_0 = (1/4\pi)\mathrm{Fr}^2\,\mathrm{dyn}^{-1}\,\mathrm{cm}^{-2}$, and $C = r$, where Fr represents the electrostatic unit of charge, the franklin.

Ohm's Law

Experiment shows that electrical charge flows along a conductor at a rate which is proportional to drop in potential. Thus

$$\Delta V = R \frac{dQ}{dt} = RI \tag{3.28}$$

where the *resistance* (R) is characteristic of the conductor. Equation (3.28) is the well-known Ohm's law; it is a special form of Eq. (2.68).

3.16   Origin and Distribution of Ions

For many purposes air acts as a good insulator; but when careful measurements are made, air turns out to have characteristic conductivity which varies with time, location, and especially with height. The conducting property of air has been known since the late eighteenth century when Coulomb called attention to the slow discharge of a charged body through air, but an adequate explanation was proposed only at the beginning of the twentieth century by Elster and Geitel in Germany and C. T. R. Wilson in England. They attributed the conductivity to the movement of positive and negative ions of molecular size through the air under an electrical field. The investigation of the source of these ions led to the discovery of cosmic rays by Hess in 1911 and to the development of cosmic-ray physics.

Primary cosmic rays are particles of very great energy, mostly protons, which enter the earth's atmosphere from all directions and produce other high energy particles by colliding with neutral air molecules. These products of collision are called secondary cosmic rays, and they are responsible for production of ions by collision with air molecules. The energy of cosmic radiation is comparable in magnitude to the light received from stars and is distributed over an enormous range. The primaries appear to have maximum total energy in the neighborhood of 10^{10} electron volts (eV) with occasional particles of 10^{12}, 10^{14}, up to as much as 10^{19} eV. The latter figure is approximately four joules, and it should be understood that each measurement of cosmic-ray energy is likely to be an underestimate. The lower energy particles are absorbed in the upper atmosphere, while the highest energy particles penetrate all the way to the earth, producing multiple high energy particles each of which leaves a trail of ionized air molecules. The average number of ion pairs produced at sea level by cosmic rays is about 1.5 cm^{-3} s^{-1}; ion production increases with latitude due to the effect of the earth's magnetic field in deflecting the charged primary rays, thereby shielding the region of the geomagnetic equator more effectively than the higher latitudes.

Ion production increases with height up to about three hundred per cubic centimeter per second at 13 km and decreases above that height, as shown in Fig. 3.18. Primary cosmic rays of the greatest energies may produce cascades or showers of as many as 10^8 high energy particles; on these occasions the ion density may be suddenly increased by more than an order of magnitude throughout a conical volume of diameter at the ground as large as a kilometer or more.

Near the ground ions are also produced by radioactive decay of elements in the soil and in the air; ion production by this process is variable depending on air currents, static stability, and atmospheric pressure as well as on proximity to radioactive rocks and the conditions of the surface. Over land radioactivity produces roughly 8 ion pairs $cm^{-3} s^{-1}$ on an average, but

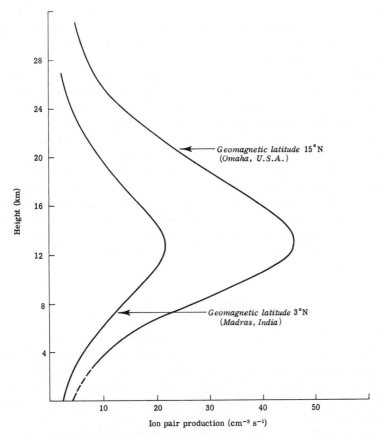

FIG. 3.18. Rate of ion pair production by cosmic rays as a function of height (after O. H. Gish, *Compendium Meteorol.*, p. 101, 1951).

radioactivity decreases to negligible intensity at a height of a few kilometers. Over the sea radioactivity is not important in producing ions.

Near the outer limit of the atmosphere cosmic rays are ineffective ionizers because only the primary particles are present in the cosmic radiation and these are few in number. Ions are produced, however, in this region by absorption of ultraviolet and x rays from the sun. At the low densities present above, say 200 km, the electrons ejected from the neutral oxygen and nitrogen atoms can remain for long periods as free electrons. Under the daytime sun's radiation electron densities of 10^6 cm^{-3} occur, and these decrease by only one order of magnitude at night. The atmosphere, therefore, contains a spherical shell of high ion density, called the *ionosphere*, which extends from a height of about 80 to above 300 km. Inside the shell, ion density produced by cosmic rays amounts to only hundreds per cubic centimeter.

Ionization, whether produced by cosmic ray or ultraviolet light, results in creation of a negative electron and a much heavier positive ion. Near sea level the electrons, with average lifetimes of 10^{-5} s, very quickly become attached to neutral molecules; the resulting negative ions as well as the positive ions have lifetimes of roughly 10^2 s. These ions may attract a group of neutral molecules but their size remains essentially molecular, and they are therefore called *small* ions. Other ions become attached to aerosol particles which are very large compared to molecules; they are called *large* or *Langevin* ions, and their average life is roughly 10^3 s.

In an electrostatic field ions are accelerated, negatives toward the center of positive potential and positives toward the center of negative potential. However, the mean free path in air at sea level is only about 10^{-7} cm, so that an ion experiences a series of accelerations interrupted by abrupt collisions; it therefore moves through the air with an average drift velocity which depends on mean free path, potential gradient, charge of the ion, and mass of the ion.

3.17   Conductivity

Small ions are constantly produced at a rate (p) which depends on the processes described above, and they are constantly destroyed by recombination with ions of the opposite sign and taken out of circulation by attachment to large particles. Recombination of each ion is proportional to ion density so that the rate of recombination is proportional to the square of the ion density. Similarly, rate of ion attachment is proportional to the density of small ions and to the density of large particles. Under conditions of equilibrium between ion production and destruction

$$p = \alpha n^2 + \beta nN \tag{3.29}$$

where n and N represent, respectively, the number of small ions and large particles per unit volume, α is called the *recombination coefficient* for small ions and has the approximate value of 1.6×10^{-6} cm^3 s^{-1}, and β is called the *combination coefficient* for small ions and large particles and has the approximate value of 3×10^{-6} cm^3 s^{-1}. The individual terms of Eq. (3.29) are only formal descriptions of the creation and destruction of ions; no additional physics is represented by the equation. However, if α, β, N, and p are known from experiment and observation, the concentration of small ions can be calculated. Near cities, where N may be 5×10^3 cm^{-3} or even larger, small ion concentration is determined mostly by attachment to large particles, but above a height of a kilometer or over the ocean recombination is the controlling process. The concentration of small ions increases with height from roughly 6×10^2 cm^{-3} at sea level to a maximum of 5×10^3 cm^{-3} at a height of 15 km. Equation (3.29) is applied in Problem 8.

The typical lifetime of small ions may be estimated in the following way. Imagine ion production within a cubic centimeter at the rate p, and allow this production to continue for a time τ until the cubic centimeter contains n ions. Therefore, $p\tau = n$. Then allow destruction of ions to proceed at the rate p. Ion concentration subsequently will remain at n, and under equilibrium conditions the average lifetime will be equal to τ and expressed by

$$\tau = n/p \tag{3.30}$$

Problem 9 provides an application of Eq. (3.30).

The *specific conductivity* of air may be defined by Ohm's law in the form

$$\lambda \equiv \frac{i}{dV/dz} \tag{3.31}$$

where i represents electrical current per unit area. Equation (3.31) is a special form of Eq. (3.28). Because the *mobility* (w) may be expressed by the identity

$$i \equiv new \frac{dV}{dz} \tag{3.32}$$

where n represents number of ions per unit volume and e charge on each ion, Eq. (3.31) may be written

$$\lambda = new \tag{3.33}$$

The mobility is defined as the average drift velocity under a potential gradient of 1 V m^{-1}. The mobility of small ions in air at sea level is roughly 1–$2 \times 10^{-4} \text{ m s}^{-1} (\text{V m}^{-1})^{-1}$, and of large ions is 10^{-8}–$10^{-6} \text{ m s}^{-1} (\text{V m}^{-1})^{-1}$. Equations (3.31)–(3.33) may be written separately for ions of different mobilities, charges and concentrations; but this refinement will be omitted here. Because of upward movement of negative ions under the atmospheric potential gradient, near the ground there are more positive than negative ions. However, the resulting positive space charge is small and is not essential to the purpose of this discussion. The mobility of small ions greatly exceeds that of large ions, so that conductivity depends mainly on the number of small ions per unit volume.

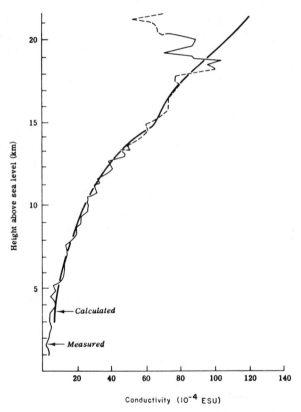

FIG. 3.19.   Specific conductivity of positive ions measured on balloon flight of Explorer II, 11 November 1935 and calculated from cosmic-ray ionization (after O. H. Gish, *Compendium Meteorol.*, p. 101, 1951). The conductivity scale can be converted to SI units by setting $1 \text{ Fr}^2 \text{ cm erg}^{-1} \text{ s}^{-1} = 1.11268 \times 10^{-10} \text{ A V}^{-1} \text{ m}^{-1}$.

One may expect ion mobility to be inversely proportional to air density; and, although it also depends on other factors, the density dependence is the only one to be considered here. Upon introducing the density dependence into Eq. (3.33), the conductivity may be expressed in terms of conductivity at a base height where the density is ρ_0 by the equation

$$\lambda = \frac{ne\rho_0}{n_0 e_0 \rho} \lambda_0 \tag{3.34}$$

Between sea level and a height of 15 km, n increases by an order of magnitude while ρ decreases by an order of magnitude. Consequently, conductivity increases by two orders of magnitude over this height range. The increase in cosmic radiation with geomagnetic latitude results in an increase in n between geomagnetic equator and high latitudes by a factor of about three. Balloon measurements of air conductivity and related calculations are shown in Fig. 3.19. Conductivity and vertical potential gradient play important roles in the processes of charge generation in clouds and lightning, which are discussed in following sections.

3.18 Charge Generation and Separation in Clouds

Clouds which extend well above the height of $0°$ C and in which there are present strong updrafts act as powerful electrostatic generators, the upward moving air carrying small positively charged particles, and the heavier falling precipitation elements carrying negative charges. This process creates a center of positive charge in the upper part of the cloud and a center of negative charge in the lower part of the cloud. This distribution, depicted in Fig. 3.20, also shows a relatively small region of positive charge near the cloud base which appears to be typical of many thunderclouds. The main centers of charge result in an electric field directed downward through the cloud, a field in the same direction as the normal "fair weather" electric field, but very much stronger. Figure 3.17 should be looked at again in this connection.

Several different plausible mechanisms have been proposed to account for generation and separation of electrical charge, and, so far as is now known each may contribute to the electrical charging of thunderstorms. However, only the polarization mechanism has been shown by numerical modeling to be capable of generating and separating amounts of charge at rates typical of observed thunderstorms. Although this constitutes a major step toward understanding electrification of thunderclouds, model calculations can be misleading and should not be accepted uncritically.

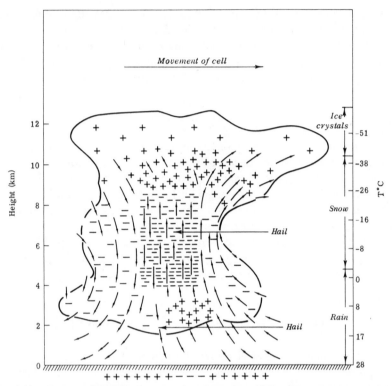

FIG. 3.20.   Cross section through very active convection cell showing temperature, vertical air velocity, and charge distribution (after L. B. Loeb, *Modern Physics for the Engineer*, p. 335. McGraw-Hill, New York, 1954).

The polarization mechanism of thundercloud electrification, illustrated schematically in Fig. 3.21, is based on the concept that a falling spherical pellet in the ambient "fair weather" electrical field will be polarized with the lower half positively charged and the upper half negatively charged. When collision and rebound occur between the falling pellet and a cloud particle, negative charge may be transferred to the pellet, leaving the droplet or ice particle charged positively. The positive charge is carried upward in the airstream, while the falling pellet carries negative charge downward. The process is self reinforcing because as charge is separated the field strength increases, thus increasing both polarization and the transfer of charge occurring at each collision. Results of model calculations shown in Fig. 3.22 indicate that the electric field grows slowly for the first 200–500 s, depending upon the cloud parameters used in the model, but then may grow rapidly to magnitudes of 3–5 kV cm^{-1} within only another 100–200 s. The

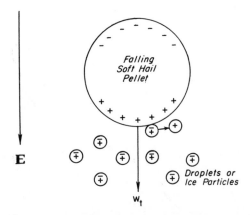

FIG. 3.21.   Schematic diagram showing polarization of spherical soft hail pellet in vertical electric field **E** and the transfer of charge to a droplet which collides and rebounds from the sphere.

model considers an arbitrary but reasonable size distribution of cloud particles, and represents the processes of growth of particles by collision and coalescence, charge transfer by collision and rebound, and accumulation of charge, as well as growth of the electric field.

The thermoelectric effect in ice also may play a role in charge generation, especially in the early stages of thunderstorm electrification. This effect results from the facts that (a) the concentration of dissociated ions in ice increases rapidly with increasing temperature (and with increasing dissolved contamination), and (b) the mobility of the positive hydrogen ion is an order of magnitude greater than the mobility of the negative hydroxyl ion. The consequence of (a) is that when a temperature gradient is created in ice, it is accompanied by a concentration gradient of both positive and negative ions. The consequence of (b) is that diffusion of positive ions is more rapid than diffusion of negative ions; and a positive space charge therefore develops in the colder region, leaving the warmer region charged negatively. Charge separation ceases when the internal electric field balances the differential ion diffusion. This occurs when the charge separation amounts to about 170 $dT/dx \, pC \, m^{-2}$ and the potential difference is about 2 ΔT mV.

The thermoelectric effect can bring about charging of cloud particles in several ways. Soft hail pellets or graupel falling through a cloud containing supercooled water droplets and ice crystals collects some of the droplets; release of the latent heat of freezing makes the surface of the hail pellets warmer than the droplets and ice crystals. Then, if some of the droplets and crystals rebound from a hail pellet after momentary contact, negative

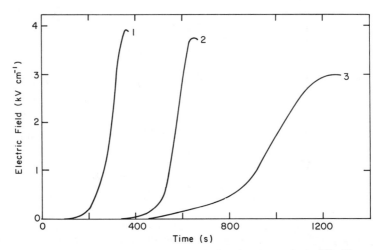

FIG. 3.22.   Growth of the thundercloud electric field for three values of liquid water content of the cloud particles (ρ_d) and the initial liquid water content of the precipitation particles [$\rho_p(0)$]. Curve 1: $\rho_d = 1$ g m^{-3}, $\rho_p(0) = 1$ g m^{-3}; Curve 2: $\rho_d = 1$ g m^{-3}, $\rho_p(0) = 0.1$ g m^{-3}; Curve 3: $\rho_d = 0.1$ g m^{-3}, $\rho_p(0) = 1$ g m^{-3}; (after A. Ziv and Z. Levin, *J. Atmos. Sci.* **31**, 1652, 1974).

charge will be transferred to the warm hail pellet while the colder particles will become positively charged. The falling hail pellet will carry negative charge downward, while the small positively charged particles will be carried upward in the rising air.

A second mechanism of thermoelectric charging may result when supercooled droplets adhere to a soft hail pellet. As the droplet begins to freeze its temperature rises to 0° C. Heat is then lost to the surrounding cold air by conduction, and an ice shell forms around the remaining liquid. As freezing proceeds toward the center of the droplet, the interior expands and the outer cold shell may be shattered as described in Section 3.9. The resulting cold ice particles will be positively charged due to the thermoelectric effect; and they will be carried away in the rising air current.

Each of these thermoelectric processes has been demonstrated in the laboratory (although important experimental differences remain to be resolved); however, cloud model calculations have indicated that charge generation by these processes may be insufficient to develop the strong electric fields of thunderstorms. Thermoelectric processes may be important in accounting for the early growth of the electric field within the cloud, before the polarization process becomes dominant.

The process or processes responsible for the positive charge near cloud base shown in Fig. 3.19 are still uncertain. One mechanism depends upon positive ions produced by point discharge or by splashing of drops at the ground surface under the main negative charge. These positive charges may be captured by falling drops in the lower part of the cloud, producing the secondary region of positive charge and creating a local strong potential gradient directed toward the primary negative charge center. In this region of strong gradient between the positive and negative centers local breakdown and the initiation of lightning may occur most readily.

3.19   The Lightning Discharge

The largest concentration of charge in a mature thunderstorm is associated with the region of strong updraft, the negative cell typically extending between heights of 4 and 9 km and the positive cell being centered above 10 km as illustrated in Fig. 3.20. The diameter of the cell is usually less than 4 km though much larger *supercells* occasionally occur. Measured electric fields in the region between the two centers have been as large as about 2000 V cm^{-1}, but local transient fields probably reach considerably higher values. Discharge by lightning may occur between the positive and negative centers within the cloud and between the cloud and the induced positive ground charge. Typically, about 20 C are discharged in a lightning flash. Following discharge an active cloud may be recharged in about 20 s.

In order to understand the phenomenon of lightning it is helpful to first understand the spark discharge in air between the plates of a charged capacitor. The spark may be initiated by a single ion pair in the space between the plates. The negative ion is accelerated toward the positive plate and the positive ion toward the negative plate, and each achieves a certain kinetic energy before it collides with a neutral particle. If this kinetic energy exceeds the ionization potential, the neutral particle yields a new ion pair, which in turn is accelerated in the electric field. In this way the number of ions grows by geometric progression until an "avalanche" is produced which provides a conducting path for discharge of the plates. The conditions required for spark discharge are that the accelerating potential gradient be so related to the mean free path between air molecules that the ions can attain kinetic energy equal to the ionization potential of the air molecule. If the potential across a capacitor at sea level is gradually increased, spark discharge occurs in dry air when the potential gradient reaches 30 kV cm^{-1}. If droplets are present, the breakdown potential is lower, and it has been estimated that the critical potential required within clouds in which droplets of 1-mm radius are present is about 10 kV cm^{-1}.

The analogy between the spark discharge and lightning is assuredly close, but the sequence of events which occurs in lightning has not been predicted by theory or explained in detail. Therefore, only a descriptive account can be given of the common lightning flash which originates in the negative cloud base and transfers negative charge to the earth. Cloud to cloud flashes develop in a similar manner, but are not directly effective in transferring charges to the ground and therefore are of less interest here.

The potential difference between convective clouds and the ground and within the clouds, as indicated by the observations available, is of the order $10-10^2$ kV m^{-1} over most of the volume. However, within the cloud the field fluctuates locally over a wide range as turbulent air motions bring the low-lying mass of positive ions closer or farther from the mass of negative ions shown in Fig. 3.20. It has been suggested that in this way the breakdown potential of about 1 MV m^{-1} may suddenly develop between the oppositely charged regions. Then there follows a lightning stroke within the cloud, which transfers negative charge downward, neutralizing the small positive charge and charging the cloud base strongly negative. Breakdown potential is not achieved in the column between cloud and ground except immediately adjacent to the negative base. Where breakdown potential is achieved negative charge advances downward and forms an ionized path of about 10 cm radius, called the *pilot leader*. This path grows toward the earth at a rate of 10^5-10^6 m s^{-1} for a distance of roughly 10^2 m length. This may be thought of as the length limited by the quantity of charge in the head of the pilot leader which can maintain breakdown potential without a resupply of electrons from the cloud base. Advance of the pilot leader produces a conducting path of $10^{13}-10^{15}$ ion pairs cm^{-1} of path. Down this path surges a negative charge which revives the potential at the head of the pilot leader, and the series of events and another extension of the pilot leader is repeated. Finally there is created a conducting path extending from the base of the cloud to within a short distance above the earth. The potential gradient in the neighborhood of sharp points connected to the earth is now high enough that the breakdown potential is reached, and a positive streamer advances from such a point to meet the pilot leader at a height of 5–50 m above the ground. When the pilot leader and streamer meet, earth and cloud are joined by a conducting path roughly 10 cm in radius, and up this path rushes the wave of ionizing potential called the *return stroke*. It advances at about 10^8 m s^{-1} and practically fully ionizes the channel. Now the negative current in the cloud base rushes earthward through the brilliantly luminous channel and discharges roughly the lowest kilometer of the cloud. Following the discharge the conductivity of the channel decreases, and concurrently the potential of the cloud base gradually recovers. After about $\frac{1}{20}$ s, the potential is sufficient for development of the *dart leader*, a new character in the drama which serves to reactivate the conducting path. The number of strokes varies widely about the average value of three to four, and each of these strokes discharges successively higher regions of the negatively charged portion of the cloud. The total charge transferred from cloud to ground also varies widely. The average value appears to be about 20 C which can be generated in a typical thunderstorm in about 20 s. Satellites have detected a relatively few (~1 in 500) lightning flashes which are more than 100 times brighter than ordinary lightning; these are believed to be flashes from the positive charge center near the top of a cumulonimbus cloud to the negative ground.†

Observations made in aircraft flying above thunderstorms show that there is an average (positive) current of nearly 1 A flowing upward from cloud to ionsphere; other observations suggest that, at least occasionally, lightning may occur between cloud and ionsphere.

3.20   The Mean Electric Field

The thunderstorms which are always active somewhere in the atmosphere may be imagined as serving as huge Van de Graaff generators providing a

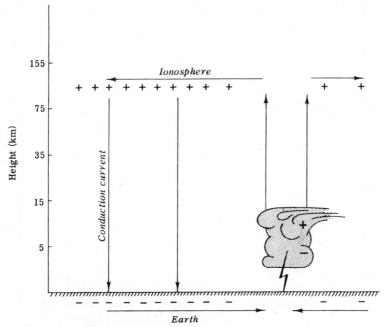

FIG. 3.23.   Charge generation by thunderstorms, the charge distribution in the earth and ionsphere and the positive conduction current.

† B. N. Turman, *J. Geophys. Res.* **82,** 2566, 1977.

potential difference between ionosphere and earth amounting to about 400 kV. Because both earth and ionosphere are excellent conductors, charge is conducted readily in the horizontal direction, and the two spherical surfaces are each at uniform potential. Therefore, positive current flows downward through the atmosphere over the entire earth except for the regions of thunderstorms; this current amounts to only about 4×10^{-12} A m^{-2} or 2×10^3 A over the whole earth. It is sufficient to discharge the earth–ionosphere potential difference in about 7.5 min if a recharging mechanism were not operating.

The vertical distribution of potential can be described using Eq. (3.31) if the current and the conductivity is known as a function of height. Evidently, since conductivity increases sharply with height, the potential gradient decreases equally sharply with height. If Eq. (3.31) is integrated between the ground and the ionosphere, the potential difference can be calculated from observations of the vertical current and the distribution of conductivity. In practice, the potential gradient near the earth is more easily measured than is the vertical current, so that Eq. (3.31) is first used

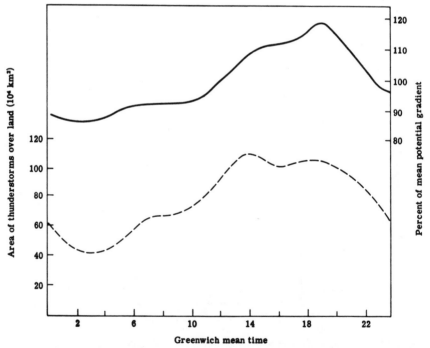

FIG. 3.24.   *Solid:* Potential gradient over the sea. *Dashed:* Area of observed thunderstorms as a function of time of day (after J. A. Chalmers, *Atmospheric Electricity*, 2nd ed., p. 204. Pergamon, New York, 1967).

to calculate the current and then is used in integrated form to calculate the potential difference. An exercise is provided in Problem 10. The generation of potential difference and the current flow between earth and ionosphere are illustrated in Fig. 3.23.

Observations of vertical potential gradient over the sea show that there occurs an average diurnal variation of about 30%. Maxima and minima occur, respectively, at the same time all over the globe. This daily variation is in close agreement with the diurnal variation of the number of observed thunderstorms as is shown in Fig. 3.24. In fact it is this evidence which first demonstrated the role played by thunderstorms in maintaining the mean vertical potential gradient of the atmosphere.

List of Symbols

		First used in Section
C	Diffusion capacitance, electrical capacitance	3.5, 3.15
D	Diffusion coefficient of water vapor in air	3.5
e	Vapor pressure, ionic charge	3.2, 3.17
$\mathbf{E}$	Electric field vector	3.15
$\mathbf{F}$	Force vector	3.15
F_D	Drag force	3.7
g	Force of gravity per unit mass	3.7
G	Gibbs function	3.2
i	Number of dissociated ions per molecule, electrical current per unit area	3.4, 3.17
I	Electrical current	3.15
L	Latent heat	3.6
m	Mass of water per drop	3.6
m_s	Molecular mass of solute	3.4
m_w	Molecular mass of water	3.4
M	Mass of solute per drop	3.4
n	Number of dissociated particles of solute, number of small ions per unit volume	3.4, 3.17
n'	Number of water molecules	3.4
N	Number of nuclei per unit volume, number of large ions per unit volume	3.5, 3.17
N_0	Avogadro's number	3.4
p_i	Internal pressure due to surface tension	3.1
p	Rate of ion pair production	3.17
q	Electrostatic test charge	3.15
Q	Electrostatic charge	3.15
r	Radius of water droplet	3.1
$\mathbf{r}$	Separation vector	3.15
R	Electrical resistance	3.15
R_w	Specific gas constant for water vapor	3.2
s	Specific entropy	3.2

		First used in Section
t	Time	3.6
T	Absolute temperature	3.2
u	Specific internal energy	3.2
V	Electrical potential	3.15
w	Updraft velocity, mobility	3.13, 3.17
w_t	Terminal fall velocity	3.7
w_d	Relative fall velocity of drops and droplets	3.7
z	Height coordinate	3.13
α	Specific volume, recombination coefficient for small ions	3.2, 3.17
β	Combination coefficient for large ions	3.17
ε	Collection efficiency, permittivity	3.7, 3.15
ε_c	Collision efficiency	3.7
η	Viscosity	3.7
λ	Thermal conductivity of air, specific electrical conductivity	3.6, 3.17
ρ	Air density	3.7
ρ_w	Water density	3.2
ρ_d	Mass of liquid water in small droplets per unit volume of air	3.7
σ	Surface tension	3.1
τ	Average lifetime of small ions	3.17

Subscripts

c	Curved surface
h	Solute
i	Ice
s	Saturation
v	Vapor
w	Water

Problems

1. Calculate the relative humidities over spherical droplets of pure water for the following radii: 10^{-1}, 10^{-2}, 10^{-3}, 10^{-4} cm if the temperature is $0°$ C and surface tension is 0.075 N m^{-1}.

2. Lord Kelvin derived Eq. (3.1) by considering the following thought experiment. A closed box contains a tray filled with water. A capillary tube is placed vertically in the tray. The air is pumped out of the box and water evaporates until equilibrium is reached at the capillary water surface and the flat water surface. Show how Eq. (3.1) can be deduced from this experiment.

3. From Eq. (3.4) find the radius which requires the greatest supersaturation for equilibrium for specified values of mass of solute, temperature, pressure, etc.

4. Show that the distance a small droplet would travel before coming to rest if shot horizontally into still air at an initial velocity v (called the penetration range) is given by

$$\lambda = \tfrac{2}{9} \frac{\rho_w r^2 v}{9\eta}$$

5. If an ice crystal of 1 μg mass falls from a cloud into an isothermal region where the temperature is $-10°$ C and the vapor pressure is 2.0×10^2 Pa (2.0 mb), how long will it fall before the crystal is reduced to 10^{-7} g? Assume that the diffusion capacitance may be expressed by $(m/4\rho_w)^{1/3}$ cm.

6. Calculate the greatest possible depth of precipitation which could fall from a cloud which extends from 70 kPa (700 mb) where the temperature is $-5°$ C to 50 kPa (500 mb) where the temperature is $-23°$ C if the average liquid water content is 0.3 g m^{-3} and if there are no updrafts within the cloud.

7. For the conditions of Problem 6 calculate the precipitation in 6 hr under steady state conditions if the updraft is 1.0 m s^{-1} at 70 kPa.

8. Calculate the small ion concentration under conditions of equilibrium between ion production and recombination for the following cases: ($\alpha = 1.6 \times 10^{-6}$ cm^3 s^{-1} $\beta = 3 \times 10^{-6}$ cm^3 s^{-1}).

(a) $N = 5 \times 10^3$ cm^{-3}; $p = 10$ cm^{-3} s^{-1}

(b) $N = 5 \times 10^2$ cm^{-3}; $p = 1.5$ cm^{-3} s^{-1}

(c) $N = 10^2$ cm^{-3}; $p = 10$ cm^{-3} s^{-1}

9. Calculate the average lifetimes of small ions for the cases given in Problem 8.

10. (a) If the vertical potential gradient near the ground is measured as 150 V m^{-1} and the conductivity is 1.5×10^{-14} A V^{-1} m^{-1}, find the vertical conduction current.

(b) Find the potential difference between the earth and the atmosphere at a height of 20 km for the following representative observations of conductivity.

Height (km)	λ (A V^{-1} m^{-1}) ($\times 10^{-14}$)	Height (km)	λ (A V^{-1} m^{-1}) ($\times 10^{-14}$)
0	2.5	12	38
2	3	14	60
4	6	16	70
6	10	18	85
8	20	20	110
10	26		

(c) If the conductivity is assumed uniform between 20 and 80 km, find the potential difference between the ground and the lower ionosphere (80 km).

Solutions

1. The ratio e_c/e_s is calculated from Eq. (3.1), noting that the exponent is small compared to unity. It follows, upon expanding in a Taylor series and retaining only two terms that

$$\frac{e_c}{e_s} = 1 + \frac{2\sigma}{\rho_w R_w Tr} = 1 + \frac{1.19 \times 10^{-7} \text{ cm}}{r}$$

Therefore,

r cm	exponent	$[(e_c/e_s) - 1]\%$
10^{-1}	1.19×10^{-6}	10^{-4}
10^{-2}	1.19×10^{-5}	0.001
10^{-3}	1.19×10^{-4}	0.012
10^{-4}	1.19×10^{-3}	0.119

These values agree well with the top curve of Fig. 3.3.

2. Water will rise in the capillary until the force of gravity acting on the column equals the upward force of surface tension lifting the column. Thus

$$\left|\frac{2\sigma}{r}\right| = \rho_w g h$$

The difference in vapor pressures between the flat surface and the curved surface in the capillary can be expressed by the hydrostatic equation (1.17) in the form

$$\frac{de}{dz} = \rho_v g = \frac{eg}{R_w T}$$

Therefore upon integrating from the flat surface $(z = 0)$ to $z = h$,

$$\ln \frac{e_c}{e_s} = -\frac{gh}{R_w T}$$

Upon substituting from the first equation above and introducing a negative sign to convert from the concave surface of the capillary column to the convex surface of a droplet

$$e_c = e_s \exp\left[\frac{2\sigma}{\rho_w R_w T r}\right]$$

3. Upon setting the derivative of Eq. (3.4) equal to zero

$$\frac{d}{dr}\left(\frac{e_h}{e_s}\right) = \frac{2\sigma}{\rho_w R_w T r^2} + \frac{9}{4}\frac{im_w M}{4\pi r^3 \rho_w m_s} = 0$$

Therefore,

$$r_{max} = \left[\frac{9}{8\pi}\frac{im_w M R_w T}{\sigma m_s}\right]^{1/2}$$

For 10^{-14} g of NaCl ($m_s = 46.5$, $i = 2$) this gives

$$r_{max} = 2.2 \times 10^{-4} \text{ cm}$$

in agreement with Fig. 3.3.

4. A droplet which suddenly encounters a nonequilibrium velocity v experiences a Stokes's force given by

$$F = 6\pi \eta r v$$

From Newton's second law this force is expressed by

$$m \frac{dv}{dt} = \frac{4}{3}\pi r^3 \rho_w \frac{dv}{dt}$$

Therefore,

$$\frac{1}{v}\frac{dv}{dt} = -\frac{9}{2}\frac{\eta}{\rho_w r^2}$$

The response time is then defined as the time required for the velocity to reduce to $1/e$ of the initial velocity. It is

$$t_r = \frac{2}{9}\frac{r^2 \rho_w}{\eta}$$

and the penetration range is

$$\lambda = vt_r = \frac{2}{9}\frac{\rho_w r^2 v}{9\eta}$$

5. Upon introducing the given expression for C, integrating Eq. (3.15), and introducing numerical values, the time required for the crystal of 10^{-9} kg to evaporate to a mass of 10^{-10} kg is 148 s. The calculation requires care and persistence.

6. The total liquid water in a vertical column of unit cross section extending through the cloud can be expressed by

$$M_l = \rho_l \Delta z = \frac{\overline{\rho_l R_m T}}{g} \int_{p_1}^{p_0} \frac{dp}{p}$$

where ρ_l represents the liquid water content. Upon substituting numerical values

$$M_l = 2.3 \text{ kg m}^{-2} = 2.3 \text{ mm (depth)}$$

7. The rate of upward transport of water vapor through the 70 kPa surface is expressed by

$$\frac{dM}{dt} = w\rho_v = \frac{wq_s p}{R_m T}$$

where q_s is the saturation specific humidity. From the pseudo-adiabatic chart $q_s = 3.8 \text{ g kg}^{-1}$ at 70 kPa and 268 K. Upon introducing numerical values

$$\frac{dM}{dt} = 3.46 \times 10^{-3} \text{ kg m}^{-2} \text{ s}^{-1}$$

and if the water vapor entering the cloud at the base is condensed and falls as precipitation, the precipitation in 6 hr is

$$M = 74.7 \text{ kg m}^{-2} \quad \text{or} \quad 74.7 \text{ mm (depth) in 6 hr}$$

8. Equation (3.29) can be solved as a quadratic equation in n. This yields a negative root, which must be discarded, and a positive root. For the given numerical values the small ion concentrations are

$$\text{(a)} \quad n = 6.1 \times 10^2 \text{ cm}^{-3}$$
$$\text{(b)} \quad n = 6.1 \times 10^2 \text{ cm}^{-3}$$
$$\text{(c)} \quad n = 2.4 \times 10^3 \text{ cm}^{-3}$$

9. (a) 61 s
    (b) 407 s
    (c) 240 s
10. (a) $2.25 \times 10^{-12} \text{ A m}^{-2}$

(b) Using Eq. (3.31) the potential difference between top and bottom of each layer may be calculated. Addition of these increments gives for the potential difference from 0 to 20 km 410 kV.

(c) The increment from 20 to 80 km is 123 kV; therefore, the difference between the ground and 80 km is 533 kV.

General References

Chalmers, *Atmospheric Electricty 2nd Ed.*, is well organized, clearly written, and provides a comprehensive account of atmospheric electricity and its relation to processes in clouds.

Fletcher, *The Physics of Rainclouds*, provides an excellent account of nucleation. His book is an excellent text for the student, but does not include the electrical aspects of cloud physics.

Halliday and Resnick, *Physics for Students of Science and Engineering*, and Sears, *Principles of Physics*, Vol. 2, give good accounts of elementary electrical principles.

Hirschfelder, Curtiss, and Bird, *Molecular Theory of Gases and Liquids*, provides a detailed and authoritative account of intermolecular forces.

Junge, "Atmospheric Chemistry" (*Advances in Geophysics, Volume 4*), presents an organized but detailed account of nuclei, their constitution, distribution, and physical processes.

Mason, *The Physics of Clouds*, in 1957 provided the first unified and comprehensive account of cloud physics. The second edition is an invaluable reference for deeper penetration of the subject.

Pruppacher and Klett, *Microphysics of Clouds and Precipitation*, provides a wealth of organized data and information and serves well as a coherent reference for the microphysics of clouds. It does not include cloud dynamics or chemistry, or the fields of application to weather modification and air quality.

Wallace and Hobbs, *Atmospheric Science, An Introductory Survey*, provides additional experimental and observational data on cloud physics, as well as descriptions of clouds and cloud systems.

CHAPTER IV

Atmospheric Motions

"Perhaps someday in the dim future it will be possible to advance the computations faster than the weather advances and at a cost less than the saving to mankind due to the information gained. But that is a dream." L. F. RICHARDSON

That was the comment made in 1922 by Lewis F. Richardson, who had carried out a numerical experiment in which he attempted to predict changes in the state of the atmosphere (its pressure, density, and velocity) using the fundamental equations of fluid mechanics and thermodynamics. He carried out much of the laborious effort of a single prediction while serving as an ambulance driver in France during the First World War and spent several more years in completing the work. The prediction failed dismally, but the causes of the failure were in large part understood by Richardson. He foresaw that it might be possible to overcome certain observational and computational difficulties, but he did not foresee development of the large, high-speed computers which has made numerical weather prediction a major research activity and a central function of modern weather services.

4.1 Atmospheric Forces

The velocity of air changes in response to forces acting on it; at the same time air velocity redistributes the mass of the atmosphere and changes the pressure field. Thus the velocity, density, and pressure fields are coupled, and we can expect to understand the atmospheric state variables or to predict them in a mathematical sense only by solving a set of simultaneous equations representing the physics of the atmosphere. The fundamental physical principles which govern atmospheric behavior have been introduced earlier; they are: Newton's second law of motion (or conservation of momentum), the conservation of mass, and the conservation of energy. In Chapter II it was shown that the latter two principles lead, respectively, to the equation of continuity (2.28a or 2.28b) and the first law of the thermodynamics (2.34 or 2.42). These will be used in discussing atmospheric motions in this chapter.

Newton's second law, which was introduced in Chapter I, is expressed economically as a vector equation (1.2); the vector form is the equivalent of three independent scalar equations written for the three velocity components, u in the x direction, v in the y direction, and w in the z direction. We shall find it advantageous in this chapter to use both the vector equation and the scalar component equations, and it proves to be easy to transform from one to the other.

Of the real forces acting on the air, the pressure and gravitational forces have been introduced in Chapter I. The pressure gradient force per unit mass may be expressed by the expansion

$$-\frac{1}{\rho}\nabla p = -\frac{1}{\rho}\left(\mathbf{i}\,\frac{\partial p}{\partial x} + \mathbf{j}\,\frac{\partial p}{\partial y} + \mathbf{k}\,\frac{\partial p}{\partial z}\right) \qquad (1.21)$$

The gravitational force per unit mass is expressed by $\mathbf{g}^*$ in Eq. (1.5).

In addition to these two forces the *frictional force* is also important in atmospheric motions. This force is most important in the *planetary boundary layer*, the layer in contact with the earth's surface about 1 km in thickness and containing about 10% of the mass of the atmosphere.

In order to formulate the frictional force it is useful to refer to the concept of stress which has been introduced in Chapter II and has been expressed by Eq. (2.73) as the product of molecular viscosity and velocity shear. Although the molecular stress is only observed under certain flow conditions, called *laminar* or nonturbulent, and the conditions in the boundary layer are usually *turbulent*, we can use this concept to illustrate how the frictional force may be formulated. Consider a fluid element in the interior of the fluid as depicted on the right side of Fig. 4.1. The fluid above the element exerts a stress *on* the fluid element in the x direction. The fluid below the

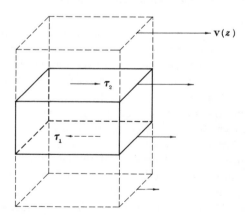

FIG. 4.1.   Vertical profile of horizontal wind velocity $\mathbf{V}(z)$ and the horizontal stresses τ_1 and τ_2 acting *on* an element of air.

element exerts an equal stress in the $-x$ direction *on* the element. Consequently, in this case there is no net stress.

More generally, however, the stress may vary with height. The net force acting on the element in the x direction is then

$$\text{Net force} = \left(\tau_x + \frac{\partial \tau_x}{\partial z} \delta z \right) \delta x \, \delta y - \tau_x \, \delta x \, \delta y$$

and the net force per unit mass becomes, upon substituting from Eq. (2.73),

$$\frac{1}{\rho} \frac{\partial \tau_x}{\partial z} = \frac{1}{\rho} \frac{\partial}{\partial z} \left(\mu \frac{\partial u}{\partial z} \right) \tag{4.1}$$

Equation (4.1) shows that there can be a net force if the shear varies in the z direction or if the viscosity varies with z.

In the case of turbulent fluids, which are further discussed in Chapter VI, it is convenient to introduce the *eddy viscosity* μ_e and the kinematic eddy viscosity $K_m \equiv \mu_e/\rho$. By analogy with Eq. (4.1) the net force per unit mass due to turbulence may now be written in the form

$$\frac{1}{\rho} \frac{\partial \tau_x}{\partial z} = \frac{\partial}{\partial z} \left(K_m \frac{\partial u}{\partial z} \right) \tag{4.2}$$

where the small variation of ρ with height in the boundary layer is neglected. Or combining the two horizontal components in vector form, the frictional force per unit mass, $\mathbf{F}_F$, may be written

$$\mathbf{F}_F = \frac{1}{\rho} \frac{\partial \tau}{\partial z} = \frac{\partial}{\partial z} \left(K_m \frac{\partial \mathbf{V}}{\partial z} \right) \tag{4.3}$$

4.2 The Coriolis Force

It was pointed out in Chapter I that Newton's second law is valid for an unaccelerated coordinate system, and consequently that if velocity is measured with respect to a point on the rotating earth, an apparent centrifugal force must be added to the real forces acting on the air. It was also shown that this apparent force could be added to the gravitational force, and the sum of these two forces is referred to as the force of gravity.

If we consider the effect of rotation, not as before on a point fixed on the earth, but on a moving point, for example, on an air parcel moving with the wind, an additional apparent force arises. To recognize this force in its simplest form, imagine a flat disk at the North Pole rotating counterclockwise with the earth as shown in Fig. 4.2. At the initial instant an object is projected at a certain speed from the Pole along a latitude line having the initial direction shown in Fig. 4.2. As seen from space, in the unaccelerated coordinate system the projectile continues in this direction, but as seen from a point on the rotating disk as the latitude line rotates through an angle $\Delta\theta$, the projectile appears to curve toward the west. This *thought* experiment can be easily verified using a phonograph turntable and a ball bearing. The curved path of the rolling ball across the turntable is easily recognized, and with a little ingenuity the curvature can be measured. The apparent force which must be invoked in this case if Newton's second law is to be valid in the rotating coordinate system is called a *Coriolis* force.

Another simple experiment demonstrates that the Coriolis force is associated with velocity in the azimuthal direction as well as with velocity in the radial direction. A projectile fired in the azimuthal direction as shown in Fig. 4.3 follows a straight line as seen in the unaccelerated coordinate system. As seen from a point on the rotating disk as the earth rotates through the angle $\Delta\theta$, the projectile curves to the right. In this case the centrifugal force acts in the same direction as the Coriolis force and complicates the problem somewhat. The point to be noted here is that the Coriolis force acts to the right of the direction of motion for azimuthal motion as well as for radial motion. Because the direction of rotation of the earth as seen from above the South Pole is clockwise rather than counterclockwise as in the Northern Hemisphere, the Coriolis force in the Southern Hemisphere acts to the left of the direction of the motion. The solutions to Problems 1 and 2 provide mathematical expressions for the Coriolis forces in the cases illustrated in Figs. 4.2 and 4.3.

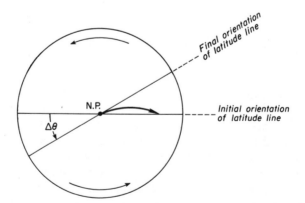

FIG. 4.2.   Disk at the North Pole rotating with the earth. The path followed by a projectile while the earth rotates through an angle $\Delta\theta$ as seen from a point on the earth is shown by the curved arrow.

It can be readily recognized that, except for motion parallel to the earth's rotational axis, all motions with respect to a point fixed on the earth give rise to Coriolis forces. Although they may be developed separately for the three velocity components, it is far simpler to develop the Coriolis force using vector methods and then to resolve the resulting vector into components. To do this we first express the velocity as measured in an unaccelerated (or absolute) coordinate system by

$$\mathbf{V}_a = \mathbf{V} + \mathbf{\Omega} \times \mathbf{r} \tag{4.4}$$

where $\mathbf{V}$ is the velocity as measured with respect to the rotating system (the relative velocity), $\mathbf{\Omega}$ is the angular frequency of the earth's rotation (directed parallel to the earth's axis outward from the North Pole), and $\mathbf{r}$ is the position vector from the center of the earth. The *vector* or *cross* product is discussed in Appendix I.B from which it may be recognized that $\mathbf{\Omega} \times \mathbf{r}$ represents the linear velocity of a point in the rotating system with respect to the absolute system. Since $\mathbf{V}$ represents $d\mathbf{r}/dt$, the relative time derivative of the position vector, and $\mathbf{V}_a$ represents $d_a\mathbf{r}/dt$, the absolute time derivative of the position vector, Eq. (4.4) may be written

$$\frac{d_a\mathbf{r}}{dt} = \frac{d\mathbf{r}}{dt} + \mathbf{\Omega} \times \mathbf{r} = \left(\frac{d}{dt} + \mathbf{\Omega} \times \right)\mathbf{r}$$

This shows that the absolute time derivative is expressible as

$$\frac{d_a}{dt} = \left(\frac{d}{dt} + \mathbf{\Omega} \times \right) \tag{4.5}$$

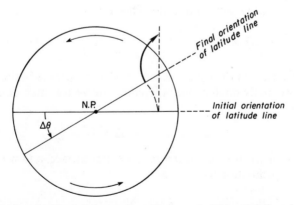

FIG. 4.3. Disk at the North Pole rotating with the earth. The path followed by a projectile while the earth rotates through an angle $\Delta\theta$ is shown as seen from an unaccelerated coordinate system (straight dashed line) and as seen from a point on the earth (curved arrow).

Now if the operator on the right side is applied to the absolute velocity given by Eq. (4.4)

$$\frac{d_a\mathbf{V}_a}{dt} = \left(\frac{d}{dt} + \mathbf{\Omega}\times\right)(\mathbf{V} + \mathbf{\Omega}\times\mathbf{r})$$

$$= \frac{d\mathbf{V}}{dt} + \mathbf{\Omega}\times\mathbf{V} + \mathbf{\Omega}\times\frac{d\mathbf{r}}{dt} + \mathbf{\Omega}\times(\mathbf{\Omega}\times\mathbf{r})$$

$$= \frac{d\mathbf{V}}{dt} + 2\mathbf{\Omega}\times\mathbf{V} + \mathbf{\Omega}\times(\mathbf{\Omega}\times\mathbf{r}) \qquad (4.6)$$

Equation (4.6) shows that the acceleration in unaccelerated coordinates represents the sum of the relative acceleration $d\mathbf{V}/dt$, the Coriolis acceleration or force $2\mathbf{\Omega}\times\mathbf{V}$, and the centrifugal acceleration or force $\mathbf{\Omega}\times(\mathbf{\Omega}\times\mathbf{r})$. We may recognize that the centrifugal force is directed normal to the earth's axis, and the Coriolis force is directed at right angles to the velocity. The vector Coriolis force is largest for velocity normal to the earth's axis and vanishes for velocity parallel to the earth's axis. Problem 3 illustrates the effect of direction on the Coriolis force, and Problem 4 extends the concept to a nonatmospheric application.

4.3 The Equations of Motion

The vector equation of motion can now be written by introducing the vector forms of the pressure, gravitational, and frictional forces [Eqs. (1.21), (1.4), and (4.3)] into Newton's second law [Eq. (1.2)] and expanding the absolute acceleration using Eq. (4.6). This gives

$$\frac{d\mathbf{V}}{dt} + 2\mathbf{\Omega}\times\mathbf{V} = -\frac{1}{\rho}\nabla p + \frac{1}{\rho}\frac{\partial\mathbf{\tau}}{\partial z} + \mathbf{g}^* - \mathbf{\Omega}\times(\mathbf{\Omega}\times\mathbf{r}) \qquad (4.7)\dagger$$

The last two terms on the right depend only on properties of the solid earth which are essentially unchanging; therefore, the vector force of gravity may be defined by

$$\mathbf{g} \equiv \mathbf{g}^* - \mathbf{\Omega}\times(\mathbf{\Omega}\times\mathbf{r}) \qquad (4.8)$$

The direction of $\mathbf{g}$ is normal to the geopotential surface, and we may define the vertical coordinate z by

$$\mathbf{g} \equiv -\mathbf{k}g \qquad (4.9)$$

† Note that the frictional force in Eq. (4.7) is limited to the horizontal components.

where **k** is the outward directed unit vector normal to the geopotential surface and g is the scalar acceleration of gravity. Equation (4.7) may now be written

$$\frac{d\mathbf{V}}{dt} + 2\mathbf{\Omega} \times \mathbf{V} = -\frac{1}{\rho}\nabla p + \frac{1}{\rho}\frac{\partial \tau}{\partial z} + \mathbf{g} \qquad (4.10)$$

When Eq. (4.10) is applied to the *free* atmosphere, i.e., the portion of the atmosphere above the planetary boundary layer, the friction term $1/\rho(\partial \tau/\partial z)$ can usually be neglected.

For many applications Eq. (4.10) is resolved into components in a convenient coordinate system. Here we will resolve each of the terms into Cartesian components in which, as shown in Fig. 4.4 at a specified point on the earth, the x axis is along a latitude line toward the east, the y axis is along a longitude line toward the north, and the z axis is vertically upward. In this system the x, y plane is tangent to the earth at the origin of coordinates. The equations also may be resolved into components in a spherical coordinate system; this more complicated step will be unnecessary for the topics discussed in this chapter.

The velocity is expressible as the vector sum of the three component velocity vectors; therefore, the vector acceleration may be resolved into components by the equation

$$\frac{d\mathbf{V}}{dt} = \mathbf{i}\frac{du}{dt} + \mathbf{j}\frac{dv}{dt} + \mathbf{k}\frac{dw}{dt} \qquad (4.11)$$

where u, v, and w represent the velocity components in the x, y, and z directions, respectively. The pressure and gravity forces are resolved into components by Eqs. (1.21) and (4.9), respectively. The friction force is assumed

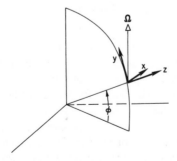

FIG. 4.4.   Cartesian coordinate system at a point on the earth at latitude ϕ.

to be horizontal; therefore, the two components are given by

$$\frac{1}{\rho}\frac{\partial \tau}{\partial z} = \frac{1}{\rho}\left(\mathbf{i}\frac{\partial \tau_x}{\partial z} + \mathbf{j}\frac{\partial \tau_y}{\partial z}\right) \qquad (4.12)$$

To resolve the Coriolis force into components we first note that $\mathbf{\Omega}$ may be resolved into components as follows

$$\mathbf{\Omega} = \Omega(\mathbf{j}\cos\phi + \mathbf{k}\sin\phi) \qquad (4.13)$$

There is no component of $\mathbf{\Omega}$ in the x direction. The Coriolis force may be expanded into components by expanding the vector product $2\mathbf{\Omega} \times \mathbf{V}$ as a determinant, as follows

$$2\mathbf{\Omega} \times \mathbf{V} = 2\begin{vmatrix} \mathbf{i} & \mathbf{j} & \mathbf{k} \\ 0 & \Omega\cos\phi & \Omega\sin\phi \\ u & v & w \end{vmatrix}$$

$$= \mathbf{i}2\Omega(w\cos\phi - v\sin\phi) + \mathbf{j}2\Omega u\sin\phi - \mathbf{k}2\Omega u\cos\phi \quad (4.14)$$

This expansion may be verified using the vector products of the unit vectors as reviewed in Appendix I.B.

Finally, by collecting the appropriate components from Eqs. (1.21), (4.9), (4.11), (4.12), and (4.14) the three scalar equations of motion may be written

$$\frac{du}{dt} + 2\Omega(w\cos\phi - v\sin\phi) = -\frac{1}{\rho}\frac{\partial p}{\partial x} + \frac{1}{\rho}\frac{\partial \tau_x}{\partial z}$$

$$\frac{dv}{dt} + 2\Omega u\sin\phi = -\frac{1}{\rho}\frac{\partial p}{\partial y} + \frac{1}{\rho}\frac{\partial \tau_y}{\partial z} \qquad (4.15)$$

$$\frac{dw}{dt} - 2\Omega u\cos\phi = -\frac{1}{\rho}\frac{\partial p}{\partial z} - g$$

For the synoptic and planetary scales of motion to which these equations are usually applied, the vertical velocity w is small compared to the horizontal velocity v or u. Typically, w is of the order of 1 cm s^{-1}, while v or u is of the order 10 m s^{-1}. Similarly, for these scales of motion the vertical acceleration dw/dt is small compared to the acceleration of gravity and the vertical pressure gradient force. Also, the Coriolis term in the third equation is very much smaller than g or $(1/\rho)\partial p/\partial z$. Upon making these slight approximations and replacing $2\Omega\sin\phi$ by the symbol f (called the *Coriolis parameter*), Eq. (4.15) may be written

$$\frac{du}{dt} - fv = -\frac{1}{\rho}\frac{\partial p}{\partial x} + \frac{1}{\rho}\frac{\partial \tau_x}{\partial z} \tag{4.16}$$

$$\frac{dv}{dt} + fu = -\frac{1}{\rho}\frac{\partial p}{\partial y} + \frac{1}{\rho}\frac{\partial \tau_y}{\partial z} \tag{4.17}$$

$$g = -\frac{1}{\rho}\frac{\partial p}{\partial z} \tag{1.18}$$

The Coriolis parameter is a slowly varying function of the south–north direction. It will be treated as constant in the following discussion. Note that in Eqs. (4.16) and (4.17) the orientation of the x axis is not limited to the west–east direction, but can be chosen arbitrarily so long as it is horizontal.

The implication of Eq. (1.18) is that pressure at a point is determined uniquely by the vertical distribution of density (and g, of course). However, the density and velocity fields are coupled through the horizontal equations of motion [Eqs. (4.16) and (4.17)]. Note that although the vertical velocity does not appear explicitly in the horizontal equations, it may be made explicit by expanding the individual derivatives as follows (see Appendix I.D):

$$\frac{du}{dt} = \frac{\partial u}{\partial t} + u\frac{\partial u}{\partial x} + v\frac{\partial u}{\partial y} + w\frac{\partial u}{\partial z} \tag{4.18}$$

and similarly for dv/dt. Although dw/dt was neglected in the third equation of motion and w was neglected compared to v in the first equation of motion, it does not follow necessarily that w can be neglected in expansions of the individual accelerations. It turns out that for synoptic and planetary scales of motion the last term on the right of Eq. (4.18) can be neglected, but for smaller scales of motion it may be comparable to the other terms.

In Sections 4.4–4.10, which are concerned with the atmosphere above the boundary layer, the frictional term in the equations of motion will be neglected. In Section 4.11 effects of boundary layer friction will be considered.

4.4   Applications of the Horizontal Equations of Motion

The Geostrophic Wind

A rich variety of different types of flow can be described or predicted from Eqs. (4.16) and (4.17). Their characteristics can be discussed in relation to the relative magnitudes of the individual terms. A particularly simple and important relationship holds under the condition that the acceleration terms, as well as the friction terms, are small compared to the Coriolis terms. Under these conditions, the Coriolis and pressure gradient forces are balanced, and this balance is used to define the *geostrophic* velocity components by the equations

$$v_g \equiv \frac{1}{f\rho}\frac{\partial p}{\partial x} \qquad (4.19)$$

$$u_g \equiv -\frac{1}{f\rho}\frac{\partial p}{\partial y} \qquad (4.20)$$

These scalar equations can be combined into the vector equation

$$\mathbf{V}_g \equiv \frac{1}{f\rho}(\mathbf{k} \times \nabla p) \qquad (4.21)$$

Equation (4.21) shows that the geostrophic wind flows at right angles to the horizontal pressure gradient, that is, along isobars on a horizontal surface. Under balance of the Coriolis and pressure gradient forces, the wind flows along the isobars with high pressure on the right and low pressure on the left in the Northern Hemisphere as shown in Fig. 4.5 (Buys Ballot's law). An exercise is provided in Problem 5.

What is implied by the condition that the acceleration terms in Eqs. (4.16) and (4.17) be small compared to the Coriolis terms? It means simply that the time for the velocity components to change by 100% of their average absolute values must be long compared to $|1/f|$. Since f has a value of approximately $10^{-4}\,\mathrm{s}^{-1}$ in middle latitudes, the time required for one of the velocity components to change by 100% must be long compared to $10^4\,\mathrm{s}$ or about 3 hr. Air moving at the typical magnitude of about $10\,\mathrm{m\,s}^{-1}$ travels about 100 km in 3 hr, so that we may conclude that the size of *weather systems* or the distance traveled between reversals of direction must be substantially larger than about 100 km in order that the actual wind velocity be approximately equal to the geostrophic velocity.

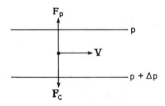

FIG. 4.5. Balance of Coriolis force ($\mathbf{F}_c$) in the Northern Hemisphere and pressure force ($\mathbf{F}_p$) for an air parcel moving at velocity $\mathbf{V}$ parallel to the isobars on a horizontal surface.

The pressure systems often shown on weather maps and the systems shown in Fig. 1.5 are much larger than 100 km, so that the winds are nearly geostrophic. One consequence is that in the Northern Hemisphere winds blow counterclockwise around large low pressure centers (cyclones) and clockwise around high pressure centers (anticyclones). In the Southern Hemisphere the opposite relation holds, flow is clockwise around low pressure and counterclockwise around high pressure. In each case the term *cyclonic* refers to flow around low pressure, and *anticyclonic* refers to flow around high pressure.

It is important to realize, however, that in general the wind velocity is not exactly geostrophic. If it were, Eqs. (4.16) and (4.17) would no longer depend on time, and weather could not change. However, even in problems concerned with prediction, it is useful to retain the concept and definition of the geostrophic wind. In fact, the horizontal equations of motion may be written in the vector form

$$\frac{d\mathbf{V}_H}{dt} + f\mathbf{k} \times (\mathbf{V} - \mathbf{V}_g) = 0 \qquad (4.22)$$

and in scalar form

$$\frac{du}{dt} - f(v - v_g) = 0 \qquad (4.23)$$

$$\frac{dv}{dt} + f(u - u_g) = 0 \qquad (4.24)$$

Equations (4.23) and (4.24) are applied to a very simple case in the solution of Problem 6. We shall return to these equations later in this section, after introducing other simplified forms of the horizontal equations of motion.

The Gradient Wind

Weather maps show that in the lower part of the troposphere low pressure and high pressure systems are often oval in form and sometimes may be nearly circular. Under these conditions the acceleration terms in Eqs. (4.16) and (4.17) or (4.23) and (4.24) may be comparable to the Coriolis or pressure gradient terms even though the speed may be nearly constant. For example, an air parcel moving in a circular path as shown in Fig. 4.6 experiences a change in the u or v velocity component from maximum negative to maximum positive in moving halfway around the circle. In many cases the time involved will be comparable to $1/f$ ($\sim 3\,hr$), and therefore the geostrophic approximation cannot be expected to hold. However, the acceleration normal

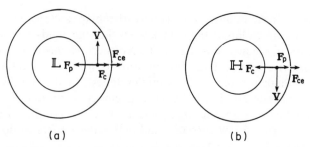

FIG. 4.6.   (a) Balance of centrifugal ($\mathbf{F}_{ce}$), Coriolis ($\mathbf{F}_c$), and pressure ($\mathbf{F}_p$) forces for flow around a low pressure center (L) in the Northern Hemisphere. (b) Balance of centrifugal, Coriolis, and pressure forces for flow around a high pressure center (H) in the Northern Hemisphere.

to the direction of flow can be expressed as centrifugal acceleration, and there may be balance of the centrifugal, Coriolis, and pressure forces. Balance of these forces in cyclonic (counterclockwise) and anticyclonic (clockwise) flow in the Northern Hemisphere is illustrated in Fig. 4.6.

In order to express Eqs. (4.16) and (4.17) in a form appropriate to the general case of curved flow, it is convenient to transform the equations to the *natural* coordinate system. In this system the direction of flow projected onto the horizontal plane at a specified point is designated as in the direction of the s axis, and the n axis is designated as perpendicular to s, to the left, as illustrated in Fig. 4.7. We now consider a Cartesian coordinate system oriented such that the s- and x axes coincide at the specified point and the n and y axes similarly coincide at this point. The horizontal wind speed, which is represented by u in the Cartesian system, is represented by V in the natural system. Since the velocity in the n direction is zero by definition, we recognize that $v = 0$ at this point. We may now transform Eqs. (4.16) and (4.17) to natural coordinates in the form

$$\frac{dV}{dt} = -\frac{1}{\rho}\frac{\partial p}{\partial s} \tag{4.25}$$

$$\frac{V^2}{r} + fV = -\frac{1}{\rho}\frac{\partial p}{\partial n} \equiv fV_g \tag{4.26}$$

where r represents the radius of curvature of the trajectory of the moving air parcel. Equation (4.25) shows that the horizontal speed changes only when there is a component of pressure gradient in the direction of flow, that is, when the pressure field does work (positive or negative) on the air parcel. Air flowing with a component toward lower pressure accelerates, while air flowing toward higher pressure decelerates. Equation (4.26) is

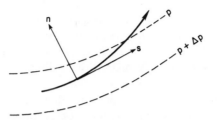

FIG. 4.7. Orientation of the s and n axes of the natural coordinate system at a specified point on the trajectory of an air parcel.

called the *gradient wind equation*; it represents balance of the centrifugal, Coriolis, and pressure gradient forces in the direction normal to the flow. It may be applied to small scale motions as well as to larger scales.

It is important to recognize that the radius of curvature of the trajectories of air parcels are in general not the same as the radius of curvature of *streamlines* of the flow, lines everywhere parallel to the direction of flow at a specified instant. If the streamlines are unchanging in time, then the moving parcel follows the streamlines, and in this case trajectories and streamlines have the same curvature. On the other hand, if the streamlines are changing, the trajectories and streamlines will have different curvatures. It follows that in applying the gradient wind equation to the case of circular isobars as shown in Fig. 4.6, the radius of curvature of an isobar can be used for r only if the pressure field is not changing in time.

The gradient wind equation may be treated as a quadratic equation in V which has the solution

$$V = -\tfrac{1}{2}fr[1 \pm (1 + 4V_g/fr)^{1/2}] \qquad (4.27)$$

V is essentially positive, but r can be positive (counterclockwise flow) or negative (clockwise flow). It follows that for counterclockwise flow in the Northern Hemisphere ($r > 0$, $f > 0$) only the negative radical may be used. On the other hand, for clockwise flow there are two possible solutions corresponding to the positive and negative radicals. Note that for both counterclockwise flow around a low pressure center and clockwise flow around a high pressure center $\partial p/\partial n$ is negative, and therefore V_g/f is positive, but for clockwise flow around a low pressure center $\partial p/\partial n$ is positive, and therefore V_g/f is negative. Negative V_g indicates that the direction of flow is opposite to the geostrophic direction. Note that balanced counterclockwise flow around a high pressure center in the Northern Hemisphere is not dynamically possible, for in this case the centrifugal, Coriolis, and pressure gradient forces would all be in the same direction.

The solutions of Eq. (4.27) may be conveniently represented by the gradient wind diagram shown in Fig. 4.8. The abscissa is conveniently taken as the dimensionless ratio of acceleration to Coriolis force per unit mass, which in this case is $V/|fr|$. This ratio, called the *Rossby number*, plays an important role in atmospheric dynamics. Similarly, the ordinate is conveniently taken as $V_g/|fr|$.

The upper branch of Fig. 4.8 represents flow in the counterclockwise direction around a low pressure center. For synoptic scale cyclones the ratio $V/|fr|$ is typically of order 0.1. When $V/|fr|$ is much greater than unity, the centrifugal force dominates the Coriolis force, and balance occurs between the centrifugal and pressure forces. This situation is encountered in small vortices such as tornadoes, water spouts, and dust devils, and these are referred to as *cyclostrophic* winds. In these cases, in which $|fr|$ is very

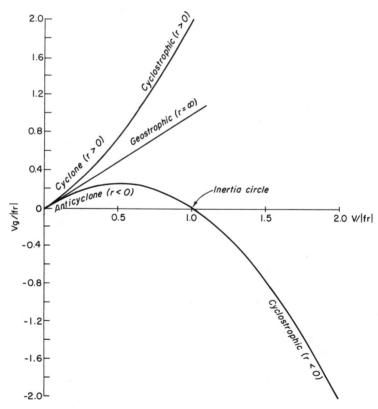

FIG. 4.8.   Gradient wind diagram for circular flow in the counterclockwise direction $(r > 0)$ and the clockwise direction $(r < 0)$ in the Northern Hemisphere.

small compared to its value in a cyclone, the upper branch would have to be extended beyond the limits of Fig. 4.8.

The lower branch in Fig. 4.8 represents flow in the clockwise direction. The portion close to the origin (Rossby number small compared to unity) represents flow in an anticyclone (central high pressure); for Rossby numbers between 0.5 and 1.0 a second possible balance occurs between centrifugal, Coriolis, and pressure gradient forces which is seldom observed in the atmsophere and is therefore referred to as the *anomalous* anticyclone. The portion of this branch below the axis ($V_g < 0$) represents clockwise flow around a low pressure center; in this case centrifugal force dominates the Coriolis force, and again we encounter the condition for the *cyclostrophic* wind. It is easy to demonstrate using water in a circular basin that cyclostrophic flow around a low pressure center may occur in either the clockwise or counterclockwise direction. To represent this case in which r is very small, Fig. 4.8 would have to be extended greatly to very large Rossby numbers.

Figure 4.8 also shows the relation of the gradient wind to the geostrophic wind. For a given pressure gradient, the geostrophic wind speed lies between the higher gradient wind speed for the anticyclone and the lower gradient wind speed for the cyclone. Figure 4.8 also shows that in an anticyclone the maximum value of the geostrophic wind is $(1/4)|fr|$, whereas there is no such limit for the cyclone. Problems 7 and 8 provide applications of the gradient wind equation to synoptic scale phenomena, while Problem 9 provides an application to a much smaller scale phenomenon.

The horizontal scales (typical distance in which a velocity component changes by 100% of its value) and the Rossby numbers corresponding to some of the important atmospheric phenomena are listed in Table 4.1.

Inertial Motions

Another simple and interesting case exists when there is no pressure gradient force but the wind speed is finite. There is then balance between

TABLE 4.1
SCALES OF SOME ATMOSPHERIC PHENOMENA

Phenomenon	Horizontal scale or radius	Rossby number $(V/|fr|)$
Planetary waves	10,000 km	10^{-2}
Cyclones, anticyclones	1,000 km	10^{-1}
Hurricanes	100 km	10^{0}
Tornadoes	<1 km	$>10^{2}$

the centrifugal and Coriolis forces as shown by Eq. (4.26), and the air must move in a circular path whose radius is given by

$$r_i = -\frac{V}{f} = -\frac{V}{2\Omega \sin \phi} \tag{4.28}$$

This is the radius of the *inertia circle*. This solution of the gradient wind equation is represented by the point on the gradient wind diagram (Fig. 4.8) where $V/|fr|$ equals unity. The negative sign in Eq. (4.28) indicates that in the Northern Hemisphere flow in the inertial circle is clockwise; this reflects the fact that, as discussed earlier, an object moving relative to the earth's surface experiences a Coriolis force which tends to result in curvature toward the right. The radius of the inertia circle is directly proportional to wind speed and inversely proportional to sine of the latitude. At the poles the radius of the inertia circle for a 10 m s^{-1} wind speed is about 70 km. The period of the inertia circle is easily found to be

$$\tau_i = \frac{\pi}{\Omega \sin \phi} \tag{4.29}$$

The period given by Eq. (4.29) is half a pendulum day, that is, half the period of rotation of a Foucault pendulum.† The period is 12 hr at the poles, 17 hr at 45° latitude, and 24 hr at 30° latitude.

Inertia circles are rarely, if ever, observed in the atmosphere because the condition of negligible horizontal pressure gradient is seldom realized over periods as long as 10 or 20 hr. In the ocean, where the pressure field changes less rapidly than in the atmosphere, evidence of inertia circles has been found.

Inertia circles may be visualized as superimposed on more complicated flow fields, and the concept of the resulting *inertial motions* is important in understanding atmospheric dynamics. Consider, for example, the case of uniform flow parallel to the x axis and parallel to isobars on a horizontal plane. Just prior to time $t = 0$ there is a change in the magnitude of the pressure gradient, so that the initial wind field is not in geostrophic balance. It is assumed that thereafter the pressure field does not change with time. The problem is to determine the response of the velocity field to the changed pressure distribution. If f is treated as a constant following the motion

---

† The Foucault pendulum demonstrates the rotation of the earth as it oscillates in a plane fixed with respect to the stars while the plane of the horizon rotates under it. Foucault pendulums can be seen in the National Academy of Sciences in Washington, D.C., the Franklin Institute in Philadelphia, the South Kensington Museum of Sciences in London, and in other institutions.

and v_g is set equal to zero, Eqs. (4.23) and (4.24) can be transformed to the two second-order differential equations

$$\frac{d^2u}{dt^2} + f^2(u - u_g) = 0$$

$$\frac{d^2v}{dt^2} + f^2v = 0 \tag{4.30}$$

Initial conditions are $v = 0$ and $u = u_0$ at $t = 0$.

General solutions for Eqs. (4.30) may be written

$$u = u_g + A \sin ft + B \cos ft$$
$$v = A' \sin ft + B' \cos ft \tag{4.31}$$

where A, B, A', and B' are constants to be determined from the initial conditions. Upon setting $t = 0$, we find that $B' = 0$ and $B = u_0 - u_g$, and from Eqs. (4.23) and (4.24) that $A = 0$ and $A' = u_g - u_0$. Therefore, the particular solutions appropriate to the initial conditions are

$$u = u_g + (u_0 - u_g) \cos ft$$
$$v = -(u_0 - u_g) \sin ft \tag{4.32}$$

Since $u \equiv dx/dt$ and $v \equiv dy/dt$, Eqs. (4.32) may be integrated to yield

$$x - x_0 = u_g t + \frac{u_0 - u_g}{f} \sin ft \tag{4.33}$$

$$y - y_0 = \frac{u_0 - u_g}{f}(1 - \cos ft) \tag{4.34}$$

These are the parametric equations of the trajectories of air parcels under the given conditions.

A trajectory given by Eqs. (4.33) and (4.34) may be visualized as a point whose average displacement is at constant speed in the x direction but which oscillates in the y direction with an amplitude proportional to $u_0 - u_g$ and a period equal to the period of the inertia circle. An oscillation of the same amplitude also occurs in the x displacement. The wavelength of the oscillations is $2\pi u_g/f$. If the initial velocity is greater than geostrophic ($u_0 > u_g$), the parcel will move initially to the right, toward higher pressure. This will result in the parcel's slowing down until it is moving at less than the geostrophic speed; the pressure gradient force will then accelerate the parcel toward lower pressure. As its speed becomes supergeostrophic, it will curve to the right and will again be directed toward higher pressure to begin another cycle.

Although this example is unrealistic, especially in treating the pressure field as independent of the velocity field, it illustrates the nature of inertial motions. These motions may occur wherever the forces acting on the air are locally unbalanced, for example, in association with intense cyclones or jet streams.

4.5   The Equations of Motion on a Constant Pressure Surface

We have recognized that for motions of the horizontal scale of cyclones and anticyclones, which we call the *synoptic* scale (roughly 1000 km) or for the still larger scale of the general circulation which we call the *planetary* scale (roughly 10,000 km), the hydrostatic equation [Eq. (1.18)] represents accurately the relationship between pressure and height. Using this equation it is possible to replace height with pressure as the vertical coordinate, and it turns out that this simplifies some of the equations. The resulting equations are said to be expressed in *pressure coordinates.*

To express the horizontal pressure gradient in a form appropriate to the use of pressure as the vertical coordinate consider the vertical cross section shown in Fig. 4.9. The pressure gradient in the x direction at height z may be represented by

$$\frac{\Delta p}{\Delta x} = \rho g \frac{\Delta z_{\mathrm{p}}}{\Delta x}$$

where Δz_{p} designates the change in height of the pressure surface corre-

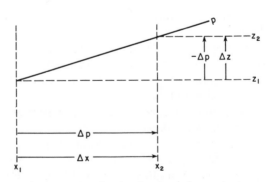

FIG. 4.9.   Vertical cross section showing the slope of a constant pressure surface p.

sponding to the displacement Δx. Therefore, upon taking limits the pressure gradient force in the x direction is

$$-\frac{1}{\rho}\frac{\partial p}{\partial x} = -g\left(\frac{\partial z}{\partial x}\right)_p = -\left(\frac{\partial \Phi}{\partial x}\right)_p \tag{4.35}$$

The geopotential (Φ) is defined by Eq. (1.10). It follows that the pressure gradient force in the y direction is

$$-\frac{1}{\rho}\frac{\partial p}{\partial y} = -g\left(\frac{\partial z}{\partial y}\right)_p = -\left(\frac{\partial \Phi}{\partial y}\right)_p \tag{4.36}$$

These relations show that by applying the equations of motion to constant pressure surfaces, the number of dependent variables is reduced by one, and this proves to be a valuable simplification. Consequently, for many theoretical and practical applications in meteorology the equations of motion are expressed in pressure coordinates.

The acceleration terms in the equations of motion implicitly involve the vertical coordinate, for the x equation the acceleration may be expanded in the form

$$\frac{du}{dt} = \left(\frac{\partial u}{\partial t}\right)_p + u\left(\frac{\partial u}{\partial x}\right)_p + v\left(\frac{\partial u}{\partial y}\right)_p + \frac{\partial u}{\partial p}\frac{dp}{dt} \tag{4.37}$$

where the subscript indicates that the derivative is taken at constant p. The meaning of these terms may be illustrated using Fig. 4.9 from which it is evident that

$$\left(\frac{\partial u}{\partial x}\right)_p \approx \frac{u_{x_2 z_2} - u_{x_1 z_1}}{x_2 - x_1}$$

where the subscripts identify the coordinates at which values of u are determined. The term dp/dt represents the rate of pressure change experienced by a moving parcel; it can be expanded in the form

$$\frac{dp}{dt} = \frac{\partial p}{\partial t} + u\frac{\partial p}{\partial x} + v\frac{\partial p}{\partial y} + w\frac{\partial p}{\partial z}$$

Upon introducing the hydrostatic equation, the last term on the right becomes $-w\rho g$. This term dominates the right side since its magnitude can be shown to be an order of magnitude or more larger than the other terms. Thus dp/dt is approximately proportional to vertical velocity; it is represented by the symbol ω.

Equations (4.16) and (4.17) can now be written in pressure coordinates in the form

$$\frac{du}{dt} - fv = -\left(\frac{\partial \Phi}{\partial x}\right)_p \tag{4.38}$$

$$\frac{dv}{dt} + fu = -\left(\frac{\partial \Phi}{\partial y}\right)_p \tag{4.39}$$

and the hydrostatic equation can be transformed to pressure coordinates in the form

$$\frac{\partial \Phi}{\partial p} = -\frac{1}{\rho} \tag{4.40}$$

The equations of motion can also be applied to other surfaces, in particular to surfaces of constant potential temperature. These equations are useful especially in studying flow under adiabatic conditions, for under these conditions air parcels adhere to potential temperature surfaces. Problem 10 requires the development of the horizontal pressure force for surfaces of constant potential temperature (*isentropic* coordinates).

4.6   Variation with Height of the Geostrophic Wind

Equation (4.40) may be integrated from pressure surface p_0 to surface p_1 with the result

$$\Phi_1 - \Phi_0 = \overline{R_m T_v} \ln \frac{p_0}{p_1} \tag{4.41}$$

where the virtual temperature as defined in Section 2.19 has been introduced through the equation of state and the bar represents an average over the layer. This *hypsometric equation* shows that the vertical separation of two pressure surfaces is directly proportional to the average virtual temperature of the air in the vertical column between them. It is obvious, therefore, that horizontal gradient of the mean virtual temperature is reflected in a change with height of the slope of pressure surfaces and in change in the geostrophic wind velocity. This result may be developed mathematically by differentiating the geostrophic equations [Eqs. (4.38) and (4.39) with acceleration terms omitted] with respect to p and differentiating Eq. (4.40) first with respect to

x at constant p and then with respect to y at constant p. The geopotential (Φ) can then be eliminated with the result

$$\frac{\partial v_g}{\partial p} = \frac{1}{f\rho^2}\left(\frac{\partial \rho}{\partial x}\right)_p = -\frac{R_m}{fp}\left(\frac{\partial T_v}{\partial x}\right)_p \tag{4.42a}$$

$$\frac{\partial u_g}{\partial p} = -\frac{1}{f\rho^2}\left(\frac{\partial \rho}{\partial y}\right)_p = \frac{R_m}{fp}\left(\frac{\partial T_v}{\partial y}\right)_p \tag{4.42b}$$

Upon combining these scalar equations

$$\frac{\partial \mathbf{V}_g}{\partial p} = -\frac{R_m}{fp}(\mathbf{k} \times \nabla_p T_v) \tag{4.43}$$

If Eq. (4.43) is multiplied by dp and integrated between pressure surfaces p_0 and p_1, the vector difference between these two surfaces is

$$\mathbf{V}_{g_1} - \mathbf{V}_{g_0} \equiv \mathbf{V}_{th} = \frac{R_m}{f}(\mathbf{k} \times \overline{\nabla_p T_v})\ln\frac{p_0}{p_1} \tag{4.44}$$

Equation (4.44) is called the *thermal wind equation*. It states that the vector difference of the geostrophic winds at two pressure surfaces is directed parallel to the mean virtual isotherms for the layer, with low temperature to the left in the Northern Hemisphere. Note the analogy of this statement to Buys Ballot's law of the geostrophic wind. Problem 11 requires developing an alternate form of the thermal wind equation, expressed in terms of the thickness of the layer between p_0 and p_1.

One consequence of Eq. (4.44) is that in most of the middle latitude troposphere, where temperature increases toward the equator, the geostrophic wind, which is usually from west to east, increases with height. This conclusion applies to the real wind as well as to the geostrophic wind.

Another consequence of Eq. (4.44) can be recognized with the aid of Fig. 4.10. In the case shown, cold air is on the left; warm air on the right; therefore, the thermal wind vector $\mathbf{V}_{th}$ is directed into the page with low temperature on the left. The geostrophic vector at the lower surface ($\mathbf{V}_0$) in Fig. 4.10 is from the cold air toward the warm air. On the upper surface the projection of the sum of $\mathbf{V}_0$ and $\mathbf{V}_{th}$ yields the geostrophic vector $\mathbf{V}_1$. Obviously, the direction of the geostrophic wind has turned in a counterclockwise direction with height. We call this *backing* of the wind with height, and we recognize that backing with height corresponds to the horizontal transport of cold air. We call this cold air *advection* and represent it quantitatively by $-\mathbf{V} \cdot \nabla_p T_v$. If the wind were directed from warm air to cold, the geostrophic

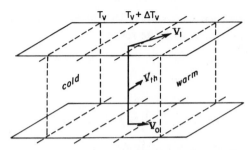

FIG. 4.10.   Three-dimensional sketch showing the relation between the thermal wind vector $\mathbf{V}_{th}$ and the geostrophic vectors at top ($\mathbf{V}_1$) and at bottom ($\mathbf{V}_0$) of layer in the case of cold advection.

wind would turn with height in the clockwise sense (*veer* with height). Therefore, by observing the direction of turning of the wind with height (from differential cloud motions, for example) one can determine whether there is warm advection or cold advection.

4.7   The Circulation Theorems

Vortices are observed in fluid flows on a variety of scales, and these vortices often persist for long periods as discrete identifiable elements. Examples are found in the wake of a canoe paddle pulled through the water, and in tobacco smoke in a closed room, as well as in tornadoes and synoptic scale cyclones and anticyclones. The relatively long lives of these vortices reveal an important property of fluid flow which can be expressed mathematically by formulating the equations of motion in a special way.

We first define a scalar quantity known as the *circulation* by the equation

$$C \equiv \oint \mathbf{V} \cdot d\mathbf{s} \tag{4.45}$$

where $d\mathbf{s}$ represents the differential of distance taken along a closed curve. The line integral is taken in the counterclockwise direction. In the special case of circular flow at uniform speed, the circulation around a circle tangent to the flow is just the speed times the circumference of the circle. In the case of uniform rectilinear flow, the circulation around any closed path is zero, and in general the circulation may be positive or negative depending upon the flow field and the path chosen. Now choose an arbitrary closed curve within the fluid, form the scalar product of the vector equation of motion in unaccelerated or absolute coordinates [Eqs. (4.6) and (4.7)] and the differential element of distance along the curve, and integrate around the curve (neglecting friction) with the result

$$\oint \frac{d_a \mathbf{V}_a}{dt} \cdot d\mathbf{s} = - \oint \frac{\nabla p}{\rho} \cdot d\mathbf{s} + \oint \mathbf{g}_a^* \cdot d\mathbf{s} \qquad (4.46)$$

The final line integral vanishes because $\mathbf{g}_a^*$ is a conservative vector (depends only on position). The first term on the right of (4.46) is equivalent to $\oint dp/\rho$. Alternatively, it may be transformed as follows, using Stokes's theorem (developed in Appendix I.F).

$$- \oint \frac{\nabla p}{\rho} \cdot d\mathbf{s} = - \int\int \nabla \times \frac{\nabla p}{\rho} \cdot d\boldsymbol{\sigma} = \int\int \frac{1}{\rho^2} \nabla \rho \times \nabla p \cdot d\boldsymbol{\sigma} \qquad (4.47)$$

where $d\boldsymbol{\sigma}$ represents the differential element of surface area.

The left side of Eq. (4.46) may be expanded as follows:

$$\frac{d_a}{dt} \oint \mathbf{V}_a \cdot d\mathbf{s} = \oint \frac{d_a \mathbf{V}_a}{dt} \cdot d\mathbf{s} + \oint \mathbf{V}_a \cdot d\frac{d_a \mathbf{s}}{dt} = \oint \frac{d_a \mathbf{V}_a}{dt} \cdot d\mathbf{s} + \frac{1}{2} \oint d(V_a^2)$$

The last term vanishes because V_a^2 is a function of position only. Then upon utilizing Eq. (4.45)

$$\frac{d_a C_a}{dt} = \oint \frac{d_a \mathbf{V}_a}{dt} \cdot d\mathbf{s}$$

where C_a is the absolute circulation. Upon noting that the individual derivative of a scalar in the absolute system is the same as the individual derivative in the relative system, Eq. (4.46) can now be written

$$\frac{dC_a}{dt} = - \oint \frac{dp}{\rho} \qquad (4.48)$$

Equation (4.48) is known as the *Kelvin circulation theorem*; it is a simple but powerful predictive equation. If density is a function of pressure only the line integral on the right is zero, and the absolute circulation is conserved following the moving contour. A fluid in which density depends only on pressure is called a *barotropic* fluid.

The *Bjerknes circulation theorem* can be derived by expanding C_a as follows:

$$C_a = \oint (\mathbf{V} + \boldsymbol{\Omega} \times \mathbf{r}) \cdot d\mathbf{s} = C + \oint \boldsymbol{\Omega} \times \mathbf{r} \cdot d\mathbf{s}$$

$$= C + \oint \boldsymbol{\Omega} \cdot \mathbf{r} \times d\mathbf{s}$$

But Ω is a constant vector and $\oint \mathbf{r} \times d\mathbf{s} = 2\mathbf{A}$ where $\mathbf{A}$ is the vector normal to the plane of the contour and equal to the area enclosed by the contour. Therefore, Eq. (4.48) becomes, upon utilizing Eq. (4.47)

$$\frac{d}{dt}(C + 2\Omega \cdot \mathbf{A}) = -\oint \frac{dp}{\rho} = \int\int \frac{1}{\rho^2} \nabla\rho \times \nabla p \cdot d\sigma \qquad (4.49)$$

where the vector $\rho^{-2} \nabla\rho \times \nabla p$ is called the *baroclinity vector*. The term $2\Omega \cdot \mathbf{A}$ may be visualized as the earth's vorticity times the projection of the area enclosed by the contour on the plane of the equator. It therefore may be represented by $2\Omega F$ where F represents the projection of $\mathbf{A}$ on the plane of the equator. The Bjerknes circulation theorem may now be written

$$\frac{d(C + 2\Omega F)}{dt} = -\oint \frac{dp}{\rho} \qquad (4.50)$$

Equation (4.50) is a remarkably concise statement of the relationship between the velocity, pressure, and density fields and can be used to predict the state of a set of air parcels forming a particular closed curve at a particular instant. A few simple applications of the Bjerknes circulation theorem are described in the following paragraphs.

First, consider a vertical plane with a closed contour defined by two isobars and two vertical lines as shown in Fig. 4.11. Upon introducing the equation of state, the right side of Eq. (4.50) becomes

$$\oint R_m T_v \left(\frac{dp}{p}\right) = R_m \oint T_v \, d(\ln p)$$

Now integrate around the contour beginning at point (1). Pressure is constant along the lines from (1) to (2) and (3) to (4). Along the other two sides

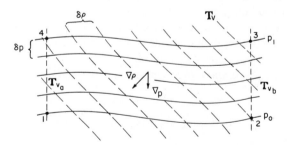

FIG. 4.11.   A closed contour lying in a vertical plane defined by the isobars p_0 and p_1. The mean virtual temperature of the vertical column between points (1) and (4) is $\overline{T}_{va}$ and the mean virtual temperature of the column between (2) and (3) is $\overline{T}_{vb}$.

integration is performed by taking the mean virtual temperature outside the integral. Then the integral becomes

$$\oint \frac{dp}{\rho} = R_m \overline{T}_{vb} \ln \frac{p_1}{p_0} + R_m \overline{T}_{va} \ln \frac{p_0}{p_1} = R_m (\overline{T}_{va} - \overline{T}_{vb}) \ln \frac{p_0}{p_1}$$

So the baroclinic term in the Bjerknes circulation theorem depends directly on the difference in mean virtual temperatures of the two vertical columns weighted by the natural logarithm of the ratio of pressures at bottom and top. If change in F can be neglected, Eq. (4.50) reduces in this case to

$$\frac{dC}{dt} = -R_m (\overline{T}_{va} - \overline{T}_{vb}) \ln \frac{p_0}{p_1} \qquad (4.51)$$

showing that circulation will become more positive (counterclockwise) if $\overline{T}_{vb} > \overline{T}_{va}$ and vice versa. A good example is the change in circulation in a room heated on one side. Another example is the sea breeze which develops when air over land is heated relative to air over the cool sea, but in this case the circulation change is more complicated because the elapsed time is often long enough that the Coriolis force may produce a significant change in the orientation of the contour and therefore in F. It should also be recognized that prediction requires integration of Eq. (4.50) or (4.51) over a time interval, and that $(\overline{T}_{va} - \overline{T}_{vb})$ in general will be influenced by the velocity field and therefore will not be constant. Problem 12 requires calculation of the baroclinic term for a sea breeze case.

The baroclinic term also may be evaluated by a simple graphical technique using the form on the far right of Eq. (4.49). The cross product $\nabla \rho \times \nabla p$ represents a vector normal to the plane of the diagram shown in Fig. 4.11 and proportional to the area of the parallelogram formed by the lines of constant ρ and p. These areas may be visualized as intersections with the plane of the diagram of tubes or *solenoids* formed by the surfaces of constant ρ and constant p. Upon forming the scalar product with the vector representing the differential area, the result is the product $\delta \rho\, \delta p$, and the area integral is evaluated from

$$\int\int \frac{1}{\rho^2} \nabla \rho \times \nabla p \cdot d\boldsymbol{\sigma} = \sum_i \frac{1}{\rho^2} \delta_i \rho\, \delta_i p = N_{\rho, p} \frac{\delta \rho\, \delta p}{\rho^2}$$

where $N_{\rho, p}$ represents the number of $\delta \rho\, \delta p$ solenoids contained within the bounding curve. The baroclinic term in the circulation theorems is sometimes referred to as the *solenoidal* term.

Next consider a contour lying in a pressure surface; clearly, the baroclinic term vanishes and change in circulation is directly related to change in area projected onto the plane of the equator. For the example of a circular cyclonic vortex and a concentric contour centered at latitude ϕ (assuming that the pressure surface is approximately horizontal), wind speed will increase in the cyclonic sense if F decreases, that is, if the area of the contour decreases, or the fluid making up the vortex moves toward the equator, or tilts with respect to the horizontal so that the contour area projected onto the plane of the equator decreases. Problem 13 requires the derivation of an equation describing these effects.

If a horizontal contour is chosen, the baroclinic term in general will not be zero as in the previous example, but it turns out in most cases of synoptic scale to be an order of magnitude smaller than the effect of change in F.

The Bjerknes circulation theorem is especially useful in cases in which the flow is contained, as in a closed pipe, or by the closed room of the example given here. In most cases in the atmosphere the choice of contour is somewhat arbitrary; the result is that although the rate of change of circulation for the particular contour may be accurately determined, it may be difficult to relate this to the velocity field which is usually the desired product.

4.8   The Vorticity Equation

It turns out to be possible to develop a general form of the equations of motion which retains the simplicity of the Bjerknes circulation theorem but applies to the velocity field at each point rather than to an arbitrarily chosen contour. In place of the circulation, we introduce the vector *vorticity*, defined by $\mathbf{V} \times \mathbf{V}$. Note that we may recognize from Stokes's theorem that circulation is related to the magnitude of vorticity by

$$|\mathbf{V} \times \mathbf{V}| = \lim_{\sigma \to 0} \frac{C}{\sigma} \qquad (4.52)$$

An exercise is provided by Problem 14. The vorticity may be expanded in Cartesian coordinates as follows

$$\mathbf{V} \times \mathbf{V} = \begin{vmatrix} \mathbf{i} & \mathbf{j} & \mathbf{k} \\ (\partial/\partial x) & (\partial/\partial y) & (\partial/\partial z) \\ u & v & w \end{vmatrix}$$

$$= \mathbf{i}\left(\frac{\partial w}{\partial y} - \frac{\partial v}{\partial z}\right) + \mathbf{j}\left(\frac{\partial y}{\partial z} - \frac{\partial w}{\partial x}\right) + \mathbf{k}\left(\frac{\partial v}{\partial x} - \frac{\partial u}{\partial y}\right) \qquad (4.53)$$

$$\equiv \mathbf{i}\xi + \mathbf{j}\eta + \mathbf{k}\zeta$$

Alternatively, the vertical component of vorticity can be expressed in natural coordinates by

$$\zeta = \frac{V}{r} - \frac{\partial V}{\partial n}$$

The vorticity describes the rotation of the fluid at each point. A component of the vorticity is positive if a small pin wheel placed in the fluid rotates in a counterclockwise sense about the relevant axis, and is negative if the pin wheel rotates in a clockwise sense. We shall be concerned in the following with the vertical component of the vorticity, designated by ζ.

To derive the vorticity equation we differentiate Eq. (4.38) with respect to y and Eq. (4.39) with respect to x, both at constant p, and subtract. The acceleration terms are then expanded using Eq. (4.37) and a similar equation for dv/dt. This yields

$$\frac{d\zeta_p}{dt} + v\frac{\partial f}{\partial y} + (\zeta_p + f)\left(\frac{\partial u}{\partial x} + \frac{\partial v}{\partial y}\right)_p + \frac{\partial v}{\partial p}\frac{\partial \omega}{\partial x} - \frac{\partial u}{\partial p}\frac{\partial \omega}{\partial y} = 0 \quad (4.54)$$

where $\zeta_p \equiv (\partial v/\partial x - \partial u/\partial y)_p$. The second term arises because the rotation of the earth about a vertical axis varies toward the north (in the y direction). The term $v\,\partial f/\partial y$ is equivalent to df/dt, so that Eq. (4.54) may be written

$$\frac{d}{dt}(\zeta_p + f) + (\zeta_p + f)\mathbf{V}_p\cdot\mathbf{V} = \frac{\partial u}{\partial p}\frac{\partial \omega}{\partial y} - \frac{\partial v}{\partial p}\frac{\partial \omega}{\partial x} \quad (4.55)$$

where $(\partial u/\partial x + \partial v/\partial y)_p$ is represented by $\mathbf{V}_p\cdot\mathbf{V}$ and is called the *horizontal divergence* evaluated at constant p. Equation (4.55) is one form of the *vorticity equation*.

It is instructive to compare Eq. (4.55) with the Bjerknes circulation theorem [Eq. (4.50)]. If a contour is chosen lying in a pressure surface, the baroclinic term in Eq. (4.50) does not appear. The time rate of change of vorticity translates directly into the time rate of change of circulation divided by the area enclosed by the contour, and the remaining terms in the vorticity equation represent ways in which F (area enclosed by the contour projected on the plane of the equator) may change with time. The first term df/dt represents the effect of motion toward the north or south, the divergence term represents the effect of change in area enclosed by the contour through horizontal divergence or convergence, and the terms on the right side, which are called "tilting" terms, represent the effect of change in orientation of the contour plane.

4.9   Potential Vorticity and Its Conservation

The terms on the left side of Eq. (4.55) can be combined into a single term. To do this we develop a form of the equation of continuity [Eq. (2.28)] appropriate to the use of pressure coordinates. Consider a differential element of air contained between two pressure surfaces as shown in Fig. 4.12. Under hydrostatic conditions the mass may be expressed by

$$\delta M = \rho \, \delta A \, \delta z = -\frac{\delta A \, \delta p}{g}$$

Upon differentiating with respect to time following the motion (the total or individual derivative) and recognizing that mass must be conserved following the motion

$$\frac{1}{\delta M}\frac{d(\delta M)}{dt} = \frac{1}{\delta A}\frac{d(\delta A)}{dt} + \frac{1}{\delta p}\frac{d(\delta p)}{dt} = 0 \qquad (4.56)$$

And, by introducing $\delta x \, \delta y$ for δA and interchanging the order of the differential operators

$$\frac{1}{\delta A}\frac{d(\delta A)}{dt} = \frac{\delta(dx/dt)}{dx} + \frac{\delta(dy/dt)}{\delta y} = \frac{-\delta(dp/dt)}{\delta p} \qquad (4.57)$$

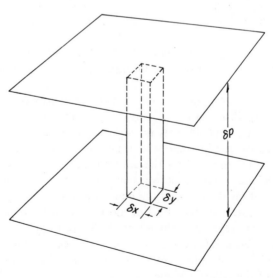

FIG. 4.12.   An element of air of mass δM contained between isobaric surfaces.

Finally, upon taking the limit as δx, δy, and δp approach zero, and replacing dp/dt by the symbol ω, we may write the two equations

$$\mathbf{V}_p \cdot \mathbf{V} \equiv \left(\frac{\partial u}{\partial x} + \frac{\partial v}{\partial y} \right)_p = -\frac{\partial \omega}{\partial p} \tag{4.58}$$

$$\frac{1}{\delta p} \frac{d(\delta p)}{dt} = -\mathbf{V}_p \cdot \mathbf{V} \tag{4.59}$$

Equation (4.58) is the equation of continuity in pressure coordinates. Note that it has only three dependent variables, whereas the general form [Eq. (2.28)] has four dependent variables. Equation (4.59) may now be introduced into Eq. (4.55) with the result

$$\frac{\delta p}{\zeta_p + f} \frac{d}{dt} \left(\frac{\zeta_p + f}{\delta p} \right) = \frac{\partial u}{\partial p} \frac{\partial \omega}{\partial y} - \frac{\partial v}{\partial p} \frac{\partial \omega}{\partial x} \tag{4.60}$$

This shows that $(\zeta_p + f)/|\delta p|$, which is called the *potential vorticity*, changes following the motion only through the "tilting" terms. For the synoptic and planetary scales of motion, these terms are usually at least an order of magnitude smaller than the individual terms in Eq. (4.55), so that Eq. (4.60) can be further simplified to yield

$$\frac{d}{dt} \left(\frac{\zeta_p + f}{|\delta p|} \right) \simeq 0 \tag{4.61}$$

Thus the potential vorticity is at least approximately conserved following the motion and may therefore be used to identify or "tag" air parcels. Other quasi-conservative properties, such as potential temperature or absolute humidity, may also be used as tags in appropriate situations

If we assume that potential temperature is conserved following the motion, we may express the potential temperature difference between the top and bottom of the column by $\delta\theta$ and introduce this into the bracket in Eq. (4.61) without affecting the validity of the equation. Then upon taking the limit as $\delta p \to 0$, Eq. (4.61) becomes

$$\frac{d}{dt} \left[(\zeta_p + f) \left| \frac{\partial \theta}{\partial p} \right| \right] \simeq 0 \tag{4.62}$$

The bracket contains an alternate form of the potential vorticity which relates to individual air parcels rather than to columns of finite height. The term $\partial\theta/\partial p$ is a measure of the static stability of the air and can be readily determined from a temperature sounding plotted on a pseudo-adiabatic chart.

Equations (4.61) and (4.62) are so simple that they can be used to explain heuristically a number of atmospheric flow fields. For example, consider flow from the west in the Northern Hemisphere over a north–south mountain barrier as shown in Fig. 4.13. We imagine a uniform current extending without change in the north–south direction. The air close to the ground is forced to flow parallel to the earth's surface, while at some height the air flow is more nearly horizontal. Therefore, as a column of mass δM and pressure increment $|\delta p|$ ascends the mountain, $|\delta p|$ decreases while δM is conserved. We recognize from Eq. (4.61) that since f is unchanging in westerly flow, ζ_p must decrease; that is, the column must curve in an anticyclonic sense. This effect continues until $|\delta p|$ reaches its minimum value at the top of the mountain. At this longitude the air is flowing from the northwest, so that f is decreasing along the trajectory. As a consequence, the maximum anticyclonic curvature is reached before the column reaches the crest of the mountain. As the column descends the mountain $|\delta p|$ increases, while f continues to decrease; both these effects require the vorticity to become more positive, and the column therefore curves rather sharply in a cyclonic sense. The trajectory reaches its most southerly latitude in the lee of the mountain and then turns toward the north.

As the column moves away from the mountain slope $|\delta p|$ no longer changes, but here the flow has a southerly component so that f is increasing and vorticity must decrease. Consequently, the column curves again in an anticyclonic sense; in this way a series of waves may be produced. The troughs are regions of low pressure while the ridges are regions of high pressure.

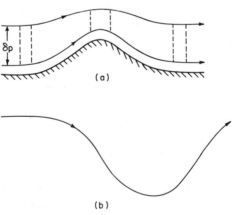

FIG. 4.13.   Flow from the west crossing a north–south mountain in the Northern Hemisphere. (a) Vertical cross section. (b) The trajectory in the horizontal plane followed by air parcels which conserve potential vorticity.

Observations of wind flow over the Rocky Mountains often reveal stream-line patterns similar to that shown in Fig. 4.13, and cyclones sometimes form in the pressure trough on the eastern slope of the mountains.

A second application of conservation of potential vorticity concerns broad, uniform flow over an extensive level surface. If we consider a column extending from the earth's surface nearly to the top of the atmosphere (a pressure surface in the lower stratosphere is high enough), the individual rate of change of $|\delta p|$ is small and can be neglected. What remains of Eq. (4.61) describes flow at constant absolute vorticity. As an aid in visualizing the flow, it may be helpful to fix attention on an air parcel at an intermediate height. If at some point the flow is disturbed so that the column moves along a straight line with a northward component as shown in Fig. 4.14, it must generate anticyclonic relative vorticity and begin to curve to the right. Maximum anticyclonic curvature is reached at the point of highest latitude, and as the column begins to move to the south, the relative vorticity must become less negative. It reaches zero relative vorticity at the initial latitude and then must begin to curve cyclonically as the column reaches lower latitudes. It is easy to recognize that the column oscillates periodically about its initial latitude. Constant absolute vorticity trajectories of this kind are similar to the streamlines and the trajectories observed in weather systems in middle and high latitudes, reflecting the fact that much of the physics of synoptic scale motions is represented by Eq. (4.61).

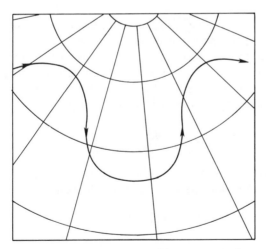

Fig. 4.14.   Trajectory followed by air moving with constant absolute vorticity in a broad uniform current.

4.10   Rossby Waves

The application just discussed can be developed more rigorously and at the same time, some important properties of synoptic scale motions can be developed. The development which follows requires an understanding of the concepts and general properties of wave motion. These topics are introduced in Chapter VII, and if the reader is not already acquainted with them, it will be helpful to read Sections 7.1–7.4 before going further with this section.

A highly simplified mathematical model of the atmosphere utilizing Eq. (4.55) will be considered. Assume that air is flowing from the west at a speed $\bar{u}$ which is uniform in x and y and constant in t. The velocity field is subjected to a small sinusoidal disturbance; the velocity components may now be represented by $\bar{u} + u'$ and v'. Next we choose a pressure surface in the atmosphere on which $\partial \omega / \partial p$ or $\mathbf{V}_p \cdot \mathbf{V}$ can be neglected. In the real atmosphere this condition is approximately valid in the middle troposphere. If as before we neglect the tilting terms, the vorticity equation reduces to

$$\frac{d}{dt}\left(\frac{\partial v'}{\partial x} - \frac{\partial u'}{\partial y}\right)_p + v'\frac{\partial f}{\partial y} = 0$$

This can be expanded to

$$\left[\frac{\partial}{\partial t} + (\bar{u} + u')\frac{\partial}{\partial x} + v'\frac{\partial}{\partial y} + \omega'\frac{\partial}{\partial p}\right]\left(\frac{\partial v'}{\partial x} - \frac{\partial u'}{\partial y}\right)_p + v'\frac{\partial f}{\partial y} = 0$$

Because f depends only on y, $\partial f / \partial y$ can be evaluated for the latitude where the equation is to be applied. It is customary to represent $\partial f / \partial y$ by the symbol β and to treat β as a constant. The approximation involved is referred to as the β-*plane approximation*. Because $\bar{u} \gg u', v'$, the products of primed terms can be neglected compared to first-order primed terms. For instance, $u'\, \partial^2 v' / \partial x^2$ can be neglected compared to $(\partial / \partial t)(\partial v' / \partial x)$. Therefore

$$\left(\frac{\partial}{\partial t} + \bar{u}\frac{\partial}{\partial x}\right)\left(\frac{\partial v'}{\partial x} - \frac{\partial u'}{\partial y}\right)_p + v'\beta = 0 \qquad (4.63)$$

Under the condition that the horizontal velocity divergence vanishes there exists a stream function (ψ) which satisfies the equations

$$v' \equiv \frac{\partial \psi}{\partial x}$$

$$u' \equiv -\frac{\partial \psi}{\partial y} \qquad (4.64)$$

Problem 15 requires a proof of this statement. Upon introducing Eq. (4.64) into (4.63), the following differential equation in one dependent variable results.

$$\left(\frac{\partial}{\partial t} + \bar{u}\frac{\partial}{\partial x}\right)\left(\frac{\partial^2\psi}{\partial x^2} + \frac{\partial^2\psi}{\partial y^2}\right)_{\mathrm{p}} + \beta\left(\frac{\partial\psi}{\partial x}\right)_{\mathrm{p}} = 0 \qquad (4.65)$$

We assume a solution of the form

$$\psi = A e^{ik(x-ct)+imy} \qquad (4.66)$$

Equation (4.66) describes a wave which is periodic in x and y moving in the x direction at speed c. The coefficients k and m are, respectively, the wave numbers in the x and y directions. The wavelengths in the x and y directions are related to the wave numbers by

$$L_x \equiv \frac{2\pi}{k} \qquad L_y \equiv \frac{2\pi}{m} \qquad (4.67)$$

Upon substituting Eq. (4.66) in (4.65)

$$(-ikc + ik\bar{u})(k^2 + m^2) - ik\beta = 0$$

and therefore the *frequency equation* which relates frequency to wave number is

$$c = \bar{u} - \beta/(k^2 + m^2) \qquad (4.68)$$

Equation (4.68) describes the speed of *Rossby* or *Rossby–Haurwitz* waves as a function of zonal wind speed, latitude, and wave numbers in the zonal and meridional directions. These waves propagate westward relative to the zonal wind speed; their westward speed increases with increasing wavelength, and they may be stationary for zonal wind speed given by

$$\bar{u}_s = \beta/(k^2 + m^2) \qquad (4.69)$$

Table 4.2 shows the stationary zonal wind speed ($\bar{u}_s$) calculated from Eq. (4.69) as a function of zonal and meridional wavelength at latitude 45°.

TABLE 4.2
STATIONARY ZONAL WIND SPEED ($\bar{u}_s$) CALCULATED
FROM EQ. (4.69) FOR LATITUDE 45°

L_y (km)	L_x (km)	$\bar{u}_s$ (m s^{-1})
∞	2000	16.4
∞	3000	36.9
∞	4000	65.8
2000	2000	8.2
3000	3000	18.5
4000	4000	32.9

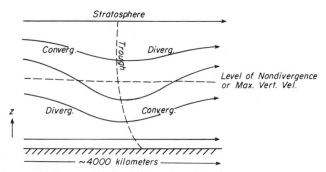

Fig. 4.15.   Streamlines projected onto an east–west vertical cross section through a region of a pressure trough. Vertical scale exaggerated by a factor of several hundred.

The theory developed here has been based on the assumption that the horizontal velocity divergence can be neglected in the vorticity equation. In general, this is by no means a valid assumption, and it must be asked how the theory can be applied to the atmosphere. In synoptic scale disturbances there is marked convergence in the lower troposphere on the east side of pressure troughs, and this is reflected in the rising air which produces the clouds and rain which is characteristic of the east sides of pressure troughs. Conversely, low level divergence and sinking air and consequent clear skies are characteristic of the west sides of pressure troughs. In the upper troposphere the distribution of divergence is reversed from that in the lower troposphere. It is easy to recognize that this must be so if flow in the stratosphere is essentially horizontal and unaffected by disturbances in the troposphere. Consequently there is an intermediate region in the mid troposphere (about 55 kPa or 550 mb) called the level of nondivergence, and it is here that the preceding theory can be applied for synoptic scale systems. The relationships described here are indicated in Fig. 4.15. The full vertical structure can be developed using more complex mathematical models. These involve simultaneous solution of the vorticity and thermal energy equations and are beyond the scope of this book.

4.11   Frictionally Driven Secondary Circulation

So far we have neglected the friction force in the planetary boundary layer and its effects. The magnitude of the friction term may be determined from observations of the wind velocity and pressure field under conditions of negligible acceleration. In this case, as shown in Fig. 4.16, balance of the pressure force $\mathbf{F}_p$, the Coriolis force $\mathbf{F}_c$ and friction force $\mathbf{F}_F$ is expressed by

$$\mathbf{F}_c + \mathbf{F}_F + \mathbf{F}_p = 0$$

where it is assumed that the friction force acts in the opposite direction from the wind. The friction force is related to the Coriolis force and the wind speed and direction by

$$F_F = F_c \tan \theta = f V \tan \theta \qquad (4.70)$$

The angle θ between the wind direction and the isobar direction near the earth's surface is commonly in the range $10°–45°$, so that the friction force and Coriolis force are of the same order of magnitude. Consequently, mass is transported toward lower pressure within the planetary boundary layer. This mass transport has important kinematic and dynamic consequences. This can be recognized easily in the case of a closed low pressure region where cross isobar flow necessarily results in horizontal velocity convergence and therefore in upward motion at the top of the planetary boundary layer. This result is not critically dependent on the exact formulation of the friction force.

When we choose the x axis in the direction of the geostrophic wind, the mass transport toward low pressure may be expressed by

$$M = \int_0^h \rho v \, dz$$

where M is the cross isobaric mass transport per unit length of isobar and h is the height of the boundary layer. Because of the choice of coordinates, $\partial p / \partial x = 0$, and if there are no accelerations, Eq. (4.16) may be integrated to

$$M = \int_0^h \rho v \, dz = \frac{\tau_x(z = 0)}{f} \qquad (4.71)$$

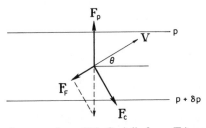

FIG. 4.16.   Balance of pressure force ($\mathbf{F}_p$), Coriolis force ($\mathbf{F}_c$), and friction force ($\mathbf{F}_F$).

which simply states that the cross isobaric flow is equal to the component of the surface stress parallel to the isobars divided by the Coriolis parameter.

By integrating Eq. (4.71) along a closed isobar and applying Gauss's theorem to the left-hand side and Stokes's theorem to the right-hand side we find

$$\mathbf{V} \cdot \mathbf{M} = -\frac{1}{f}\mathbf{V} \times \boldsymbol{\tau}_0 \cdot \mathbf{k} \tag{4.72}$$

where $\mathbf{M}$ is the vector transport of mass. Conservation of mass requires that if the local rate of change of density is neglected

$$\mathbf{V} \cdot \mathbf{M} = (\rho w)_0 - (\rho w)_h = -\rho w_h \tag{4.73}$$

because $w_0 = 0$. Substituting this result in Eq. (4.72) yields the vertical velocity at the top of the boundary layer in the form

$$w_h = \frac{1}{f\rho_h}\mathbf{V} \times \boldsymbol{\tau}_0 \cdot \mathbf{k} \tag{4.74}$$

which states that the vertical velocity at the top of the boundary layer is proportional to the curl of the surface stress. This stress is usually parameterized by the wind near the surface as we shall see in Chapter VI, and therefore can easily be derived when observations are available.

If observations near the surface are not available, then the stress may be related to the geostrophic wind speed, the height of the boundary layer, the static stability of the boundary layer, and the roughness of the surface. These relations are explored in some detail in Chapter VI. Here we shall restrict outselves to the simple dimensional argument that the mass flux (M) is proportional to the geostrophic wind speed (v_g) and the height of the boundary layer (h). Thus

$$M = c_1 \bar{\rho} v_g h \tag{4.75}$$

where c_1 represents the ratio of cross isobar mass flux to geostrophic mass flux; it is treated here as constant although it turns out to be a function of static stability and surface roughness. As before, Eq. (4.75) is integrated along a closed contour, and Gauss' and Stokes' theorems are introduced. Upon substituting Eq. (4.73), we find

$$w_h = \frac{\bar{\rho}}{\rho} c_1 h \zeta_g \tag{4.76}$$

where $\zeta_g \equiv \mathbf{k} \cdot \mathbf{V} \times \mathbf{v}_g$, the geostrophic vorticity. Equation (4.76) shows that upward motion should accompany positive geostrophic vorticity and downward motion should accompany negative geostrophic vorticity. For the region of a well developed middle latitude cyclonic vortex we will choose a value of geostrophic vorticity of $6 \times 10^{-5} \, \text{s}^{-1}$, a depth of the boundary layer of 1 km, and $c_1 \simeq 0.17$, corresponding to cross isobar flow at half the geostrophic speed at an angle of $20°$ to the geostrophic direction. This gives a vertical velocity of about $1 \, \text{cm s}^{-1}$. The vorticity chosen here corresponds to a situation of solid rotation in which the tangential speed 330 km from the center of rotation is $10 \, \text{m s}^{-1}$.

The flow field discussed here is referred to as a *secondary* circulation, whereas the vortex flow is the *primary* circulation. The specific process described in this case is called *Ekman pumping*. Ekman pumping contributes significantly to horizontal convergence and upward motion in low pressure (cyclonic) regions, horizontal divergence and downward motion in high pressure (anticyclonic) regions. It is responsible in part for the fact that clouds and precipitation occur in regions of low pressure at the ground, while clear skies occur in regions of high pressure at the ground.

Ekman pumping also has an important effect on the vorticity of the system in which it occurs. The effect on the free atmosphere above the boundary (or Ekman) layer may be discussed qualitatively with the aid of Fig. 4.17 and the equation of conservation of potential vorticity [Eq. (4.61)]. If we assume that there exists a height in the free atmosphere where the vertical velocity is zero, then upward flow at h as indicated schematically in Fig. 4.17 requires that the vertical extent of elemental columns in the free atmosphere must decrease with time following the motion. Consequently, if in this case we do not consider change in f, ζ must decrease following the motion. In this way vorticity of the air which enters the system through upward motion at the top of the Ekman layer, and which is already less than the vorticity of

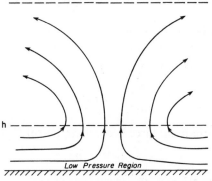

FIG. 4.17.   Vertical cross section through a cyclonic vortex showing schematically streamlines of secondary flow in the Ekman layer and in the free atmosphere.

the air which it replaces, is further "spun down" as the air diverges in the free atmosphere. An alternate way to look at the process is to recognize that the diverging air flows toward high pressure and therefore must slow down. Problem 18 requires a mathematical treatment of the spin down process.

An especially simple example of Ekman pumping and the spin down of vorticity can be observed when a cup of tea containing tea leaves is stirred in a circular fashion. When quasi-equilibrium is reached, the centrifugal force (which takes the place of the Coriolis force in the atmosphere) is approximately balanced by the radial pressure forces except close to the surface of the cup. Within the bottom boundary layer the tea is slowed by viscous drag, and here the centrifugal force cannot balance the pressure force. Therefore, just as in the atmosphere, cross isobar flow occurs toward lower pressure, and the tea leaves are swept together at the center of the bottom of the cup. The convergence occurring within the boundary layer results in upward flow at the top of the boundary layer, and since the upper surface of the tea is not sensibly affected, continuity requires slow radial outflow and reduced rotation throughout the rest of the fluid.

4.12   Vertical Convection

In Section 2.14 the static stability was defined as the downward acceleration experienced by a parcel per unit geopotential displacement from equilibrium. It was shown that for unsaturated air the static stability is given by Eq. (2.51) $[s_z = (1/\theta) \, \partial\theta/\partial z]$. Under saturated conditions the rising air is heated through release of the latent heat of condensation or cooled through evaporation in sinking air, and the static stability in this case is expressed by Eq. (2.103) $[s_z = (1/\theta_e) \, \partial\theta_e/\partial z]$ where θ_e represents the equivalent potential temperature.

If we consider a parcel displaced vertically in an environment at rest, the acceleration due to buoyancy force discussed in Section 2.14 can be expressed by the differential equation

$$\frac{d^2z}{dt^2} + (gs_z)z = 0 \tag{4.77}$$

assuming that no frictional force acts on the parcel. The general solution is

$$z = Ae^{iNt} + Be^{-iNt} \tag{4.78}$$

where $N^2 \equiv gs_z$. If $N^2 > 0$, Eq. (4.78) describes a vertical oscillation with angular frequency N and period $2\pi/N$. N is called the *Brunt–Väisälä* frequency. The case of vertical oscillation will be discussed further in Section 4.14.

If $N^2 < 0$, it is convenient to replace iN by N^* and to write the general solution in the form

$$z = Ae^{N^*t} + Be^{-N^*t} \qquad (4.79)$$

where N^* is real. For initial conditions, $z = z(0)$ and $w = 0$, $A = B$, and the solution becomes

$$z = z(0) \cosh N^*t \qquad (4.80)$$

Equation (4.80) describes displacement which increases with time without limit; for an initial upward displacement the parcel rises at an increasing rate, and for an initial downward displacement it sinks at an increasing rate. An exercise is provided by Problem 19. The equation is valid only for the initial stage of convection after which turbulence develops and entrainment of air from outside the parcel introduces a frictional force. Through these processes vertical convection may lead to efficient vertical transfer of momentum, heat, and water vapor.

Vertical convection occurs frequently in the atmosphere and is most often made visible by the presence of cumulus clouds. Upward convection usually occurs in columns or plumes having diameters which range from a few hundred meters to as much as 20 km or more in very severe storms. An example is shown in Fig. 3.17. Vertical convection below the clouds is governed by the unsaturated criterion, while convection within the clouds is governed by the saturated criterion. Although convection cells are continuous through the cloud base, the vertical acceleration may change abruptly as the air enters the cloud and condensation begins.

There are several important factors not accounted for in this simple treatment of convection. First, in deriving the expression for static stability the pressures of the parcel and the environment were assumed to be the same. This is a valid assumption only for small departures from hydrostatic equilibrium (small vertical accelerations). Second, when vertical convection occurs, there must be compensating general subsidence in the area between ascending columns; consequently, the density of the environment is likely to be decreased by the subsidence, and the vertical acceleration is reduced. Since the proportion of horizontal area occupied by ascending columns is usually small, this effect is of minor importance. Third and most important, entrainment of air from the environment results in decreasing the vertical acceleration through both introduction of a frictional force and dilution of the rising moist air. Finally, evaporation occurring at the top and sides of a cumulus cloud results in local cooling and a tendency for downdrafts to occur along the cloud surfaces. These downdrafts help to maintain a sharp definition of the cloud boundaries. For all the reasons cited, the vertical velocities in convective cells and the heights reached by cumulus clouds are usually less than calculation using the theory of this section would indicate.

4.13 Waves on an Interface

If a stone is dropped into a body of water the surface is disturbed and a wave will travel outward away from the initial disturbance. Gravity acting on the upward displaced water produces a horizontal pressure gradient which accelerates water particles, and although the details are not intuitively obvious, the wave motion with which everyone is familiar results.

Waves may also be produced on an interface separating two fluids of different densities. In fact, waves on a water surface are really waves on the air–water interface. In this case the ratio of densities is about 10^{-3}. In other cases in the atmosphere and ocean, waves may form on interfaces separating fluids differing by one part in a hundred or less. These are referred to as shearing-gravitational or Helmholtz waves. Waves may be present without net transfer of bulk properties such as momentum, heat, and water vapor. However, if the displaced fluid mixes with its environment or if the waves are unstable so that they "break," transfer does occur. Transfer processes are discussed in Chapter VI.

In the following discussion we will assume that the two fluids are homogeneous and incompressible and that the earth's rotation can be neglected. Therefore, the motion may be considered to be limited to the x, z plane. The heavier fluid is assumed to lie below the lighter, and the lower fluid is assumed to be at rest in the undisturbed state while the upper fluid flows with undisturbed speed $\Delta \bar{u}$, as illustrated in Fig. 4.18. If we now consider a small amplitude disturbance of the boundary, we may distinguish between the undisturbed state (barred letters) and the perturbation state (primed letters). The total pressure is expressed by $\bar{p} + p'$, where $\bar{p}$ is constant in x and t and varies hydrostatically in z, and p' may vary in $x, z,$ and t. The total velocity in the x direction in the upper fluid is $\Delta \bar{u} + u'_2$. In developing the

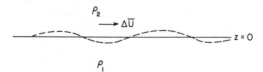

FIG. 4.18. A horizontal undisturbed boundary (solid line) between two incompressible homogeneous fluids of density ρ_1 and ρ_2 and the position of the disturbed boundary (dashed line) at a particular instant.

equations which govern the perturbations we will neglect, as in Section 4.10, the nonlinear terms in the perturbation quantities compared to linear terms. The equations of motion and of continuity may now be written

$$\frac{\partial u_2'}{\partial t} + \Delta \bar{u} \frac{\partial u_2'}{\partial x} = -\frac{1}{\rho} \frac{\partial p_2'}{\partial x}$$

$$\frac{\partial w_2'}{\partial t} + \Delta \bar{u} \frac{\partial w_2'}{\partial x} = -\frac{1}{\rho} \frac{\partial p_2'}{\partial z} \tag{4.81}$$

$$\frac{\partial u_2'}{\partial x} + \frac{\partial w_2'}{\partial z} = 0$$

Since these are linear equations, we expect the solutions to be exponential in each of the independent variables. We choose on the basis of observation the following solutions representing a wave of wave number k traveling with speed c in the x direction.

$$u_2' = A e^{\gamma z} e^{ik(x-ct)}$$

$$w_2' = C e^{\gamma z} e^{ik(x-ct)}$$

$$p_2' = D e^{\gamma z} e^{ik(x-ct)}$$

Upon substituting these solutions in Eqs. (4.81), the following algebraic equations result

$$A = \frac{D}{\rho(c - \Delta\bar{u})} \qquad D = \frac{ik(c - \Delta\bar{u})}{\gamma} \rho C \qquad A = \frac{i\gamma}{k} C \tag{4.82}$$

And upon eliminating A, C, and D, we find that $\gamma = \pm k$. It then follows that the general solutions may be written for the upper fluid

$$u_2' = (A_2 e^{kz} + A_2^* e^{-kz}) e^{ik(x-ct)}$$

$$w_2' = (C_2 e^{kz} + C_2^* e^{-kz}) e^{ik(x-ct)} \tag{4.83}$$

$$p_2' = (D_2 e^{kz} + D_2^* e^{-kz}) e^{ik(x-ct)}$$

Now if we apply the condition that the disturbance cannot be infinite in amplitude as $z \to \infty$, we recognize that A_2, C_2, and D_2 must each equal zero. And, A_2^* and D_2^* may be eliminated through the use of Eq. (4.82). Therefore, the solutions become

$$u_2' = -iC_2^* e^{-kz} e^{ik(x-ct)}$$

$$w_2' = C_2^* e^{-kz} e^{ik(x-ct)} \tag{4.84}$$

$$p_2' = -iC_2^* \rho_2 (c - \Delta\bar{u}) e^{-kz} e^{ik(x-ct)}$$

In the lower fluid the general solutions are identical in form to Eq. (4.83) with the subscript 1 replacing the 2. In this case the requirement that the amplitude remain noninfinite as $z \to -\infty$ requires that A_1^*, C_1^*, and D_1^* each equal zero. The solutions for the lower layer may then be written

$$u_1' = iC_1 e^{kz} e^{ik(x-ct)}$$

$$w_1' = C_1 e^{kz} e^{ik(x-ct)} \qquad (4.85)$$

$$p_1' = iC_1 \rho_1 c e^{kz} e^{ik(x-ct)}$$

In order to eliminate the remaining two arbitrary constants C_1 and C_2^*, two boundary conditions are required. The flow must satisfy two conditions at the boundary: (1) a *kinematic* condition that the shape of the boundary must be the same as defined within either fluid, that is, the two fluids must move so that no voids appear between them and they must not simultaneously occupy the same space, and (2) a *dynamic* condition that the pressure at adjacent points on either side of the boundary must be the same. Recognizing that fluid parcels which are situated on the boundary in either fluid remain on the boundary, the kinematic and dynamic conditions can be expressed in combined form by the equations

$$\left[\frac{d}{dt}(p_1 - p_2)\right] = 0 \qquad \begin{array}{l} \text{for a parcel on the boundary} \\ \text{in the upper fluid} \end{array}$$

$$\qquad\qquad\qquad\qquad\qquad\qquad\qquad\qquad\qquad (4.86)$$

$$\left[\frac{d}{dt}(p_1 - p_2)\right] = 0 \qquad \begin{array}{l} \text{for a parcel on the boundary} \\ \text{in the lower fluid} \end{array}$$

Equation (4.86) can be expanded upon recalling that the total pressure p_1 is equivalent to $\bar{p}_1 + p_1'$ and that $\bar{p}_1$ varies only with z (and similarly for p_2). If nonlinear terms in the perturbation quantities are neglected as before, the following equation represents the boundary condition for the upper fluid

$$\left[\frac{\partial}{\partial t}(p_1' - p_2') + \Delta \bar{u}\frac{\partial}{\partial x}(p_1' - p_2') + w_2'\frac{\partial}{\partial z}(\bar{p}_1 - \bar{p}_2)\right] = 0 \qquad \text{at} \quad z = 0$$
$$(4.87)$$

Evaluation of this equation at $z = 0$ rather than at the disturbed boundary introduces an approximation of the same order as neglect of the nonlinear terms. For the lower fluid the corresponding equation is

$$\left[\frac{\partial}{\partial t}(p_1' - p_2') + w_1'\frac{\partial}{\partial z}(\bar{p}_1 - \bar{p}_2)\right] = 0 \qquad \text{at} \quad z = 0 \qquad (4.88)$$

Finally, upon introducing the hydrostatic equation in the last terms of Eqs. (4.87) and (4.88) and substituting the perturbation solutions we find that

$$\frac{C_1}{C_2^*} = \frac{g\,\Delta\rho - \rho_2 k(c - \Delta\bar{u})^2}{\rho_1 kc(c - \Delta\bar{u})} = \frac{\rho_2 kc(c - \Delta\bar{u})}{g\,\Delta\rho - \rho_1 kc^2}$$

where $\Delta\bar{u} \equiv \bar{u}_2 - \bar{u}_1$, and $\Delta\rho \equiv \rho_1 - \rho_2$. This leads to the quadratic frequency equation

$$c^2(\rho_1 + \rho_2) - 2\rho_2\,\Delta\bar{u}c + \rho_2\,\Delta\bar{u}^2 - g\,\Delta\rho/k = 0$$

which has the solutions

$$c = \frac{\rho_2\,\Delta\bar{u}}{\rho_1 + \rho_2} \pm \left[\frac{g\,\Delta\rho}{k(\rho_1 + \rho_2)} - \frac{\rho_1\rho_2\,\Delta\bar{u}^2}{(\rho_1 + \rho_2)^2}\right]^{1/2} \tag{4.89}$$

The first or *convective* term on the right of Eq. (4.89) represents the weighted average speed of the undisturbed current. The second or *dynamic* term shows that waves may propagate at two speeds which depend on the magnitudes of the density and velocity discontinuities. For the case of zero shear and for heavier fluid lying below lighter (positive $\Delta\rho$) the wave speed is proportional to the square root of the density discontinuity and to the square root of wavelength. For the case of a free surface where $\rho_2 = 0$, the wave speed becomes $\pm(g/k)^{1/2}$ which is called the deep water wave speed. The effect of the upper layer is to decrease the wave speed from the free surface case. It should be noted that Eq. (4.89) was derived for the case of large depth or, alternatively, short wavelengths. In considering atmospheric gravity waves on an inversion surface, say at the top of the boundary layer, it may be necessary to apply different boundary conditions and to derive a different frequency equation.

Equation (4.89) shows that if the vertical shear exceeds a critical value given by

$$\Delta\bar{u}_c \equiv \left(\frac{g}{k}\frac{\Delta\rho(\rho_1 + \rho_2)}{\rho_1\rho_2}\right)^{1/2}$$

the second term becomes imaginary. This means that the amplitude of the perturbations given by Eqs. (4.84) and (4.85) may increase exponentially with time. Therefore, for shear exceeding $\Delta\bar{u}_c$ instability exists, and we recognize that for any finite shear of either sign there is a critical wavelength below which all waves are unstable. An exercise is provided in Problem 20.

In the real atmosphere discontinuities of density and wind velocity seldom, if ever, occur; however, strong vertical gradients are often observed. The theory of instability of waves in a region of strong gradients has been comprehensively developed, but requires more extensive treatment than is appropriate here. However, an instructive relationship can be easily developed from the preceding equation by considering a shear layer for which $k \approx \Delta z^{-1}$ and assuming $\rho_1 \approx \rho_2$. Then upon dividing by Δz and passing to the limit, we find that

$$\left(\frac{\partial u}{\partial z}\right)_c = \left(-\frac{2g}{\rho}\frac{\partial \rho}{\partial z}\right)^{1/2}$$

Therefore

$$-\frac{g}{\rho}\frac{\partial \rho/\partial z}{(\partial u/\partial z)_c^2} = \frac{1}{2}$$

The ratio on the left is referred to as the critical *Richardson number* and is designated Ri_c. The significance of the Richardson number is that it identifies the transition from laminar to turbulent flow.

This derivation of Ri_c is obviously inexact, and consequently the result is not quantitatively correct. A theory including compressibility and linear variation of temperature and velocity was first developed by G. I. Taylor.[†] He deduced that $Ri_c = 0.25$, and this value has been confirmed by later more precise work. The Richardson number is further discussed in Section 6.7.

4.14 Acoustic-Gravity Waves

In Section 4.12 it was shown that in a stable environment an air parcel displaced vertically from equilibrium oscillates with the Brunt–Väisälä angular frequency given for adiabatic change of state by

$$N = (gs_z)^{1/2} = \left(\frac{g}{\theta}\frac{\partial \theta}{\partial z}\right)^{1/2}$$

For unsaturated air in an isothermal environment, the period of oscillation is easily found to be about six minutes, and the period increases as stability decreases.

The Brunt–Väisälä frequency was derived using the vertical equation of motion and the hydrostatic equation assuming the pressures of the parcel and the environment are equal. Obviously, this cannot be generally valid, and if equality of pressures is not assumed the problem becomes a fluid

† G. I. Taylor, *Proc. Roy. Soc. A*, **132**, 499, 1931. The work reported in this paper was done in 1914 and formed part of the essay for which the Adams Prize was awarded in 1915.

dynamical problem involving the equations of motion in the x and z direction, the equation of continuity, and the equation of conservation of energy for adiabatic change of state. We will again consider only frequencies high enough that the Coriolis force and motion in the y direction can be neglected. Equations (4.15) and (2.28) in this case reduce to

$$\frac{\partial u}{\partial t} + u\frac{\partial u}{\partial x} + w\frac{\partial u}{\partial z} = -\frac{1}{\rho}\frac{\partial p}{\partial x}$$

$$\frac{\partial w}{\partial t} + u\frac{\partial w}{\partial x} + w\frac{\partial w}{\partial z} = -\frac{1}{\rho}\frac{\partial p}{\partial z} - g \qquad (4.90)$$

$$\frac{\partial \rho}{\partial t} = -\frac{\partial(\rho u)}{\partial x} - \frac{\partial(\rho w)}{\partial z}$$

Conservation of energy will be expressed for adiabatic change of state by $d\theta/dt = 0$, where θ represents potential temperature. Upon expanding

$$\frac{\partial \theta}{\partial t} + u\frac{\partial \theta}{\partial x} + w\frac{\partial \theta}{\partial z} = 0 \qquad (4.91)$$

We will assume that the atmosphere is motionless in the undisturbed state and that the undisturbed pressure, density, and potential temperature vary only with height. If, as before, nonlinear terms in the disturbed quantities (indicated by primes) are neglected, Eqs. (4.90) and (4.91) become

$$\bar{\rho}\frac{\partial u'}{\partial t} = -\frac{\partial p'}{\partial x}$$

$$\bar{\rho}\frac{\partial w'}{\partial t} = -\frac{\partial p'}{\partial z} - \rho'g$$

$$\frac{\partial \rho'}{\partial t} = -\frac{\partial(\bar{\rho}u')}{\partial x} - \frac{\partial(\bar{\rho}w')}{\partial z} \qquad (4.92)$$

$$\frac{\partial \theta'}{\partial t} + w'\frac{\partial \bar{\theta}}{\partial z} = 0$$

By combining the equation of state [Eq. (2.10)] and the definition of potential temperature [Eq. (2.47)], we find that

$$\rho' = \frac{p'}{c_s^2} - \frac{\bar{\rho}}{\bar{\theta}}\theta' \qquad (4.93)$$

where $c_s^2 \equiv (c_p/c_v)R_m T$, the square of the speed of sound.

By combining the second and fourth equations of Eqs. (4.92) together with Eq. (4.93) and treating c_s^2 as independent of height, we find the following relation between the disturbed pressure and vertical velocity.

$$\frac{\partial^2 p'}{\partial t \, \partial z} + \frac{g}{c_s^2} \frac{\partial p'}{\partial t} = -\bar{\rho}\left(\frac{\partial^2 w'}{\partial t^2} + N^2 w'\right) \tag{4.94}$$

Then by combining the first, third, and fourth equations of Eqs. (4.92) a second relation between p' and w' may be written

$$\frac{1}{c_s^2} \frac{\partial^2 p'}{\partial t^2} - \frac{\partial^2 p'}{\partial x^2} = -\bar{\rho}\left(\frac{\partial^2 w'}{\partial t \, \partial z} + \frac{N^2}{g} \frac{\partial w'}{\partial t}\right) \tag{4.95}$$

We next reduce these partial differential equations to ordinary differential equations by introducing the solutions

$$p' = P(z)e^{i(kx + vt)}$$

$$\bar{\rho}w' = W(z)e^{i(kx + vt)}$$

Equations (4.94) and (4.95) then become

$$iv\left(\frac{dP}{dz} + \frac{g}{c^2} P\right) = W(v^2 - N^2)$$

$$\left(k^2 - \frac{v^2}{c_s^2}\right)P = -iv\left(\frac{dW}{dz} + \frac{N^2}{g} W\right) \tag{4.96}$$

If we assume exponential solutions of the form

$$P = P*e^{(\gamma + in)z}$$

$$W = N*e^{(\gamma + in)z} \tag{4.97}$$

we find by substituting into Eqs. (4.96) that

$$\gamma = -\frac{1}{2}\left(\frac{N^2}{g} + \frac{g}{c_s^2}\right) \tag{4.98}$$

and the frequency equation relating the wave numbers k and n to the frequency v is

$$v^2\left[n^2 + \frac{1}{4}\left(\frac{g}{c_s^2} - \frac{N^2}{g}\right)^2\right] = \left(k^2 - \frac{v^2}{c_s^2}\right)(N^2 - v^2) \tag{4.99}$$

Equation (4.99) can be written as a quadratic equation in v^2 which has the solutions

$$v_a^2 = \frac{c_s^2 \lambda^2}{2} \left\{ 1 + \left[1 - \frac{4k^2 N^2}{c_s^2 \lambda^4} \right]^{1/2} \right\} \tag{4.100}$$

$$v_g^2 = \frac{c_s^2 \lambda^2}{2} \left\{ 1 - \left[1 - \frac{4k^2 N^2}{c_s^2 \lambda^4} \right]^{1/2} \right\} \tag{4.101}$$

where

$$\lambda^2 \equiv k^2 + n^2 + \tfrac{1}{4}(g/c_s^2 + (N^2/g))^2$$

The higher frequency waves, represented by Eq. (4.100), are called *acoustic* waves, while the lower frequency waves, represented by Eq. (4.101), are called *gravity* waves.

The second term in the radical in Eqs. (4.100) and (4.101) is small compared to unity, so that the following are good approximations.

$$v_a^2 = c_s^2 [k^2 + n^2 + \tfrac{1}{4}(g/c_s^2 + (N^2/g))^2] \tag{4.102}$$

$$v_g^2 = \frac{k^2 N^2}{k^2 + n^2 + \tfrac{1}{4}(g/c_s^2 + (N^2/g))^2} \tag{4.103}$$

We may recognize from Eq. (4.97) that n^2 determines the character of the height dependence of the solutions. If n^2 is positive the waves are periodic in z, if n^2 is negative the waves vary exponentially in z. In the former case $(n^2 > 0)$ the waves are called *internal*, in the latter case $(n^2 < 0)$ they are *external*. In this discussion n^2 has been treated as independent of other parameters of the problem. In a more complete treatment, n can be determined explicitly.

The acoustic-gravity wave solutions can be conveniently represented on a wave number–frequency diagram as illustrated in Fig. 4.19. It is easily recognized from Eq. (4.99) that for $v^2 = N^2$, $n^2 < 0$, and the waves are external. Also, for $v^2 = kc_s^2$, $n^2 < 0$. Equations (4.102) and (4.103) show that for $n^2 = 0$, v_a^2 is asymptotic to $c_s^2 k^2$ for large values of k^2, while v_g^2 is asymptotic to N^2 for large values of k^2. Thus the Brunt–Väisälä frequency is correct for short gravity waves; as wavelength increases, frequency decreases toward zero at very long wavelengths.

Other characteristics of acoustic-gravity waves can be easily deduced from Eqs. (4.102) and (4.103) with the aid of Fig. 4.19. (1) Acoustic waves of all lengths have periods less than a critical value. For very long waves $(k \sim 0, n \sim 0)$ the period is about 5 min. These long waves move faster than the "speed of sound" c_s. As wavelength decreases, the speed approaches

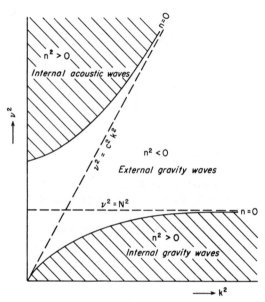

FIG. 4.19.   Wave number frequency diagram for the acoustic-gravity wave. The relations given by Eqs. (4.102) and (4.103) are shown but not to scale.

c_s, and the critical limit decreases, so that for waves 1 km in length the critical period is about 3 s. (2) External gravity waves must have periods within a particular range. For a wavelength of 1 km the period lies between about 3 s and 6 min. (3) Internal gravity waves have periods greater than the Brunt–Väisälä period. The period increases with wavelength and increases as vertical wavelength decreases.

The solutions for the pressure and vertical (velocity) perturbations in the case of internal waves are

$$p' = P^* e^{-|\gamma|z} e^{i(kx + nz + vt)}$$
$$\bar{\rho}w' = W^* e^{-|\gamma|z} e^{i(kx + nz + vt)}$$

(4.104)

The solutions represent waves having velocity components in the x and z directions, and their direction of propagation depends upon the values of k, n, and v. Problem 21 provides an exercise and Problem 22 requires extending the theory developed here.

Acoustic-gravity waves are ubiquitous in the atmosphere, but their effects on weather are somewhat subtle and not easy to detect. Large amplitude waves of this type produced, for example, by the 1883 explosion of Mt. Krakatoa or by nuclear explosions have provided measurements of the temperature and winds in the stratosphere and the thermosphere. For many meteorological purposes the gravity waves are important, but the acoustic waves constitute unwanted "noise." It turns out that acoustic waves can be eliminated by neglecting the local rate of change of density in the equation of continuity without significantly affecting the gravity wave solutions.

List of Symbols

		First used in Section
c	wave speed in x direction	4.10
c_1	dimensionless constant	4.11
c_s	speed of sound	4.14
C	circulation	4.7
f	Coriolis parameter $[2\Omega \sin \phi]$	4.3
$\mathbf{F}$	vector force	4.1
g	scalar acceleration of gravity	4.3
$\mathbf{g}$	vector acceleration of gravity	4.3
$\mathbf{g}^*$	vector gravitational acceleration	4.3
h	height of the boundary layer	4.11
$\mathbf{i}, \mathbf{j}, \mathbf{k}$	unit vectors in x, y, z directions, respectively	4.1
k, m	wave numbers in x and y directions, respectively	4.10
K_m	kinematic eddy viscosity	4.1
M	mass transport	4.11
n	wave number in z direction	4.14
N	Brunt–Väisälä angular frequency $[(g/\theta)\, \partial\theta/\partial z]^{1/2}$	4.12
p	pressure	4.1
r	radius of curvature of trajectory	4.4
$\mathbf{r}$	position vector from center of earth	4.2
R_m	gas constant for dry air	4.6
s_z	static stability $[(g/\theta)\, \partial\theta/\partial z]$	4.12
t	time	4.2
T_v	virtual temperature	4.6
u, v, w	velocity components in x, y, z directions, respectively	4.3
V	scalar horizontal velocity	4.4
$\mathbf{V}$	vector velocity	4.2
x, y, z	Cartesian coordinates	4.1
β	$\partial f/\partial y$	4.10
ζ, η, ξ	components of vorticity in the x, y, z direction, respectively	4.7
θ	potential temperature, cross isobar angle	4.9, 4.11
μ	viscosity	4.1
ρ	density	4.1
τ	stress	4.1
ϕ	latitude	4.3
Φ	geopotential	4.5
ω	individual rate of pressure change $[dp/dt]$	4.5
Ω	scalar angular frequency of earth	4.3
$\mathbf{\Omega}$	vector angular frequency of earth	4.2

Subscripts

a   absolute, acoustic
c   Coriolis, critical
e   eddy
F   friction
g   geostrophic, gravity
h   height of the boundary layer
H   horizontal
i   inertial
p   pressure
Primes   disturbed values
Overbar   undisturbed or average value
s   sound

Problems

1. If a projectile is fired horizontally from the North Pole at a speed V toward a target at a short distance S, show that the distance the projectile would be deflected by the Coriolis force is expressed by $S' = \Omega S^2/V$. Find the acceleration due to the Coriolis force.

2. If the projectile of Problem 1 is fired toward the east from a point at a distance r from the North Pole, as shown in Fig. 4.3, show that the distance the projectile would be deflected by the centrifugal and Coriolis forces is expressed by

$$S' = \frac{S^2}{2r}\left(\frac{\Omega r}{V} + 1\right)^2$$

where S is the distance between the target and the point of firing and $S \ll r$. Find the acceleration and compare with the answer in Problem 1.

3. An archer standing on the equator shoots an arrow toward the north a horizontal distance of 200 m in 4 s and another arrow an equal distance at the same speed toward the east. Assuming conditions are identical except for direction and that the northward-directed arrow strikes the center of the target, by how much and in what direction would the eastward-directed arrow be deflected by the Coriolis force?

4. A 200 ton locomotive (1 ton equals 10^3 kg) of the Alaskan Railroad (latitude 65°N) travels along a straight track at 30 m s^{-1}. What lateral force is exerted on the rails? Find the angle of tilt of the roadbed which would balance this lateral force and comment on whether it is feasible to account for this effect in railroad design.

5. (a) Find the geostrophic wind speed and direction at 30°N where pressure on a horizontal surface increases at 1 mPa m^{-1} [1.0 mb (100 km)$^{-1}$] toward the northeast and the pressure and temperature are, respectively, 85 kPa (850 mb) and 0° C.

(b) If the situation described in part (a) occurred at 70°S, what would be the geostrophic speed and direction?

6. Calculate the speed which an air parcel would achieve after 1 hr if it started from rest near the equator in a region in which the pressure decreased eastward at 1 mPa m^{-1}. The pressure is 90 kPa (900 mb) and the temperature is 20° C.

7. Estimate by calculation the distance and direction from Miami (latitude 26°N) to the center of a slowly moving hurricane if the pressure increases toward the southwest at 20 mPa m^{-1} (1 mb in 5 km) and the wind speed at the top of the boundary layer is 40 m s^{-1} from the northwest. Assume circular trajectories and use 1.10 kg m^{-3} for the density of air.

8. Consider a circular anticyclone at 60°N, in which the geostrophic wind speed between radii of 100 and 1000 km is 5 m s^{-1}. If the geostrophic speed at all radii remains constant as the system moves slowly southward, what are the limiting latitudes for balance of the pressure, centrifugal, and Coriolis forces at radii of 200 and 300 km? What is the gradient wind speed corresponding to these points?

9. If the wind speed in a tornado between center and radius r is given by

$$V = ar$$

show that the pressure distribution is given by

$$p = p_0 \exp\left[\frac{-a^2(r_0^2 - r^2)}{2R_m T}\right]$$

where a represents a constant, p_0 the pressure at radius r_0, and T the temperature (assumed constant). Find the central pressure if at a distance of 200 m from the center the pressure is 100 kPa, wind speed is 40 m s^{-1} and temperature is 20° C.

10. Show that the pressure gradient force in the horizontal direction can be expressed by

$$-\nabla_\theta(c_p T + \Phi)$$

where the subscript θ indicates that the gradient is taken at constant potential temperature.

11. Show that the thermal wind equation [Eq. (4.44)] also can be expressed in the form

$$\mathbf{V}_{g_1} - \mathbf{V}_{g_0} = \frac{1}{f} \mathbf{k} \times \nabla_p(\Phi_1 - \Phi_0)$$

12. Calculate the velocity of the sea breeze after 1 hr under the following conditions if the initial state was motionless. The mean virtual temperature in a column over land extending from 100 kPa (1000 mb) to 90 kPa (900 mb) is 15° C, and the corresponding temperature over the sea is 10° C. The return flow is assumed to occur along the 90 kPa surface, and the ascending and descending columns are assumed to be 25 km apart.

13. Show that the time rate of change of azimuthal wind speed in a uniform circular vortex lying in a pressure surface may be expressed by

$$\frac{dV}{dt} = -\left(\frac{V}{r} + f\right)V_r - \Omega r \cos\phi\left(\frac{v}{R} + \frac{d\alpha}{dt}\right)$$

where r is the distance of the air parcel from the center of the vortex, R is the radius of the earth, v the mean south to north velocity, V_r the uniform radial velocity, and α the inclination of the contour with respect to the pressure surface.

14. Find the circulation around the following circular vortices and determine the vorticity at interior points. (a) The tangential velocity varies as $V = ar$ where a is constant, (b) The tangential velocity varies as $V = k/r$ where k is constant.

15. Show that if the horizontal velocity divergence is zero, there exists a stream function ψ which satisfies Eqs. (4.64).

16. Find the time required for the mass of air contained in the boundary layer of a cylindrical hurricane to be lifted by Ekman pumping out of the boundary layer under the following conditions. The boundary layer is 1 km deep, and the wind in the boundary layer at a radius of 300 km blows at 30 m s^{-1} at a cross isobar angle of 20° toward lower pressure.

17. (a) If horizontal stress decreases linearly with height to zero at the top of the boundary layer and if turning of the wind with height is neglected, show that the mean kinematic eddy viscosity is related to the surface stress by

$$\tau_0 = 2\rho K_m\left(\frac{V_h - V_0}{h}\right)$$

where K_m is the kinematic eddy viscosity. Assume that V increases linearly with height.

(b)   Show also that the surface stress can be represented by

$$\tau_0 = \rho h f \bar{V} \tan \theta$$

18.   For the process of spin down of a circular vortex as described in Section 4.11 show that the vertically averaged geostrophic vorticity can be expressed as a function of time by the equation

$$\bar{\zeta_g} = \bar{\zeta_g}(t = 0)\exp[-fc_1 t]$$

19.   Suppose that an air parcel in an environment of static stability equal to -10^{-5} m^{-1} is given an initial upward velocity from its equilibrium position of 1 m s^{-1}. Calculate its vertical speed when it reaches a height of 1 km from its equilibrium position.

20.   Suppose that a vertical shear of 1 m s^{-1} exists across an inversion surface representing a discontinuity of 1° C. If the inversion is subjected to a disturbance of 50 m wavelength and maximum vertical displacement of 1 m, estimate the time required for the maximum of the oscillating vertical velocity of the wave to reach 10 m s^{-1} according to the Helmholtz theory. Use a temperature of 273 K.

21.   Find the horizontal and vertical phase speeds for a gravity wave whose length in the x direction is 5 km and whose length in the z direction is 2 km if the undisturbed atmosphere is isothermal and $T = 273$ K.

22.   In the case of two-dimensional flow over a series of sinusoidal mountains and valleys the wavelength of the mountains prescribes the value of the wave number in the x direction. By allowing the coordinate system to move with the undisturbed wind, the theory of gravity waves as developed in the chapter can be applied to this case. Find the range of wind speeds for which the gravity wave set up by flow over the ridges decreases exponentially in amplitude with height.

Solutions

1.   From Fig. 4.2 the constant radial velocity is $V = S/t$. The magnitude of the lateral deflection is equal to the displacement in time t of a point at radius S with respect to the fixed coordinate system. This is expressed by

$$S' = \Omega S t$$

$$S' = \Omega S^2/V$$

The Coriolis acceleration is found by differentiating twice with respect to t and noting that $V = dS/dt$.

$$\frac{d^2 S'}{dt^2} = 2\Omega V$$

2. From Fig. 4.3 the side of the right triangle at right angles to r can be expressed by $(\Omega r + V)t$. The Pythagorean theorem then gives

$$(r + S')^2 = r^2 + (\Omega r + V)^2 t^2$$

Since $Vt = S$, it follows that

$$2rS' + S'^2 = S^2[(\Omega r/V) + 1]^2$$

and upon neglecting S' compared to $2r$

$$S' = \frac{S^2}{2r}\left(\frac{\Omega r}{V} + 1\right)^2$$

The acceleration is found to be

$$\frac{d^2 S'}{dt^2} = \Omega^2 r + 2\Omega V + \frac{V^2}{r}$$

The second term is the Coriolis acceleration and is identical to the Coriolis acceleration found in Problem 1. The other two terms are centrifugal accelerations associated with rotation of the earth and the relative velocity of the projectile.

3. 5.8 cm upward.

4. 790 N, 4.0×10^{-4} rad.

5. (a) $V_g = 12.6\ \text{m s}^{-1}$ toward the NW.

    (b) $V_g = 6.7\ \text{m s}^{-1}$ toward the SE.

6. $V = 3.36\ \text{m s}^{-1}$ toward the east.

7. $r = 102\ \text{km}$ toward the NE.

8. $\phi = 43°, 27°,\ V_{gr} = 10\ \text{m s}^{-1},\ 10\ \text{m s}^{-1}$.

9. For a tornado in which Coriolis force is small compared to centrifugal force and pressure increases with radius, Eq. (4.34) may be written

$$\frac{V^2}{r} = \frac{1}{\rho}\frac{\partial p}{\partial r}$$

This may be expressed

$$a^2 r\, dr = R_m T\, d(\ln p)$$

Integrating from r to r_0 yields

$$p = p_0 \exp\left[-\frac{a^2(r_0^2 - r^2)}{2R_m T}\right]$$

From the values given at $r = 200\ \text{m}$, $a = 0.2\ \text{s}^{-1}$. Upon setting $r = 0$ and substituting numerical values, the central pressure is calculated to be

$$p = 82.8\ \text{kPa}$$

10. A figure similar to Fig. 4.9 can be used to develop the gradient of pressure at constant θ in the form

$$\mathbf{V}_\theta p = \mathbf{V}p + \mathbf{V}_\theta z\frac{\partial p}{\partial z}$$

The definition of potential temperature [Eq. (2.47)] yields

$$\frac{\nabla\theta}{\theta} = \frac{\nabla T}{T} - \frac{R_m}{c_p}\frac{\nabla p}{p}$$

And upon applying this at constant θ and substituting above

$$-\frac{1}{\rho}\nabla p = -\nabla_\theta(c_p T + \Phi)$$

11. Upon taking the gradient of the hypsometric equation [Eq. (4.41)] at constant p

$$\nabla_p(\Phi_1 - \Phi_0) = R_m\nabla_p T \ln\frac{p_0}{p_1}$$

And substitution of this equation into Eq. (4.44) gives

$$\mathbf{V}_{g_1} - \mathbf{V}_{g_0} = \frac{1}{f}\mathbf{k} \times \nabla_p(\Phi_1 - \Phi_0)$$

12. $V = 11\ \mathrm{m\,s^{-1}}$.

13. In Eq. (4.50) express the circulation by $2\pi rV$ and the projection of the area of the contour on the plane of the equator by $\pi r^2 \sin\phi$. The baroclinic term does not appear because the contour lies in the pressure surface. This leads directly to the required result.

14. (a) $C = 2\pi a r^2$, $\zeta = 2a$.

(b) $C = 2\pi k$.

Choose a contour which does not enclose the center; since circulation is a constant independent of radius, the circulation around the contour is zero. This is easily shown for a contour consisting of arcs of two concentric circles and two radii. It follows from Eq. (4.52) that the vorticity is zero everywhere but at $r = 0$ where it is infinite.

15. Set Eq. (4.58) equal to zero and substitute Eq. (4.64).

16. 14.6×10^3 s.

17. (a) The average stress (stress at a height of half the boundary layer h) can be expressed from Eq. (4.2), and upon introducing the kinematic eddy viscosity K_m

$$\bar{\tau} = \rho K_m\frac{\partial V}{\partial z} \approx \rho\bar{K}_m\frac{V_h - V_0}{h}$$

Since $\tau_0 = 2\bar{\tau}$

$$\tau_0 = 2\rho\bar{K}_m\frac{V_h - V_0}{h}$$

(b) The frictional force acting on the boundary layer may be expressed by Eq. (4.3) and by Eq. (4.70). Upon equating the two and integrating

$$\int_{\tau_0}^{0} d\tau = -\int_0^h fV\rho\tan\theta\,dz$$

The negative sign is introduced to convert the stress exerted by the layer to stress exerted on the layer. Therefore

$$\tau_0 = \rho h f\bar{V}\tan\theta$$

18. The vorticity equation [Eq. (4.55)] can be further simplified by neglecting the tilting terms, the df/dy term, and neglecting ζ compared to f. Then if ζ is replaced by ζ_g

$$\frac{d\zeta_g}{dt} = -f\frac{\partial\omega}{\partial p} \approx -f\frac{\partial\omega}{\partial z}\frac{\partial z}{\partial p} \approx fg\rho\frac{\partial w}{\partial z}\left(-\frac{1}{\rho g}\right) = -f\frac{\partial w}{\partial z}$$

Upon replacing $\partial w/\partial z$ by w_h/h and substituting Eq. (4.76)

$$\frac{d\zeta_g}{\zeta_g} = -fc_1 \, dt$$

$$\zeta_g = \zeta_g(t = 0)\exp(-fc_1 t)$$

19. Equation (4.79) may be differentiated with respect to time. Then if the initial conditions $w = w_0$ and $z = 0$ at $t = 0$, are introduced

$$A = -B = w_0/2N^*$$

It follows that

$$z = (w_0/N^*)\sinh N^* t$$

$$w = w_0 \cosh N^* t$$

Introducing the identity $\cosh^2 x - \sinh^2 x = 1$ leads to

$$w^2 = w_0^2 - N^{*2}z^2$$

$$w = 9.95 \text{ m s}^{-1}$$

20. At $z = 0$, $x = 0$ Eq. (4.84) yields $w_2' = C_2^* e^{-ikct}$ and $t = -(1/ikc)\ln(w_2'/C_2^*)$. After integrating the vertical velocity, the vertical displacement can be expressed by

$$z = -(w_2'/ikc) + C_2^*/ikc$$

If the initial vertical velocity is zero, the initial displacement is $z_0 = C_2^*/ikc$. Since c is complex, as follows from Eq. (4.89), we write $c = c_r \pm ic_i$, and the real part of the initial displacement becomes

$$z_0 = \frac{C_2^* kc_i}{k^2(c_r^2 + c_i^2)}$$

The time may now be expressed

$$t = -\frac{1}{ikc}\ln\left(\frac{w_2' kc_i}{k^2 z_0(c_r^2 + c_i^2)}\right)$$

Taking the real part, we find

$$t = \frac{kc_i}{k^2(c_r^2 + c_i^2)}\ln\left(\frac{w_2' kc_i}{k^2 z_0(c_r^2 + c_i^2)}\right)$$

The numerical values given yield $kc_r = .0628 \text{ s}^{-1}$, $kc_i = .0411 \text{ s}^{-1}$, $w_2' = 10 \text{ m s}^{-1}$, $z_0 = 1 \text{ m}$, and these values result in

$$t = 31.3 \text{ s}$$

21. The phase speeds in the x and z directions are, respectively

$$c_{gx} = v_g/k \qquad \text{and} \qquad c_{gz} = v_g/n$$

Then if Eq. (4.103) is substituted for v_g, the values given result in

$$c_{gx} = 5.5 \text{ m s}^{-1}$$

$$c_{gz} = 2.2 \text{ m s}^{-1}$$

22. In the moving coordinate system the frequency will be given by $v_g = k\bar{u}$. Upon solving Eq. (4.103) for n^2 and neglecting small terms for convenience

$$n^2 = k^2\left(\frac{N^2}{v_g^2} - 1\right)$$

Exponentially decreasing solutions occur for $n^2 < 0$. Therefore, for this condition

$$N^2 < v_g^2 = k^2\bar{u}^2$$

$$\bar{u} > N/k = NL/2\pi$$

General References

Dutton, *The Ceaseless Wind*, concentrates on mathematical methods in developing the basic relationships of atmospheric flow.

Gossard and Hooke, *Waves in the Atmosphere*, presents a comprehensive, well organized and readable account of acoustic-gravity waves in the atmosphere.

Haltiner and Martin, *Dynamical and Physical Meteorology*, as well as other older texts, represents a "classical" presentation of the subject. The treatment is more extensive than has been attempted here.

Hess, *Introduction to Theoretical Meteorology*, presents basic concepts and theory of atmospheric motions without the use of vectors.

Holton, *An Introduction to Dynamic Meteorology*, gives a more detailed and extensive treatment of the subject than is attempted here. It is notably clear, especially in relating the theory of large scale motions to observations.

Morel, ed., *Dynamic Meteorology*, the first half of this book, consisting of lectures by N. Phillips, Charney, and Lilly, constitutes a comprehensive account of major topics in dynamic meteorology on an advanced level.

Pedlosky, *Geophysical Fluid Dynamics*, is a notably lucid and unified account of the dynamics of the atmosphere and ocean on an advanced level.

Wallace and Hobbs, *Atmospheric Science: An Introductory Survey*, provides descriptions of atmospheric motion systems which should be useful as a supplement to this chapter.

CHAPTER V

Solar and Terrestrial Radiation

"Common sense is a docile thing.
It sooner or later learns the ways of science." HENRY MARGENAU

PART I: PRINCIPLES OF RADIATIVE TRANSFER

The atmospheric motions discussed in Chapter IV are driven by a constant flow of energy. In this chapter and in Chapter VI the sources of atmospheric energy and some of the major processes of energy transfer are examined.

The fundamental energy source for the atmosphere is the sun, with only trivial amounts provided by the stars, by the full moon, by heat transfer from within the earth, or by energy released through human actions. For every 100 units of energy provided by the sun, the stars provide only about 10^{-5} units, the full moon 0.01, the earth's interior 0.005, and human actions 0.01. The sun's energy is transmitted to the earth through the intervening space as electromagnetic radiation. Electromagnetic waves travel through vacuum at 2.998×10^8 m s^{-1}, so radiation from the sun reaches the earth at a distance of about 150×10^6 km in 500 s (8 min). Radiation emitted by the sun varies in wavelength from about 10^{-14} to 10^{10} m and in frequency from about 10^{22} to 10^{-2} s^{-1}, and the totality of all wavelengths is called the electromagnetic spectrum. Waves at the short end of the spectrum are comparable to the size of an atomic nucleus, while waves at the long end of the spectrum are about $\frac{1}{10}$ of the distance of the earth from the sun in length. The electromagnetic spectrum includes the many categories of rays or waves shown in Fig. 5.1. The names of some of these categories make clear that electromagnetic radiation is important not only in transferring energy to the earth and within the atmosphere, but also in communication and remote sensing. This latter aspect is discussed in Chapter VII.

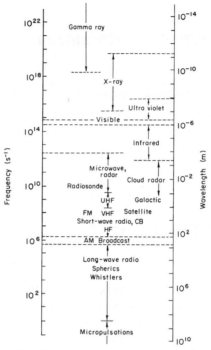

FIG. 5.1.   The electromagnetic spectrum.

5.1   Definition and Concepts

Radiant energy is energy in transit. The amount of radiant energy passing an area per unit time is called the *radiant flux*, and the radiant flux per unit area the *irradiance*† is represented by

$$E = \frac{d^2Q}{dA\,dt}$$

where Q represents radiant energy. The irradiance is usually expressed in watts per square meter.

The radiant energy per unit time coming from a specific direction and passing through unit area perpendicular to that direction is called the *radiance*.‡ The radiance L and irradiance are related by

$$L = \frac{dE}{\cos\theta\,d\omega} \tag{5.1}$$

---

† Irradiance is also referred to as *radiant flux density*.
‡ Radiance is also referred to as *intensity*.

where $d\omega$ represents the differential of solid angle and θ the angle between the beam of radiation and the direction normal to the surface (usually horizontal) on which the radiance is measured. The definition of solid angle is given in Appendix I.I. Radiant energy propagated in a single direction is called *parallel beam* radiation, and in this case $d\omega$ vanishes in Eq. (5.1). Therefore, radiance is not defined in the special case of parallel beam radiation.

Equation (5.1) can also be written in the integral form

$$E = \int_0^{2\pi} L \cos \theta \, d\omega \tag{5.2}$$

Figure 5.2 illustrates the effect on irradiance of orientation of the radiant beam with respect to the surface. Radiation whose radiance is independent of direction is called *isotropic radiation*. In this case because $d\omega = 2\pi \sin \theta \, d\theta$, as shown in Appendix I.I [Eq. (I.2)], Eq. (5.2) may be integrated to yield

$$E = \pi L \tag{5.3}$$

This relation is used frequently in following sections. In cases where radiation is not isotropic integration of Eq. (5.2), usually by numerical means, leads to more complicated results.

Since the radiant energy is distributed over various wavelengths and also because the absorptive or emissive properties of materials are a function of wavelength (λ), it is useful to define *monochromatic irradiance* (E_λ) and *monochromatic* or *spectral radiance* (L_λ) by the following relations

$$E = \int_0^\infty E_\lambda \, d\lambda \quad \text{and} \quad L = \int_0^\infty L_\lambda \, d\lambda \tag{5.4a}$$

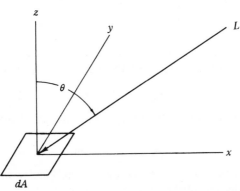

FIG. 5.2.   Radiation of radiance L at zenith angle θ incident on differential area dA.

Irradiance and radiance are also often expressed by integrals over frequency (v) in the forms

$$E = \int_0^\infty E_v \, dv \quad \text{and} \quad L = \int_0^\infty L_v \, dv \tag{5.4b}$$

The transformation from wavelength to frequency follows from

$$v = c/\lambda \tag{5.5}$$

where c represents the speed of light. The frequency is the number of waves that pass a given point in one second. From Eqs. (5.4) and (5.5) it follows that

$$E_\lambda = (c/\lambda^2)E_v \quad \text{and} \quad L_\lambda = (c/\lambda^2)L_v \tag{5.6}$$

Certain properties of radiation may be understood as depending on the characteristics of a continuous series of waves; other properties require that radiation be treated as a series of discrete particles or *quanta*. Some of the properties associated with waves are discussed in Chapter VII, while some of the properties associated with quanta are discussed in Sections 5.3 and 5.5.

5.2   Absorption and Emission of Radiation

Radiation is absorbed in passing through matter, and the fraction absorbed is a specific characteristic of the material The ratio of the absorbed to the incident radiation at a certain wavelength is called the *spectral* or *monochromatic absorptance* (a_λ) and is usually a function of the wavelength. A body with spectral absorptance equal to unity for all wavelengths is called a *black body*. Perfect black bodies do not exist in nature, but they may be very closely approximated especially in the infrared or long wave range. Of the incident radiation that is not absorbed part is reflected and part is transmitted. The ratio of the reflected to the incident radiation at a certain wavelength is called the *spectral reflectance* (r_λ), and the ratio of the transmitted to the incident radiation at a certain wavelength is called the *spectral transmittance* (τ_λ). The three ratios are related by

$$a_\lambda + r_\lambda + \tau_\lambda = 1$$

For a black body $r_\lambda = \tau_\lambda = 0$ and $a_\lambda = 1$ for all wavelengths.

Kirchhoff's Law

A molecule which absorbs radiation of a particular wavelength also is able to emit radiation of the same wavelength. The rate at which emission takes place is a function of the temperature and the wavelength. This is a fundamental property of matter which leads to a statement of Kirchhoff's law.

Consider two parallel plates, infinite in extent, separated by empty space and perfectly insulated on the outside. When the temperature of both surfaces is the same, the radiative transfer of energy must be the same in all directions, otherwise one of the plates would heat up at the cost of the other plate, in violation of the second law of thermodynamics. Assume now that wall 1 has an absorptance $a_1 = 1$ for all wavelengths and that wall 2 is capable of emitting the maximum amount of radiation $E^*(T)$ at the temperature T, i.e., its *emissivity* (ε) equals unity for all wavelengths. Emissivity is here defined as the ratio of the emitted radiation to the maximum possible emitted radiation at that temperature. With this information we can now write the balance equations of radiative transfer. The irradiance from wall 1 to wall 2 is $E_1 = \varepsilon_1 E^*(T)$, where ε_1 remains to be determined. And from wall 2 to wall 1 we have $E_2 = E^*(T) + r_2 E_1$. But since $E_1 = E_2$, we find that $\varepsilon_1 = 1 + r_2\varepsilon_1$. This is only possible when $\varepsilon_1 = 1$ and $r_2 = 1 - a_2 = 0$. Thus we find that for a black surface

$$a = \varepsilon = 1$$

$E^*(T)$ is called the black-body irradiance. It follows also that the radiance emitted by a black body is independent of its composition and of direction.

Consider now the same system with the exception that the absorptance of wall 2 is less than unity. The equilibrium is now

$$E_1 = E^*(T) = E_2 = \varepsilon_2 E^*(T) + r_2 E^*(T)$$

or

$$\varepsilon_2 + (1 - a_2) = 1, \quad \text{i.e.,} \quad \varepsilon_2 = a_2$$

Therefore, $\varepsilon = a$ regardless of whether the surface is black or not. This is Kirchhoff's law for absorptances and emissivities independent of wavelength. The law can be generalized for spectral absorptances and emissivities by considering the same system as before with wall 2 black for all wavelengths except for a small interval at wavelength λ. Because E^* is uniquely distributed over the spectrum, $E^*_\lambda(T)$ is the maximum possible radiation at wavelength λ for temperature T. An argument similar to that above can now be used to derive Kirchhoff's law in the form

$$\varepsilon_\lambda = a_\lambda \tag{5.7}$$

Note that although the assumptions of thermal and radiative equilibrium are useful in deriving Eq. (5.7), they do not restrict validity of Kirchhoff's law. Thermodynamical equilibrium, as defined in Section 2.4, is required.

Schwarzschild's Equation

Kirchhoff's law may be applied to media that partially transmit and partially absorb radiation. Consider a thin slab with thickness dx, as depicted in Fig. 5.3, and incident radiation with radiance L_λ perpendicular to this slab. The radiance will be diminished by absorption and enhanced by emission in this layer. The amount absorbed will be proportional to the thickness of the layer, the density (ρ_c) of the absorbing constituent, and to the radiance of the incident radiation. This may be expressed by

$$(dL_\lambda)_{abs} = -k_\lambda \rho_c L_\lambda \, dx \equiv -L_\lambda \, du_\lambda$$

where k_λ represents the *absorption coefficient* of the layer and du_λ represents the increment of *optical thickness* defined by

$$u_\lambda \equiv \int_0^x k_\lambda \rho_c \, dx \tag{5.8}$$

It is convenient also for future use to introduce the *density weighted height*, defined by

$$Z \equiv \int_0^z \rho_c \, dz \tag{5.9}$$

Because the absorptance of this layer must be equal to the emissivity, the contribution by emission to the radiance is given by

$$(dL_\lambda)_{em} = L_\lambda^* \, du_\lambda$$

where L_λ^* represents the black-body spectral radiance. The total change in radiance dL_λ is the sum of the absorbed and emitted contributions

$$dL_\lambda = -(L_\lambda - L_\lambda^*) \, du_\lambda \tag{5.10}$$

This equation, known as Schwarzschild's equation, can be integrated if the temperature and the optical thickness along the path are known. This will be discussed further in Section 5.14.

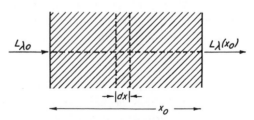

FIG. 5.3.   Depletion of radiance in traversing a homogeneous absorbing layer.

Of special interest is the case that L_λ^* is negligible. Eq. (5.10) can then simply be integrated to

$$L_\lambda = L_{\lambda 0} e^{-u_\lambda}$$ (5.11)

which is known as *Beer's law*. If the total optical thickness of the layer is represented by $u_{\lambda 0}$, then the spectral transmittance may be expressed by

$$\tau_\lambda \equiv L_\lambda(x_0)/L_{\lambda 0} = e^{-u_{\lambda 0}}$$ (5.12a)

Similarly, the spectral absorptance may be expressed by

$$a_\lambda = 1 - e^{-u_{\lambda 0}}$$ (5.12b)

Because the transmitted and reflected components are simply additive, when reflection occurs the reflected radiation must be subtracted before Beer's law is applied. An example is provided by Problem 1.

5.3 Theory of Black-Body Radiation†

The theory of the energy distribution of black-body radiation was developed by Planck and was first published in 1901. He postulated that energy can be emitted or absorbed only in discrete units or photons defined by

$$u = hv$$ (5.13)

where the constant of proportionality h has the value $(6.626176 \pm 0.000036) \times 10^{-34}$ J s and is known as *Planck's constant*. Although the Planck postulate was introduced arbitrarily, some intuitive support may be added if one considers each atom to be analogous to a mechanical oscillator (tuning fork) which may oscillate at only discrete characteristic frequencies. Einstein subsequently showed that the energy of the photon is expressible by mc^2, where m represents the equivalent mass of the photon. Therefore, the momentum of the photon is given by

$$p = hv/c$$ (5.14)

Consider now a system filled with photons which move in all directions with the speed c. To find the distribution of the photons in phase space, or more specifically, in the frequency interval Δv, it is necessary to consider the most probable distribution of photons. This is analogous to the calculation of most probable distribution of molecular velocities which was dis-

---

† This section contains advanced material which is not essential for understanding the following sections and chapters.

cussed in Section 2.15. Equations (2.57), (2.58), and (2.60) are still valid in this case. However, photons are created and destroyed at each interaction, so that there is no constraint on the number N, and Eq. (2.59) is no longer valid.

Lagrange's method of undetermined multipliers applied to Eqs. (2.58) and (2.60) now yields for the average number of photons per cell in the interval Δv_i

$$N_i = (e^{\beta u_i} - 1)^{-1} \tag{5.15}$$

Combining Eq. (5.15) with Eqs. (2.64) and (5.13) yields the Bose–Einstein distribution in the form

$$N_i = (e^{hv_i/kT} - 1)^{-1}$$

The fact that β has the same value in this case as is indicated by Eq. (2.64) can be checked by substituting Eq. (5.15) back into Eq. (2.58), with the result

$$\delta \ln W = \sum \beta u_i \, \delta N_i = \beta \, \delta u$$

Upon combining this result with Eq. (2.66)

$$\delta s = (k/Nm) \, \delta \ln W = k\beta \, \delta u$$

But for a system with constant volume

$$ds = T \, du$$

which leads again to the conclusion that

$$\beta = 1/kT \tag{2.64}$$

To find the number of photons per frequency interval it may be recognized from Eq. (5.14) that a shell of phase volume corresponding to Δv is represented by $4\pi h^3 V v^2 \Delta v/c^3$, where V represents the total volume in ordinary space. Upon dividing this volume in phase space by H, the volume of a single cell, the number of cells contained in the increment Δv is represented by

$$\Delta n = \frac{4\pi V h^3 v^2 \, \Delta v}{c^3 H}$$

This is the number of cells for photons of a single transverse degree of freedom; since photons exhibit two transverse degrees of freedom, the total number of cells is twice that given here. The number of photons in Δv is then

$$\Delta N = 2\bar{N}_i \, \Delta n = \frac{8\pi V h^3 v^2 \, \Delta v}{c^3 (e^{hv/kT} - 1)H}$$

and the energy density per unit frequency interval is

$$\frac{hv\,\Delta N}{V\,\Delta v} \equiv u_v = \frac{8\pi hv^3 h^3}{c^3(e^{hv/kT} - 1)H} \tag{5.16}$$

Although Planck had originally intended to allow the cell size to become zero, it is clear from Eq. (5.16) that this is not possible. The smallest possible cell size is determined by Heisenberg's uncertainty principle, which states that if the position of a photon is known to within Δx, the momentum cannot be known more accurately than within $h/\Delta x$. Therefore, if the precise position of the photon is known, its momentum is completely unknown, and if the momentum is known precisely, its position is completely uncertain. So, clearly, h^3 is the smallest possible cell size in which a photon can be specified.

Because for black-body radiation the photons travel in all directions with speed c

$$L_v^* = u_v c/4\pi$$

Upon combining this with Eq. (5.16) Planck's law of black-body radiance per unit frequency interval may be written in the form

$$L_v^* = \frac{2hv^3}{c^2(e^{hv/kT} - 1)} \tag{5.17}$$

This law has been experimentally verified to a high degree of accuracy.

5.4   Characteristics of Black-Body Radiation

If the radiance is expressed in terms of wavelength, Eq. (5.17) may be written [see Eq. (5.6)]

$$L_\lambda^* = 2hc^2\lambda^{-5}(c^{hc/k\lambda T} - 1)^{-1} \tag{5.18}$$

Figure 5.4 shows that the black-body radiance increases markedly with temperature and that the wavelength of maximum radiance decreases with increasing temperature.

The total radiance of a black body is found by integrating Eq. (5.18) over all wavelengths. Upon substituting $x \equiv hc/\lambda kT$, the radiance is given by

$$L^* = \int_0^\infty L_\lambda^* \, d\lambda = \frac{2k^4 T^4}{c^2 h^3} \int_0^\infty \frac{x^3 \, dx}{e^x - 1}$$

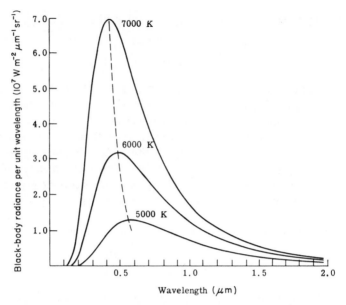

FIG. 5.4.   Black-body radiance per unit wavelength calculated from Eq. (5.18) for temperatures 5000, 6000, and 7000 K.

The integral $\int_0^\infty x^3(e^x - 1)^{-1}\,dx$ can be shown to have the value $\pi^4/15$, so that

$$L^* = \frac{2\pi^4 k^4}{15c^2h^3}\,T^4 \equiv bT^4$$

This is the *Stefan–Boltzmann law,* according to which the radiance emitted by a black body varies as the fourth power of the absolute temperature.

Since black-body radiance is independent of the direction of emission, Eq. (5.3) shows that the irradiance emitted by a black body is

$$E^* = \pi bT^4 \equiv \sigma T^4 \tag{5.19}$$

which is another form of the Stefan–Boltzmann law. The constant σ, which has the value $5.67 \times 10^{-8}\ \mathrm{W\ m^{-2}\ K^{-4}}$, is called the Stefan–Boltzmann constant.

The wavelength of maximum radiance for black-body radiation may be found by differentiating Planck's law with respect to the wavelength, equating to zero, and solving for the wavelength as required in Problem 2. The result is *Wien's displacement law*, which may be written

$$\lambda_m = \alpha/T \tag{5.20}$$

where $\alpha = 2897.8 \, \mu m \, K$. This relation makes it possible to compute the temperature of a black body by measuring the wavelength of maximum spectral radiance. In Problem 3 it is required to obtain the relation between $L^*_{\lambda \, max}$ and the temperature. If the body is only approximately black, computation using Eq. (5.20) yields an approximate temperature called the *color temperature* of the body.

5.5 The Line Spectrum

When inspected in detail with high resolution, the absorption and emission spectra of a particular material consist of large numbers of individual and characteristic spectral lines. The relationship of spectral lines to atomic structure can be understood on an elementary basis by use of the model of the hydrogen atom proposed by Niels Bohr in 1913. He postulated that the electron revolves in a circular orbit, its motion being governed by the balance between the electrostatic and the apparent centrifugal force, with the restriction that its angular momentum is a multiple of $h/2\pi$. Consequently, transition from one orbit to another orbit cannot take place continuously, but occurs in jumps each of which is accompanied by a change in energy. The energy expressed by Eq. (5.13) is emitted by the atom as a photon of electromagnetic radiation, and the atom may absorb a photon of the same energy. The energy of the quantum jump is analogous to the change in kinetic plus potential energy required to shift an earth satellite from one orbit to another, as expressed by Eq. (1.14).

Each quantum jump between fixed energy levels results in emission or absorption of characteristic frequency or wavelength. These quanta appear in the spectrum as emission or absorption lines. For the simple hydrogen atom the line spectrum is correspondingly simple, whereas the spectra of water vapor and carbon dioxide are considerably more complex.

Spectral lines have a characteristic shape suggestive of the Maxwellian distribution function. The finite width of the line is, in part, a result of molecular collisions; the interaction between atoms when they are very close together causes a slight change in the energy levels of the electrons and in the energy of an absorbed photon. This effect is called pressure broadening of the line. A less important source of line broadening is the Doppler shift associated with the velocity distribution of the molecules. Those molecules moving toward the approaching photons experience a shift toward higher frequency, and vice versa. It is therefore clear that line width should increase with temperature, as well as with pressure. The Doppler shift is discussed in Section 7.14.

The shape of a single spectral line in the lowest 80 km of the atmosphere may be expressed by

$$k_v = \frac{k_1}{\pi} \frac{\alpha}{(v - v_0)^2 + \alpha^2} \tag{5.21}$$

where α represents the half-width of the line, and the *normalized line intensity* is expressed by

$$k_1 = \int_{-\infty}^{\infty} k_v \, dv$$

Equation (5.21) follows from the discussion of refractive index in Section 7.6. The spectral transmittance or absorptance may be calculated by substituting Eq. (5.21) into the appropriate forms from Sections 5.1 and 5.2. The absorptance is plotted in Fig. 5.5 for three values of the optical thickness (u) expressed in the dimensionless form $u/2\pi\alpha \equiv X$. For small values of X the absorption is proportional to k_1. For large value of X the absorptance in the center of the line is unity, and further increase in optical thickness has no effect. However, in the line wings the absorptance continues to increase with further increase in optical thickness.

The actual absorption spectrum of the atmosphere contains innumerable lines some of which are organized into absorption bands of limited spectral interval. In this case the wings of the various lines overlap, and the combined

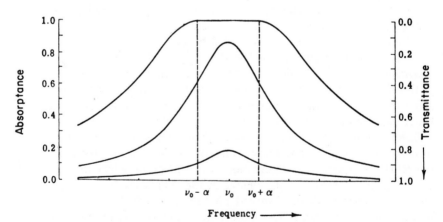

FIG. 5.5. Absorptance for a spectral line calculated from Eq. (5.21) for $u/2\pi\alpha$ equal to 0.1, 1.0, and 5 (after W. M. Elsasser, *Meteorol. Monographs* **4** (23), 5, 1960).

absorption has to be taken into account. For small absorptance the individual line absorptances can be added directly to give the combined absorptance. For strong absorptance the combined effect is more complicated and is always less than the sum of the individual line absorptances.

PART II: RADIATION OUTSIDE THE ATMOSPHERE

The energy source for nearly all the physical processes discussed in earlier chapters—expansion of air, precipitation, generation of ionospheric potential, atmospheric motions—is, of course, the sun. This statement is equally true for many phenomena outside the normal scope of atmospheric physics: biological processes, formation of soils, and the oxidation of automobile tires, for example. In each of these examples energy originally incident on the top of the atmosphere in the form of radiation has undergone a series of transformations, finally culminating in the phenomenon of special interest. The link between solar energy and atmospheric phenomena is particularly close, so that it is of the greatest importance to understand how the energy transformations occur.

5.6 The Sun

In spite of its importance to virtually all processes on earth, the sun is by no means unique. Among the billions of stars in our galaxy it is about average in mass and in size, and our galaxy is, of course, only one among millions of galaxies. The sun's importance results from its closeness; it is 150×10^6 km from the earth, whereas the next closest star is 3×10^5 times as far away.

The sun is a gaseous sphere with a diameter of 1.42×10^6 km and a surface temperature of about 6×10^3 K. The temperature increases toward deeper layers until temperatures high enough to sustain nuclear reactions are reached. The source of solar energy is believed to be fusion of four hydrogen atoms to form one helium atom, and the slight decrease in mass which occurs in this reaction accounts for the energy released in the solar interior. This energy is transferred by radiation and convection to the surface, and is then emitted as both electromagnetic and particulate radiation.

Each square centimeter of the solar surface emits on the average with a power of about 6.2 kw. This irradiance is maintained by the generation within the solar interior of only 0.265 J m^{-3} s^{-1}, whereas a corresponding value for burning coal is typically 10^9 times as large. The solar spectrum is shown in Fig. 5.6.

The sun's energy is radiated more or less uniformly in all directions, and nearly all of the energy disappears into the expanding universe. As is illustrated in Problem 4, only a minute fraction of the output of the sun is intercepted by the earth. Although there is no a priori reason to assume that the sun emits radiation at a constant rate, observations indicate that the short term variations are remarkably small (see Section 5.7).

The Solar Spectrum

The distribution of electromagnetic radiation emitted by the sun approximates black-body radiation for a temperature of about 6000 K as shown in Fig. 5.6. The similarity between the solar and black-body spectra provides the basis for estimates of the temperature of the visible surface layer of the sun. Because the sun does not radiate as a perfect black body, different comparisons are possible with different results. The Stefan–Boltzmann law may be used in combination with the solar constant, as in Problem 5, to calculate a temperature of 5733 K. Or the wavelength of maximum intensity, 0.4750 μm, combined with Wien's displacement law [Eq. (5.20)] may be used to calculate a temperature of 6100 K. Although it is not pos-

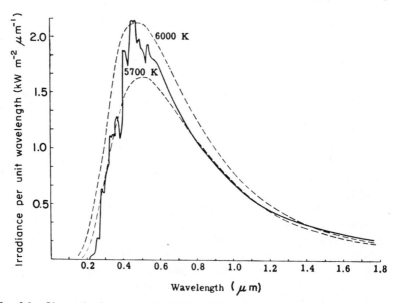

FIG. 5.6.   Observed solar spectral irradiances corrected to mean solar distance (*solid line*) and corresponding black-body spectral irradiances for temperatures of 6000 and 5700 K (*dashed lines*) (after F. S. Johnson, *J. Meteorol.* **11**, 431, 1954).

sible from the two radiation laws to determine a unique temperature for the solar surface, the uncertainty amounts to less than 10^3 degrees. This range of temperature may result from the fact that different parts of the spectrum are emitted by different layers of the sun at different temperatures.

Other Features of the Sun

Although to the eye armed with only a smoked glass, the sun seems to have a uniform brightness, inspection of the solar surface (*photosphere*) with a telescope reveals a granular structure. The granulae are bright areas with a diameter averaging between three hundred and fifteen hundred kilometers. They have a lifetime of the order of a few minutes, suggesting violent convection of tremendous scale. Figure 5.7 is a picture of the sun's surface showing clearly the granular structure. The resemblance to some types of stratocumulus clouds or to *Bénard-cells* is particularly challenging.

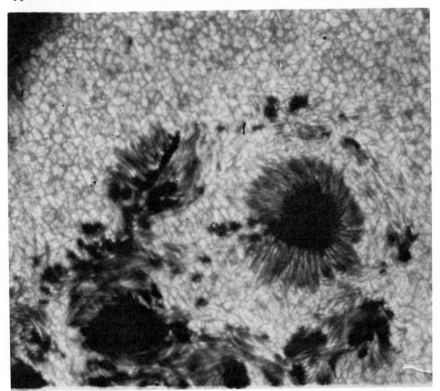

FIG. 5.7.   A portion of the surface of the photosphere at 10.01 CST 17 August 1959 taken by Project Stratoscope. The radii of the large sunspots are roughly equivalent to the diameter of the earth (published by courtesy of M. Schwarzschild).

Figure 5.7 also shows the structure of a sunspot. Clearly visible are the dark core (the umbra) and the radiant structure (the penumbra) surrounding the core. The origin and dynamics of sunspots are still uncertain although an important role must be played by magnetic fields near the surface. Because the solar gas consists of charged particles, a magnetic field exerts a force on the gas particles moving in the field. This force may prevent the development of convection and the transport of hot interior matter upward to the photosphere. This may account for the fact that sunspots appear as comparatively dark areas showing that they are cooler than the surrounding photosphere. The sunspots are much more stable than the granulae and have a lifetime varying from a few days to more than a month. Sometimes, usually in the vicinity of sunspots, very bright areas, called *flares*, become visible.

The photosphere is surrounded by a spherical shell, about 1.5×10^4 km in thickness, called the *chromosphere*. It is composed of thin gases and may be considered the solar atmosphere. The spectrum of the chromosphere is characterized by many rather sharp absorption lines, called *Fraunhofer lines* after their discoverer. Although these lines have been intensively studied as clues to properties of the chromosphere, it is likely that the region is so far from thermodynamic equilibrium that Kirchhoff's law does not apply, and therefore the Fraunhofer lines do not provide a reliable indication of temperature. The temperature is consequently uncertain, but has been estimated to vary from 8×10^3 K in the center of the chromosphere to about 20×10^3 K in the transition region between the chromosphere and the corona. Violent disturbances in the chromosphere result in large masses of ionized hydrogen being emitted from the sun. Beyond the chromosphere the *corona* extends many millions of miles into space. The weak continuous spectrum of the corona indicates that it consists of an extremely thin gas of a very high temperature (10^6 K).

The period of rotation of the sun, as determined from sunspot observations, is about 24 days at the equator, 26.5 days at 25° latitude, and is not well known near the poles.

The frequency of sunspots exhibits cycles of roughly 22 years. A cycle begins with a few sunspots appearing at about 35° latitude. Gradually the number of sunspots increases and at the same time the latitude of initial detection decreases. Five to six years after the beginning of a cycle a maximum in sunspot activity is reached. After that the number of sunspots decreases and the latitude of initial detection also decreases until at about eleven years, a minimum in activity occurs. Before the last sunspots disappear at about 5° latitude, already new sunspots have appeared at high latitudes. Although this suggests a cycle of roughly 11 years, closer observation shows that the new sunspots have a reversed sense of magnetic polarity with respect to the old ones. Therefore the total cycle of the sunspots consists of two periods, each of about 11 years.

5.7 Determination of the Solar Constant

Even though solar radiation is attenuated by scattering and absorption in passing through the atmosphere, the irradiance of solar radiation at the top of the atmosphere, the *solar constant*, may be calculated from measurements made at the earth's surface. If the atmosphere is assumed to consist of a series of homogeneous plane parallel layers, the optical thickness of the atmosphere at zenith angle θ is $u_{\lambda\infty} \sec \theta$, where $u_{\lambda\infty} \equiv \int_0^\infty \sigma_\lambda \rho \, dz$, and σ_λ is the *extinction* coefficient which represents the combined effects of absorption and scattering. This assumption introduces an error which becomes important only for zenith angles greater than 70°. The monochromatic radiance may be expressed by Beer's law [Eq. (5.11)] in the form

$$L_\lambda = L_{\lambda 0} \exp(-u_{\lambda\infty} \sec \theta)$$

or

$$\ln L_\lambda = \ln L_{\lambda 0} - u_{\lambda\infty} \sec \theta \tag{5.22}$$

where $L_{\lambda 0}$ represents the monochromatic radiance at the top of the atmosphere. Observations of L_λ are made for several zenith angles during a single day. If for each observation σ_λ has been the same, then a plot of L_λ against $\sec \theta$, as shown in Fig. 5.8, may be extrapolated to $\sec \theta = 0$, and $L_{\lambda 0}$ may be read from the ordinate. Because $\sec \theta$ cannot be less than unity, the point

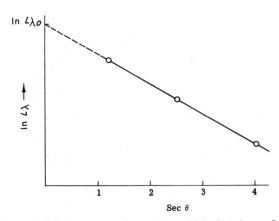

FIG. 5.8. Observed monochromatic solar radiances (L_λ) plotted as a function of zenith angle (θ).

determined in this way may be understood as representing the top of the atmosphere ($u = 0$).

A complication is introduced because the *spectrobolometer* used to measure L_λ is not capable of making absolute determinations of radiance. Instead, monochromatic radiances L_λ' are measured relative to a reference radiance. The corresponding relative radiances at the top of the atmosphere are then obtained by extrapolation as illustrated in Fig. 5.8. The relative radiances may now be plotted against wavelength, and the area under the curve expressed by the integrals

$$\int_0^\infty L_\lambda' \, d\lambda \quad \text{and} \quad \int_0^\infty L_{\lambda 0}' \, d\lambda$$

may be measured graphically. However, because the atmosphere is opaque for wavelengths shorter than 0.34 μm and for wavelengths longer than 2.5 μm, these integrals are evaluated only between 0.34 and 2.5 μm. Corrections, amounting to about 8%, are added for the omitted ranges.

In order to convert the above integrals to energy units, the total solar irradiance is measured by an instrument called the *pyrheliometer* which is illustrated in Fig. 5.9. The black surface at the far end of the instrument is exposed to solar radiation by directing the instrument toward the sun and is kept within a small temperature range by contact with circulating water. Water flow is maintained at a constant rate, so that the heat flow is proportional to the temperature difference between inflow and outflow. The radiation emitted through the opening of the instrument and other minor neglected effects are accounted for by appropriate corrections.

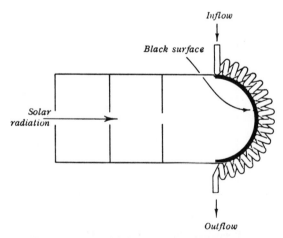

FIG. 5.9.   Essential elements of a pyrheliometer.

Now assuming that the ratio L'_λ/L_λ is independent of wavelength, the solar irradiance outside the atmosphere may be expressed by

$$E_0 = E\left(\int_0^\infty L'_{\lambda 0} \, d\lambda \middle/ \int_0^\infty L'_\lambda \, d\lambda \right) \qquad (5.23)$$

Finally, in order to determine the solar constant from E_0, it is necessary to recognize that E_0 changes as the distance of the earth from the sun changes. The relation between E_0 and the solar constant is obtained by assuming that all the solar energy which passes through a sphere of radius R_0 also passes through a sphere of radius R. Therefore, where $\bar{E}_s$ represents the solar constant

$$4\pi R_0^2 E_0 = 4\pi \bar{R}^2 \bar{E}_s \qquad (5.24)$$

The accuracy of the determination of the solar constant is limited by errors introduced in interpolation between measured wavelengths, and, particularly at short wavelengths, by correction for the ultraviolet portion of the spectrum. This portion of the spectrum is completely absorbed by the atmosphere, so that appreciable uncertainty is unavoidable by this method. Recent observations using balloons and rockets have extended the spectral range into the ultraviolet wavelengths and have reduced the uncertainty in the solar constant. Wilson and Hickey[†] report that the best estimate at present is $(1370 \pm 1)\,\mathrm{W\,m^{-2}}$ for the solar constant. The fluctuations observed by satellites over periods of days to weeks appear to be of the order of 0.1% of the solar constant or less. Over long periods there is evidence that there may be larger variations in the solar constant.

5.8 Short- and Long-Wave Radiation

The solar radiation received by the earth is partly reflected and partly absorbed. Over an extended period energy equivalent to that absorbed must be emitted to space by the earth and the atmosphere; otherwise, significant changes in temperature of the earth would be observed.

Although the irradiance absorbed by the earth and atmosphere is closely equal to that emitted, and although both distributions are very roughly black in character, the spectral curves of absorbed and emitted radiation overlap almost not at all. In Fig. 5.10 black-body radiation is plotted for a body of 6000 K corresponding to the radiation emitted by the sun and for a body of 250 K corresponding roughly to the radiation emitted by the earth

[†] R. C. Wilson and J. R. Hickey, "The Solar Output and Its Variation," (O. R. White, ed.), p. 111–116, Colo. Assoc. Univ. Press, 1977.

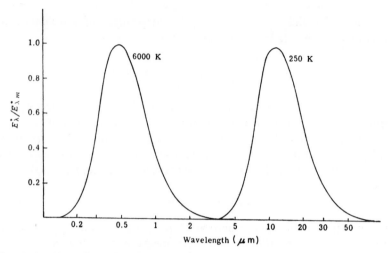

FIG. 5.10.   Normalized black-body irradiance per unit wavelength calculated from Eqs. (5.3) and (5.17) for temperatures of 6000 and 250 K.

and the atmosphere. The two curves are clearly separated into two spectral ranges above and below about 4 μm; for this reason it is customary to call the radiation received from the sun *short-wave radiation* and the radiation emitted by the earth and atmosphere *long-wave radiation*. This distinction makes it possible to treat the two types of radiation separately and thereby to reduce the general complexity of transfer of radiant energy.

5.9   Radiation Measurements from Satellites

Satellite instruments provide measurements of long-wave radiation from the earth and atmosphere as well as measurements of direct solar radiation and solar radiation reflected from the earth and atmosphere. These three separate fluxes have been determined by a method developed by Suomi in 1961. Two spherical sensors, one black and one white, are exposed to the three irradiances, and in a short time each achieves radiative equilibrium. Equilibrium for the black sensor is expressed by

$$4\pi r^2 a_b \sigma T_b^4 = \pi r^2 a_b (E_0 + E'_{sr} + E'_l)$$

where the subscript b refers to the black sensor. The absorptance a_b is assumed to be the same for short- and long-wave radiation. The irradiances

E'_{sr} (short-wave reflected radiation) and E'_1 (long-wave radiation) are defined for spherical sensors by

$$E'_{sr} = \int_0^\omega L_{sr}\, d\omega$$

and (5.25)

$$E'_1 = \int_0^\omega L_1\, d\omega$$

where ω is the solid angle by which the satellite "sees" the earth. L_{sr} and L_1 are, respectively, the radiances of reflected and emitted radiation received from the earth. Equation (5.25) express the fact that a spherical sensor is equally sensitive to radiation coming from all directions. To reconcile these equations with Eq. (5.2) is left as Problem 6.

Radiative equilibrium for the white sensor is represented by

$$4a_{wl}\sigma T_w^4 = a_{ws}(E_0 + E'_{sr}) + a_{wl}E'_1$$

where the subscript w refers to the white sensor. The absorptance of white paint at short wavelengths is small (about 0.1) and at long wavelengths is large (about 0.9) and is assumed to be the same as for the black sensor.

The two preceding equilibrium equations for the black and white sensors may be used to solve for the sum of the short-wave irradiances $(E_0 + E'_{sr})$ and the longwave irradiance (E'_1). The result is, upon introducing $a_1 = a_b = a_{wl}$ and $a_s = a_{ws}$

$$E_0 + E'_{sr} = \frac{4\sigma a_1}{a_1 - a_s}(T_b^4 - T_w^4) \qquad (5.26)$$

and

$$E'_1 = \frac{4\sigma}{a_1 - a_s}(a_1 T_w^4 - a_s T_b^4) \qquad (5.27)$$

E_0 may be measured directly by a pyrheliometer in the satellite or it may be estimated from Eq. (5.24).

In order to convert the measured irradiances into radiances emitted and reflected by the earth, it may be recalled that E'_{sr} and E'_1 depend on the solid angle ω under which the earth is seen by the satellite sensor. Figure 5.11 indicates that, where ψ represents half the plane angle subtended by the earth

$$R/(R + h) = \sin \psi$$

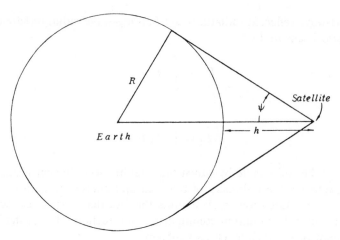

F<small>IG</small>. 5.11.   Cross section through the earth in the plane of an earth satellite.

and Appendix I.I shows that the solid angle may be expressed in terms of ψ by

$$\omega = 2\pi \int_0^\psi \sin \psi' \, d\psi' = 2\pi(1 - \cos \psi)$$

Upon combining these two equations

$$\omega = 2\pi\left(1 - \frac{(2Rh + h^2)^{1/2}}{R + h}\right) \tag{5.28}$$

Thus the solid angle depends only on the height of the satellite above the earth. As the height decreases to zero, the solid angle approaches 2π.

The radiance of the reflected solar radiation depends on the reflectance of earth and atmosphere (*albedo*) and on the zenith angle (θ) of the incoming solar radiation. If the reflected radiation is diffuse, then the irradiance of reflected radiation is equal to πL_{sr}, and this must equal the radiation received per unit area ($E_0 \cos \theta$) multiplied by the albedo. Thus the albedo (A) may be calculated from

$$\pi L_{sr} = AE_0 \cos \theta$$

Substituting this equation into (5.25) yields

$$E'_{sr} = \frac{E_0}{\pi} \int_0^\omega A \cos \theta \, d\omega$$

For a satellite at fairly low altitude, $\cos \theta$ does not vary much over the observed area, and so may be taken outside the integral. And if the average albedo is defined by

$$\bar{A} = \frac{1}{\omega} \int_0^\omega A \, d\omega$$

the reflected solar radiation may be expressed by

$$E'_{sr} = \frac{E_0}{\pi} \bar{A}\omega \cos \theta \qquad (5.29)$$

Upon introducing Eq. (5.28), the average albedo becomes

$$\bar{A} = \frac{E'_{sr}}{2E_0\{1 - [(2Rh + h^2)^{1/2}/(R + h)]\} \cos \theta} \qquad (5.30)$$

Thus determination of the average albedo requires measurements of the solar irradiance, the reflected solar irradiance, zenith angle, and height of the satellite above the earth.

The long-wave radiation emitted by the earth and atmosphere may be represented by black-body radiation at an effective temperature T_e. This means that a black body with temperature T_e radiates with radiance L_1, and the effective temperature is defined by

$$L_1 \equiv \frac{\sigma T_e^4}{\pi} \qquad (5.31)$$

Because the satellite responds to an average over the solid angle intercepted by the earth, it is convenient to introduce an average effective temperature defined by

$$\bar{T}_e^4 \equiv \frac{1}{\omega} \int_0^\omega T_e^4 \, d\omega$$

It follows from Eqs. (5.25), (5.28), and (5.31) that

$$\bar{T}_e^4 = \frac{\pi E'_1}{\sigma \omega} = \frac{E'_1}{2\sigma\{1 - [(2Rh + h^2)^{1/2}/(R + h)]\}} \qquad (5.32)$$

World-wide averages of outgoing long-wave radiation ($\sigma \bar{T}_e^4$) and $\bar{A}$ representative of a fairly short period can be obtained from a well-designed satellite observational program. If these averages are sufficiently accurate, it is possible to determine the net rate of gain or loss of energy by the earth and atmosphere. An exercise is provided in Problem 7. A satellite measurement program has been undertaken to obtain this information. An early result of this program has led to the conclusion that poleward transport of heat by ocean currents undergoes very large seasonal variations especially in the tropics.† Observations of radiation from this program should provide an important foundation for climatological research and applications. With this information we should be able to determine the balance of total energy received and emitted by the earth and atmosphere over certain time periods.

5.10 Distribution of Solar Energy Intercepted by the Earth

The flux of solar radiation per unit horizontal area at the top of the atmosphere depends strongly on zenith angle of the sun and much less strongly on the variable distance of the earth from the sun. If the zenith angle is assumed to be constant over the solid angle subtended by the sun, the irradiance on a horizontal surface may be expressed by

$$E = E_0 \cos \theta$$

The mean value of the distance of earth from the sun ($\bar{R}$) is 149.5×10^6 km. R_0 varies from 147.0×10^6 km about January 3 (perihelion) to 152.0×10^6 km about July 5 (aphelion).

The zenith angle depends on latitude, on time of day and on the tilt of the earth's axis to the rays of the sun, that is, on the celestial longitude or the date. The total energy received per unit area per day (q_0) may be calculated by integrating over the daylight hours. Thus if R_0 is assumed constant during a single day

$$q_0 = \bar{E}_s \left(\frac{\bar{R}}{R_0} \right)^2 \int_{\text{sunrise}}^{\text{sunset}} \cos \theta \, dt \qquad (5.33)$$

Equation (5.33) has been evaluated for a variety of latitudes and dates, with the results summarized in Fig. 5.12.

Figure 5.12 shows that maximum insolation occurs at summer solstice at either pole. This is a result of the long solar day (24 hr). The maximum at the Southern Hemisphere is higher than at the Northern Hemisphere because the earth is closer to the sun during the northern winter than during the northern summer. However, the reader is invited to prove in Problem 8 that the total annual insolation is the same for corresponding latitudes in Northern and Southern Hemispheres.

† A. Oort and T. Vonder Haar, *J. Phys. Oceanog.* **6,** 781, 1976.

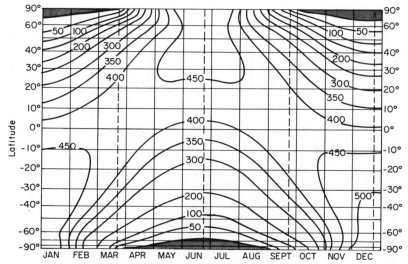

FIG. 5.12. Daily solar radiation in W m^{-2} incident on a unit horizontal surface at the top of the atmosphere as a function of latitude and date (adapted from M. Milankovitch, Mathematische Klimalehre. Handbuch der Klimatologie, Bd I, Teil A, Köppen, W. und Geiger, R., ed., Borntraeger, Berlin, 1930).

PART III: EFFECTS OF ABSORPTION AND EMISSION

5.11 Absorption

Attenuation of radiation results from absorption, scattering, and reflection. In this section the major absorbing properties of atmospheric gases will be described. Oxygen and nitrogen, which together constitute 99% of the atmosphere by volume, absorb strongly that fraction of solar radiation having wavelengths shorter than about 0.3 μm. Consequently, ultraviolet radiation from the sun is absorbed in the very high atmosphere. The very short wavelengths of great energy ionize the gas molecules. Other wavelengths in the ultraviolet are absorbed by ozone in the layer between 20 and 50 km above the earth. The photochemical processes associated with these radiations are discussed in Part IV of this chapter.

The absorption spectra for atmospheric gases are shown in Fig. 5.13. In the infrared a large fraction of the radiation is strongly absorbed by water vapor and carbon dioxide. Ozone, nitrous oxide, and methane absorb less but still appreciable energy. The very rapid change in absorptance (and therefore emissivity) with small change in wavelength which is evident in

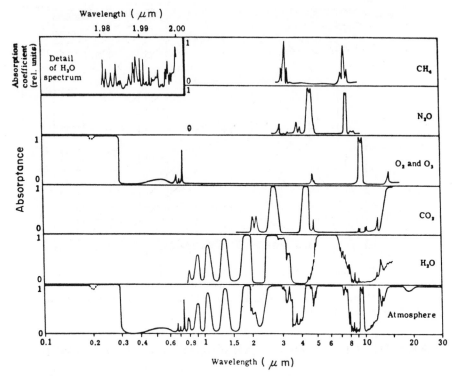

FIG. 5.13.   Absorption spectra for H_2O, CO_2, O_2, N_2O, CH_4 and the absorption spectrum of the atmosphere (after J. N. Howard, *Proc. I.R.E.* **47**, 1451, 1959; and R. M. Goody and G. D. Robinson, *Quart. J. Roy. Meteorol. Soc.* **77**, 153, 1951).

the expanded portion of Fig. 5.13 makes calculation of the total energy absorbed or transmitted by the atmosphere very difficult, and atmospheric absorption spectra are still not known with sufficient detail for highly accurate calculations. The atmosphere is evidently nearly transparent between 0.3 and 0.8 μm, but the irradiance of terrestrial and atmospheric radiation is very weak in this spectral range. Within the 8–12 μm range, except for strong absorption by ozone at 9.6 μm, terrestrial radiation escapes directly to space. This is the wavelength interval of strongest terrestrial radiation, and the radiation is said to escape through the open atmospheric "window."

5.12   The "Atmosphere Effect"

Figures 5.12 and 5.13 taken together emphasize an important characteristic of the atmosphere. The atmosphere is much more transparent for short-wave radiation than for long-wave radiation. The result is that solar energy passes through the atmosphere and is absorbed at the earth's surface. On the other hand, most of the long-wave radiation emitted by the earth's surface is absorbed by the atmosphere. Consequently, solar energy is trapped by the earth and atmosphere.

This effect is sometimes referred to as the *greenhouse effect* because in a similar manner glass covering a greenhouse transmits short-wave solar radiation but absorbs the long-wave radiation emitted by the plants and earth inside the greenhouse. However, it is not commonly recognized that, whereas the absorbing effect of the atmosphere results in temperatures well above what they would be without an atmosphere, the high temperatures in a greenhouse are not to be attributed to absorption of long-wave radiation by the glass. In 1909 R. W. Wood carried out an experiment with two small model greenhouses of which one was covered with glass and one with rock salt. Rock salt is transparent for both short- and long-wave radiation and therefore does not "trap" the radiation in the sense mentioned above. Both model greenhouses reached about the same high temperatures, proving that the effectiveness of greenhouses in growing plants is not the result of absorption of long-wave radiation by glass. Greenhouses reach much higher temperatures than the surrounding air because the glass cover of the greenhouse prevents the warm air from rising and removing heat from the greenhouse. This effect is four to five times as important as the absorption of long-wave radiation by the glass.

Trapping of radiation by the atmosphere is typical of the atmosphere and therefore correctly may be called the "*atmosphere effect*"; however, the term "*greenhouse effect*" continues to be used more widely. In order to make an estimate of this effect, the average temperature of the earth may be compared with the radiative equilibrium temperature. For an albedo of 0.35 and solar constant of 1370 W m^{-2} the equilibrium temperature is 250 K, whereas the average observed temperature near the earth's surface is 283 K, 33 K above the radiative equilibrium temperature. If all long-wave radiation is assumed to be absorbed in a single layer of the atmosphere, the result of Problem 9 shows that the average temperature of the earth's surface is 296 K. It may be concluded that the atmosphere forms a protective blanket for life on earth although, as shown in Problem 10, the "effective absorptance" is less than unity. A more detailed discussion of radiative transfer of energy is given in the following sections.

5.13 Transfer of Radiation between Two Parallel Surfaces

Radiative exchange between two parallel surfaces with negligible absorption in the intervening space provides one of the simplest models of radiation transfer. This model is approximated by the earth's surface and a cloud layer or by two cloud layers if the water vapor content in the intervening air is very low or the distance between the surfaces is small.

Assume that the radiating surfaces are *gray*, that is, that they have absorptance a_1 and a_2, respectively, which are independent of wavelength. The question is, how much energy is transferred from surface (1) to surface (2)? If E_1 represents the irradiance in the direction of surface (2) and E_2 the irradiance in direction of surface (1), then the net transfer of energy per unit area and unit time from surface (1) to surface (2) is $E_1 - E_2 = E_n$, the *net irradiance* or *net radiation*. E_1 consists of radiation emitted by surface (1) plus the radiation reflected from the same surface, so

$$E_1 = a_1 \sigma T_1^4 + (1 - a_1)E_2$$

and similarly

$$E_2 = a_2 \sigma T_2^4 + (1 - a_2)E_1$$

Solving of these equations yields

$$E_1 = \frac{1}{1 - (1 - a_1)(1 - a_2)}[a_1 \sigma T_1^4 + (1 - a_1)a_2 \sigma T_2^4]$$

$$E_2 = \frac{1}{1 - (1 - a_1)(1 - a_2)}[a_2 \sigma T_2^4 + (1 - a_2)a_1 \sigma T_1^4]$$

and

$$E_1 - E_2 = E_n = \frac{a_1 a_2}{1 - (1 - a_1)(1 - a_2)} \sigma(T_1^4 - T_2^4)$$

When $a_1 = a_2 = 1$, this equation reduces to

$$E_n = \sigma(T_1^4 - T_2^4)$$

which represents the transfer between black surfaces. In Problem 11 a possible case of transfer between the surface of the earth and a cloud layer is given. The problem quickly becomes very complicated if the energy transfer between arbitrarily shaped bodies is considered.

5.14 Transfer of Long-Wave Radiation in a Plane Stratified Atmosphere

The problem is to express the upward and downward radiation fluxes through unit horizontal areas as a function of the vertical distribution of temperature and of absorbing substance. Temperature and concentrations of absorbing gases are assumed independent of the x and y directions. The procedure to be followed is (1) to express the contribution made by an arbitrary volume element to the monochromatic irradiance at the reference level and (2) to integrate over all wavelengths the separate contributions of all volume elements above and below the reference level.

To obtain the contribution from a volume element choose a height element dz at height z above the reference level z_r, as illustrated in Fig. 5.14. The monochromatic radiance dL_λ emitted by the element in the direction θ is given by Kirchhoff's law in the form $L_\lambda^* \sec \theta \, du_\lambda$, where u_λ is the optical thickness measured in the vertical. This radiation is attenuated according to Beer's law, so that the portion of the radiance increment dL_λ that arrives at the reference level is

$$dL_\lambda \exp(-u_\lambda \sec \theta) = L_\lambda^* \sec \theta \exp(-u_\lambda \sec \theta) \, du_\lambda$$

All volume elements seen at zenith angle θ contribute an equal increment to the irradiance at dA. Remembering that $d\omega = 2\pi \sin \theta \, d\theta$ and that the relation between radiance and irradiance is given by Eq. (5.1), the contribution to irradiance at the surface made by a circular ring of the differential layer is recognized to be

$$dE_{\lambda, u_\lambda, \theta} = 2\pi L_\lambda^* \sin \theta \exp(-u_\lambda \sec \theta) \, d\theta \, du_\lambda$$

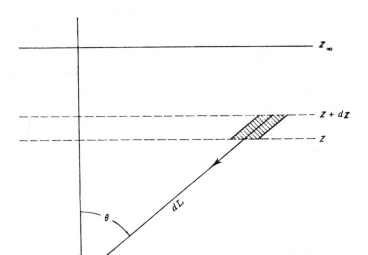

FIG. 5.14.   Contribution to irradiance $(dE_{u,\theta})$ at $z = z_r$ made by a differential element at z and zenith angle θ.

Integration with respect to θ gives the contribution to the monochromatic irradiance from the entire layer in the form

$$dE_{\lambda, u_\lambda} = 2\pi L_\lambda^* \, du_\lambda \int_0^{\pi/2} \sin \theta \exp(-u_\lambda \sec \theta) \, d\theta \qquad (5.34)$$

The integral is a function of u_λ only. If the transformation $y \equiv \sec \theta$ is made, the integral takes the form

$$\int_1^\infty \frac{e^{-u_\lambda y}}{y^2} \, dy \equiv H_2(u_\lambda) \qquad (5.35)$$

where $H_2 u_\lambda$ is called a Gold function of the second order. This function has been tabulated for various u_λ values. Now Eq. (5.34) is integrated over λ. If we then recall that $\pi L_\lambda^* = E_\lambda^*$

$$dE_u = 2 \left\{ \int_0^\infty E_\lambda^* H_2(u_\lambda) \, d\lambda \right\} du \qquad (5.36)$$

The integral on the right-hand side is a function of temperature and optical thickness and can be evaluated for each pair of values of T and u if the absorption coefficient is known.

One further integration, this time over the entire atmosphere, yields the total downward directed irradiance. A similar integration describes the upward directed irradiance. However, because evaluation of k_λ depends on difficult laboratory measurements which are limited in resolution and accuracy, evaluation of the integral is less than exact. For this reason, it is appropriate to simplify the procedure. To do this we use a thought experiment to define the spectral *flux transmittance* of a finite layer such as that shown in Fig. 5.14. We consider that isotropic radiation is incident on one face of the layer, and by analogy with Eq. (5.12a) define the flux transmittance $(\tau_{f\lambda})$ as the proportion of the incident irradiance which is transmitted through the layer. Thus, $\tau_{f\lambda} = E_\lambda / E_{\lambda 0} = E_\lambda / \pi L_{\lambda 0}$, and the transmitted irradiance (E_λ) may be calculated by integrating the transmitted radiance $[L_{\lambda 0} \exp(-u_\lambda \sec \theta)]$ over the solid angle as in Eq. (5.2). It follows that

$$\tau_{f\lambda} = 2 \int_0^{\pi/2} \exp(-u_\lambda \sec \theta) \sin \theta \cos \theta \, d\theta$$

Upon differentiating we find that

$$\frac{\partial \tau_{f\lambda}}{\partial u_\lambda} = -2 \int_0^{\pi/2} \exp(-u_\lambda \sec \theta) \sin \theta \, d\theta \qquad (5.37)$$

Equation (5.37) may be introduced into Eq. (5.34) with the result

$$dE_{\lambda, u} = -E_\lambda^* \frac{\partial \tau_{f\lambda}}{\partial u_\lambda} du_\lambda$$

The *flux emissivity* is related to $\tau_{f\lambda}$ by $\varepsilon_{f\lambda} \equiv 1 - \tau_{f\lambda}$, and therefore

$$dE_{\lambda, u} = E_\lambda^* \frac{\partial \varepsilon_{f\lambda}}{\partial u_\lambda} du_\lambda = E_\lambda^* \, d\varepsilon_{f\lambda}$$

Finally, upon integrating over wavelength and optical depth, the radiation from a cloudless atmosphere reaching a reference level is expressed formally by

$$E^\downarrow = \int_0^{\varepsilon_{f\infty}} E^* \, d\varepsilon_f \tag{5.38}$$

where $\varepsilon_{f\infty}$ represents the flux emissivity from the reference level to the top of the radiating atmosphere. Determination of ε_f as a function of height is particularly simple if ε_f is independent of temperature over the range encountered in the vertical column, that is, if the proportion of the area under the black-body envelope which represents energy absorbed in an atmospheric layer does not change significantly as the black-body envelope changes with change in temperature. This condition is at least approximately fulfilled for water vapor and carbon dioxide for the atmospheric temperature range, as may be inferred from Fig. 5.13.

The upward irradiance consists of a contribution from the atmosphere below the reference level plus a contribution from the black-body radiation emitted by the earth's surface. The latter is given by multiplying the radiation emitted by the surface by the flux transmissivity of the atmosphere below the reference level. The upward irradiance therefore can be written as

$$E^\uparrow = \int_0^{\varepsilon_{f0}} \sigma T^4 \, d\varepsilon_f + (1 - \varepsilon_{f0}) \sigma T_0^4 \tag{5.39}$$

where the index 0 refers to the earth's surface and the flux transmittance is replaced by $(1 - \varepsilon_{f0})$, following Kirchhoff's law. Black-body radiation from cloud layers may be in principle accounted for in a similar way.

Equations (5.38) and (5.39) can be evaluated when ε_f is known as a function of density weighted height [defined by Eq. (5.9)]. In the following section a discussion of the experimental determination of the flux emissivity is given. Because the temperature is an arbitrary function of the height, Eqs. (5.38) and (5.39) cannot be integrated analytically. Instead, graphical integration may be performed by dividing the atmosphere into finite layers and approximating the integrals by summation over the layers. Equations (5.38) and (5.39) may be written as summations in the form

$$E^{\downarrow} = \sigma \sum_{i=1}^{n_{\infty}} T_i^4 \, \Delta\varepsilon_{fi} \tag{5.40}$$

and

$$E^{\uparrow} = (1 - \varepsilon_{f0})\sigma T_0^4 + \sigma \sum_{i=1}^{n_0} T_i^4 \, \Delta\varepsilon_{fi} \tag{5.41}$$

where T_i represents the average temperature of the ith layer, n_{∞} represents the number of layers that are taken from the reference level to the top of the atmosphere and n_0 represents the number of layers between reference level and the earth's surface.

To evaluate the irradiance it is necessary to determine mean values of T and $\Delta\varepsilon_f$ for each layer. The flux emissivity is a function of the vapor density weighted height. Because vertical soundings are usually given with pressure as the vertical coordinate, it is convenient to transform the vertical coordinate to pressure. If only the contribution of water vapor (ρ_v) to the flux emissivity is considered

$$Z = \int_{z_r}^{z} \rho_v \, dz = \frac{1}{g} \int_{p}^{p_r} q \, dp \tag{5.42}$$

where q is the specific humidity. This procedure is to be followed in Problem 12.

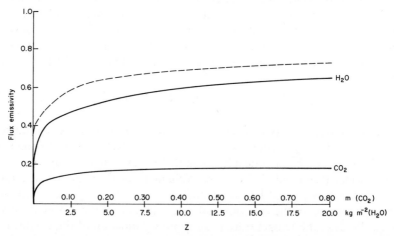

FIG. 5.15.   Experimental observations of emissivity of pure water vapor, of carbon dioxide, and of an atmospheric mixture of CO_2 (0.032%) and H_2O (mixing ratio 5 g/kg) as a function of density weighted height Z for temperature of 10° C and pressure 1013 mb (after W. M. Elsasser, *Meteorol. Monographs* **4**, 23, 1960).

The contribution to total atmospheric radiation made by carbon dioxide may be treated in a similar manner if care is taken to avoid counting twice the contributions from overlapping CO_2 and H_2O bands. Figure 5.15 illustrates the individual and combined contributions to the flux emissivities from CO_2 and H_2O. Other gases in the atmosphere emit very much less strongly and are usually neglected in these calculations.

5.15 Experimental Determination of Flux Emissivity

The flux emissivity may be derived from the results of laboratory measurements of the transmittance of the various absorbing constituents of the atmosphere. A tube is filled with a known amount of the absorbing gas at a known temperature; a cold radiation source is placed at one end of the tube, and the emitted and transmitted radiance is measured at the other end. The beam emissivity (ε_λ) is then found using Eq. (5.10) and solving for $\varepsilon_\lambda = L_\lambda / L_\lambda^*$. The beam emissivity for all wavelengths is expressed by $\varepsilon = L/L^*$. By changing the amount of absorbing gas in the tube or by using mirrors inside the tube, the optical thickness may be varied, and the emissivity may be determined as a function of optical thickness.

The beam emissivity determined as described here may be converted to flux emissivity by integrating over the hemisphere shown in Fig. 5.14. By analogy with Eq. (5.2), the flux emissivity for a particular wavelength is found to be

$$\varepsilon_f(u_\lambda) = 2 \int_0^{\pi/2} \varepsilon(u_\lambda \sec \theta) \sin \theta \cos \theta \, d\theta \tag{5.43a}$$

or

$$\varepsilon_f(u_\lambda) = 2u_\lambda^2 \int_{u_\lambda}^{\infty} \varepsilon(x) x^{-3} \, dx \tag{5.43b}$$

where u_λ or $k_\lambda Z$ represents the optical thickness of the slab for which the flux emissivity is to be determined. Equation (5.43b) provides a convenient form for numerical calculations of the integral.

In general the flux emissivity exceeds the beam emissivity for the same optical thickness. The limits are easily recognized. When optical thickness is very small, absorption by the intervening atmosphere is negligible, so that flux emissivity is directly proportional to optical thickness and $\varepsilon(u_\lambda \sec \theta) = \varepsilon(u_\lambda) \sec \theta$, and Eq. (5.43a) reduces to

$$\varepsilon_f(u_\lambda) = 2\varepsilon(u_\lambda) = \varepsilon(2u_\lambda)$$

On the other hand, if the optical thickness is very large, the radiation emitted

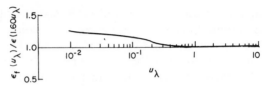

FIG. 5.16. Ratio $\varepsilon_f(u_\lambda)/\varepsilon(1.60\,u_\lambda)$ as a function of u_λ computed from Eq. (5.43).

by the slab becomes black-body radiation for all directions and $\varepsilon_f(u_\lambda) = \varepsilon(u_\lambda) \approx 1$. For a wide range of optical thickness in between the two extremes Eq. (5.43) can be approximated by

$$\varepsilon_f(u_\lambda) = \varepsilon(1.60\,u_\lambda)$$

Figure 5.16 shows that this equality holds within 20% over the range of u_λ from 10^{-2} to beyond 10. This range represents virtually all the atmospheric absorption. Therefore, the same relationship applied to flux emissivity integrated over wavelength can be used for approximate calculations.

By using this equation or Eq. (5.43) where higher accuracy is necessary, experimental observations of the beam emissivity may be converted to flux emissivity. Experimental results obtained in this way for water vapor and

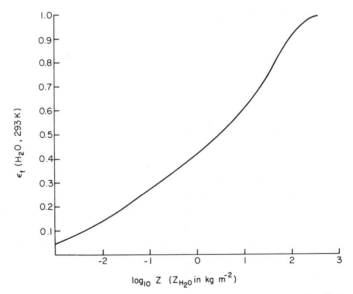

FIG. 5.17. Flux emissivity as a function of density weighted height for H_2O at 293 K (after D. O. Staley, and G. M. Jurica, *J. Appl. Meteor.* **9**, 365–372, 1970).

for carbon dioxide are shown as functions of Z in Fig. 5.15. Also shown in the figure is the flux emissivity for a typical mixture of water vapor and carbon dioxide. At very small Z (less than 10^{-4} kg m^{-2} of water vapor as indicated by Elsasser†) emissivity increases linearly with Z. Beyond 10^{-4} kg m^{-2} the emissivity increases more and more slowly with increasing thickness as a result of absorption along the path. Emissivity of unity (complete blackness) is reached as $Z \to \infty$. The Z for H_2O in the clear atmosphere ranges between about 10 to 100 kg m^{-2} (1 to 10 g cm^{-2}), and the corresponding flux emissivity of the atmosphere ranges from about 0.65 to 0.85.

A more detailed graph of ε_f for water vapor versus Z is given in Fig. 5.17.

5.16 Divergence of Net Radiation

The cooling or warming experienced by a layer of air due to radiation may be calculated from the principle of conservation of energy. If the net irradiance defined by $E_n \equiv E^\uparrow - E^\downarrow$ is smaller at the top of the layer than at the bottom, the difference must be used to warm the layer. Thus

$$c_p \rho \left(\frac{\partial T}{\partial t}\right)_R \Delta z = -E_n(z + \Delta z) + E_n(z) = -\frac{\partial E_n}{\partial z} \Delta z$$

or after introducing Eq. (5.42)

$$\left(\frac{\partial T}{\partial t}\right)_R = -\frac{1}{c_p \rho} \frac{\partial E_n}{\partial z} = -\frac{q}{c_p} \frac{\partial E_n}{\partial Z} \tag{5.44}$$

The term $\partial E_n / \partial Z$ is referred to as the *divergence of net irradiance* or simply *flux divergence*.

After substituting Eqs. (5.38) and (5.39) into Eq. (5.44) and carrying out the differentiation, the rate of radiational warming may be expressed by

$$\left(\frac{\partial T}{\partial t}\right)_R = \frac{\sigma q}{c_p} \left\{ 4 \int_0^{\varepsilon_{f\infty}} T^3 \frac{\partial T}{\partial Z} d\varepsilon_f - T_\infty^4 \left(\frac{\partial \varepsilon_f}{\partial Z}\right)_\infty + 4 \int_0^{\varepsilon_{f0}} T^3 \frac{\partial T}{\partial Z} d\varepsilon_f \right\}$$

or in finite difference form, which is of more practical use

$$\left(\frac{\partial T}{\partial t}\right)_R = \frac{\sigma q}{c_p} \left\{ 4 \sum_{i=1}^{n_\infty} T_i^3 \Delta T_i \left(\frac{\Delta \varepsilon_f}{\Delta Z}\right)_i - T_\infty^4 \left(\frac{\Delta \varepsilon_f}{\Delta Z}\right)_{n_\infty} + 4 \sum_{i=1}^{n_0} T_i^3 \Delta T_i \left(\frac{\Delta \varepsilon_f}{\Delta Z}\right)_i \right\}$$
$$\tag{5.45}$$

Calculations of radiative temperature change using Eq. (5.45) are straight-

† W. M. Elsasser, *Meteorol. Monographs* **4**, 23, 1960.

forward although somewhat tedious, particularly if detail is required. Problem 13 provides an example. Results depend, of course, on the vertical distributions of temperature and humidity, but throughout most of the troposphere, where temperature decreases with height at 4–8 K km^{-1}, the air cools by radiation at 1–2 K day^{-1} in cloudless areas. The excess of temperature over radiational equilibrium temperature indicates that internal energy is supplied to the troposphere by other processes. These are discussed in Chapter VI.

Because $\partial \varepsilon_f / \partial Z$ decreases sharply beyond density weighted path lengths corresponding to vertical distances of a few centimeters, radiational temperature change depends strongly on the temperature distribution in the immediate neighborhood of the reference point. At the base of temperature inversions, above which temperature increases with height, radiational warming at a rate of several degrees per hour may be calculated from Eq. (5.45).

Errors in radiation calculations are to be attributed to inaccuracies in determination of flux emissivity and, especially, to errors in humidity observations. The latter are very serious at low temperatures and low humidities, where the humidity element carried aloft by the radiosonde is insensitive. Flux divergence in the cloudless regions above or below cloud layers may be calculated by treating the cloud layers as black bodies at known temperatures. This is not very satisfactory, however, because the "blackness" of clouds depends on thickness and on the existing drop size spectrum, and because infrared radiation is scattered by cloud droplets in a complex manner. For these reasons it is important that radiation calculations be supplemented by direct observations.

5.17   Direct Measurement of Flux Divergence

Flux divergence may be measured by the radiometersonde, a protected flat plate radiometer which is carried aloft by a balloon. As shown in Fig. 5.18, one plate faces upward and the other downward, and each receives radiation from the appropriate hemisphere. The polyethylene cover is transparent for most of the short- and long-wave ranges of wavelengths.

In order to develop a simple relation between rate of radiational temperature change and temperature observed in the radiometersonde, the small absorptance of the polyethylene shield is neglected. Under conditions of equilibrium the upward and downward irradiances in the instrument can be derived using a modification of the method developed in Section 5.13.

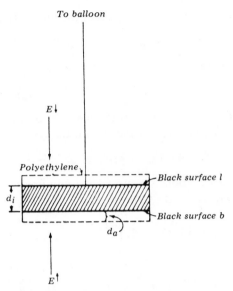

FIG. 5.18.   Essential elements of a flat plate radiometer with plates t and b shielded from ventilation by a polyethylene cover.

Considering only long-wave radiation the result for the top plate is

$$E^{\downarrow} = \sigma T_t^4 + \frac{1 - r(1 - a)}{a(1 - r)} L_t \qquad (5.46)$$

where r represents the reflectance of polyethylene, a the absorptance of the black surface, T_t temperature of the top plate, and L_t loss from top plate due to conduction to the surrounding air and to the other plate. Similarly, the equilibrium condition for the bottom plate is given by

$$E^{\uparrow} = \sigma T_b^4 + \frac{1 - r(1 - a)}{a(1 - r)} L_b \qquad (5.47)$$

where the subscript b refers to the bottom plate. Problem 14 requires the derivation of Eqs. (5.46) and (5.47). The losses are assumed to be proportional to temperature gradient. Under steady state conditions

$$L_t = (\lambda_a/d_a)(T_t - T_a) + (\lambda_i/d_i)(T_t - T_b)$$

where λ_a represents the thermal conductivity of the air trapped between the plate and the polyethylene, d_a the vertical separation of plate and polyethylene, T_a the air temperature, λ_i the thermal conductivity of insulation between the plate, and d_i the separation of the plates. In a similar manner

$$L_b = (\lambda_a/d_a)(T_b - T_a) + (\lambda_i/d_i)(T_b - T_t)$$

T_t^4 may be expanded in a binomial series about T_a with the result

$$T_t^4 = T_a^4 + 4T_a^4(T_t - T_a) + 6T_a^2(T_t - T_a)^2 + \cdots$$

For $T_t - T_a \ll T_a$, all terms beyond the first two on the right may be neglected. Similarly, T_b^4 may be expanded with the result

$$T_b^4 = T_a^4 + 4T_a^3(T_b - T_a) + 6T_a^2(T_b - T_a)^2 + \cdots$$

Finally, when these equations are substituted into Eqs. (5.46) and (5.47) and then into Eq. (5.44), the linear terms yield

$$\left(\frac{\partial T}{\partial t}\right)_R = \frac{1}{c_p\rho}\left[12\sigma T_a^2(T_t - T_b)\frac{\partial T_a}{\partial z} \right.$$
$$\left. + \left\{4\sigma T_a^3 + \left(\frac{\lambda_a}{d_a} + \frac{2\lambda_i}{d_i}\right)\frac{1 - r(1 - a)}{a(1 - r)}\right\}\frac{d}{dz}(T_t - T_b)\right] \quad (5.48)$$

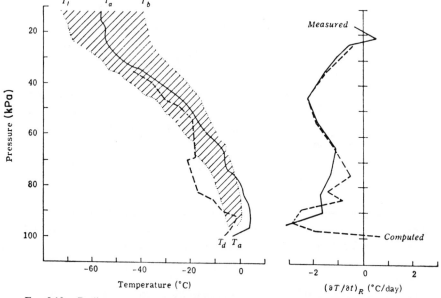

FIG. 5.19.   Radiometersonde observations of air temperature T_a, dew point T_d, radiometer temperatures T_t and T_b, rate of radiational temperature change measured using Eq. (5.48) and computed from Eq. (5.45) (after V. E. Suomi *et al. Quart. J. Roy. Meteorol. Soc.* **84**, 134, 1958).

The coefficient on the right is slowly varying with air temperature and pressure and may be determined with reasonable accuracy by calibration. The rate of radiative temperature change, therefore, is easily evaluated from observations of the rate of change with height of the temperature difference between upper and lower plates of the instrument as illustrated in Problem 15. In order to obtain the greatest accuracy, λ_i and r should be as small as possible. Observations made by a radiometersonde are shown in Fig. 5.19.

PART IV: PHOTOCHEMICAL PROCESSES

5.18   Dissociation of Oxygen

Absorption of radiation by a gas takes place when a photon strikes a molecule and the energy of the photon is transferred to the molecule. The molecule may gain so much energy in this way that it becomes unstable and splits into two new particles. Because the change in structure is produced by a photon, it is called a photochemical reaction.

An important photochemical reaction that occurs in the upper atmosphere is the dissociation of molecular oxygen by absorption of a photon. The reaction is expressed by

$$O_2 + h\nu \rightarrow O + O \tag{5.49}$$

The minimum energy necessary for this reaction corresponds to a wavelength of 0.2424 μm. The rate of dissociation of O_2 is proportional to the concentration of O_2 molecules and the amount of absorbed radiation integrated over all wavelengths less than 0.2424 μm. Therefore, for the reaction described by (5.49)

$$\left(\frac{dn_2}{dt}\right)_{\text{diss}} = -n_2 \int_0^{0.2424} \beta_{\lambda 2} k_{\lambda 2} E_{\lambda 2} \, d\lambda \equiv -J_2 n_2 \tag{5.50}$$

where n_2 is the number of O_2 molecules per cubic centimeter, β_λ is the fraction of the number of O_2 molecules dissociated per unit energy of radiation absorbed, and J_2 is the reaction rate or the inverse time constant for the dissociation process.

Equation (5.50) can be evaluated reliably because $\beta_{\lambda 2}$ and $k_{\lambda 2}$ are relatively well-known. However, $E_{\lambda 2 0}$ is uncertain by as much as a factor of 2. Future satellite observations in the ultraviolet which reduce this uncertainty may permit much more accurate calculation of the rate of dissociation. The irradiance may be expressed using Beer's law in the form

$$E_{\lambda 2} = E_{\lambda 2 0} \exp\left(-\frac{m_2}{N_0} \int_0^\infty k_{\lambda 2} n_2 \sec \theta \, dz \right) \tag{5.51}$$

where m_2 is the molecular mass of O_2 and N_0 is Avogadro's number.

Following formation of atomic oxygen as described by reaction (5.49), a series of further reactions may occur, of which the most important in the region known as the ozonosphere (height of 20–50 km) are

$$O + O_2 + M \rightarrow O_3 + M \tag{5.52}$$

and

$$O_3 + O \rightarrow 2O_2^* \tag{5.53}$$

where the asterisk indicates an excited state.

Reaction (5.52) requires a three-body collision, and the third body, represented by M, may be any particle capable of absorbing the extra energy released by the reaction. When two particles (O and O_2) collide, the O_3 formed would be unstable and dissociate again unless it were able to release its excess energy to a third particle within a very short time. Recombination of atomic oxygen also occurs through three-body collisions, but this reaction is slow enough in the ozonosphere to be neglected.

Reaction (5.52) represents the process which is chiefly responsible for the creation of ozone in the ozonosphere. Ozone is decomposed again by the reaction (5.53) and by the additional important reaction

$$O_3 + h\nu \rightarrow O_2 + O \tag{5.54}$$

In this reaction the wavelength of the photon must be less than 1.1340 μm. The equation expressing the rate of dissociation of O_3 based on reaction (5.54) is similar to Eq. (5.50) and may be written

$$\left(\frac{dn_3}{dt} \right)_{\text{diss}} = -n_3 \int_0^{1.1340} \beta_{\lambda 3} k_{\lambda 3} E_{\lambda 3} \, d\lambda \equiv -J_3 n_3 \tag{5.55}$$

The following equations may now be written for the rates of change of the concentration n_1 of O, n_2 of O_2, and n_3 of O_3.

$$\frac{dn_1}{dt} = -K_2 n_1 n_2 n_M - K_3 n_1 n_3 + 2J_2 n_2 + J_3 n_3 \tag{5.56}$$

$$\frac{dn_2}{dt} = -K_2 n_1 n_2 n_M + 2K_3 n_1 n_3 - J_2 n_2 + J_3 n_3 \tag{5.57}$$

$$\frac{dn_3}{dt} = +K_2 n_1 n_2 n_M - K_3 n_1 n_3 - J_3 n_3 \tag{5.58}$$

Here n_M is the number of M particles per unit volume, and K_2 and K_3 are the reaction coefficients of reactions (5.52) and (5.53), respectively. The distribution of the three oxygen constituents with height is now determined by the above equations together with two equations of the form (5.51), the hydrostatic equation, and the equation of state. The fact that the absorbed radiation increases the temperature, which in its turn affects the chemical reaction coefficients, complicates the calculations. It turns out that at the height where the ozone-forming reaction (5.52) occurs, little of the $E_{\lambda 2}$ radiation penetrates because most of this radiation is absorbed at greater heights. Therefore, very little dissociation of O_2 takes place, and even in the *ozonosphere* (the layer where ozone is formed and decomposed) the concentrations of O and O_3 are small compared to the concentration of O_2. Thus n_2 may be considered independent of n_1 and n_3, and if uniform temperature is assumed, n_2 may be determined by Eq. (2.21) in the form

$$n_2 = n_{20} \exp(-\Phi/R_m T) \simeq n_{20} \exp(-z/H) \qquad (2.21a)$$

where n_{20} is the number of O_2 molecules per unit volume at geopotential $\Phi = 0$ and $H \equiv R_m T/g$ is the *scale height* (height where density reaches $1/e$ of its value at reference height). Using Eqs. (5.56), (5.58), and (2.21a) we would like to find the *photochemical equilibrium* concentration of O and O_3, and we would like to determine how long it takes to achieve this equilibrium. An approximate time dependent solution of (5.56) and (5.58) may be obtained by neglecting the small terms $K_3 n_1 n_3$ and $J_2 n_2$. Thus

$$\frac{dn_1}{dt} = J_3 n_3 - K_2 n_1 n_2 n_M \qquad (5.59)$$

and

$$\frac{dn_3}{dt} = -J_3 n_3 + K_2 n_1 n_2 n_M \qquad (5.60)$$

Equations (5.59) and (5.60) can be solved for the ratio n_1/n_3.[†] We note that the equilibrium ratio, reached after infinite time, is $(n_1/n_3)_\infty \equiv r_\infty = J_3/K_2 n_2 n_M$. Upon combining the two differential equations and setting $n_1/n_3 \equiv r$

$$\int_{r_0}^{r} \frac{dr}{K_2 n_2 n_M [r^2 - (r_\infty - 1)r - r_\infty]} = -\int_0^t dt$$

The integral on the left is given in integral tables and leads to the solution

$$\frac{1}{J_3 + K_2 n_2 n_M} \ln \frac{r - r_\infty}{r + 1} \bigg|_{r_0}^{r} = -t$$

† The following derivation was provided by C. D. Leovy (private communication).

Therefore

$$r(t) = \frac{r_\infty(r_0 + 1) + (r_0 - r_\infty)e^{-(J_3 + K_2 n_2 n_M)t}}{r_0 + 1 - (r_0 - r_\infty)e^{-(J_3 + K_2 n_2 n_M)t}} \qquad (5.61)$$

Equation (5.61) indicates that the ratio n_1/n_3 varies approximately as the hyperbolic cotangent. The time constant $(J_3 + K_2 n_2 n_M)$ is no more than about 100 s within the ozone layer. Because other photochemical and dynamical processes are considerably slower than this, the concentrations of O and O_3 can be considered to be in equilibrium during the daytime. Then upon adding Eqs. (5.56) and (5.58) and utilizing the equilibrium form of Eq. (5.61)

$$\frac{d}{dt}(n_1 + n_3) = \left(1 + \frac{J_3}{K_2 n_2 n_M}\right)\frac{dn_3}{dt} = -2K_3 n_1 n_3 + 2J_2 n_2$$

$$= -\frac{2K_3 J_3}{K_2 n_2 n_M} n_3^2 + 2J_2 n_2 \qquad (5.62)$$

Equation (5.62) has a time dependent solution with an equilibrium concentration for O_3 given by

$$n_3 = \left(\frac{J_2 K_2 n_M}{K_3 J_3}\right)^{1/2} n_2 \qquad (5.63)$$

Below 20 km the ratio J_2/J_3 becomes very small, which means that the O_3 concentration becomes independent of the O_2 concentration and equilibrium is not reached. Above 20 km J_2/J_3 remains fairly constant and

$$(n_3)_e \propto n_M^{1/2} n_2 \propto e^{-(3\Phi/2R_m T)} = e^{-(3z/2H)}$$

This means that the equilibrium concentration of ozone falls off with a scale height of about 5 km. Some complexities arise because K_2 and K_3 are temperature dependent.

In general, we may say that photochemical equilibrium is approached when the major formation and decomposition processes are so rapid that the time required to form the equilibrium reservoir is short compared with diurnal variations. The photochemical reactions (5.49) and (5.54) are most active at the height where the rates of depletion of the respective irradiances $(E_{\lambda 2}$ and $E_{\lambda 3})$ are greatest. In the lower part of the ozonosphere the active radiation is almost completely absorbed, and consequently the time to restore equilibrium is very long. Therefore, as shown in Fig. 5.20, the observed concentration in the lower ozonosphere may differ markedly from the corresponding equilibrium concentration.

The simple mechanism described above does not quite account for the

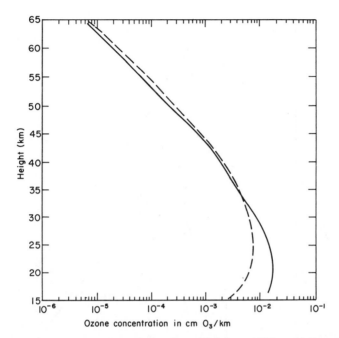

FIG. 5.20.   Ozone concentration as a function of height at 40°N on 21 December 1975.
The solid line is determined from satellite observations using limb scanning (Section 7.18).
The dashed line has been computed using a one-dimensional model which includes diffusion
in the vertical but does not take horizontal transport into account. The discrepancy between
the observed and computed values below 30 km is believed to be caused by neglecting the
poleward transport in the lower stratosphere [courtesy of J. Gille (observations) and P. Crutzen
(computation), NCAR, Boulder, Colorado, 1979].

observed ozone distribution. There are a large number of chemical reactions
involving small amounts of H, OH, HO_2, NO, NO_2, and other trace gases
which have the net effect that the equilibrium ozone concentrations are
lower than predicted by Eq. (5.63). In the next section the effects of some
trace gases are discussed in more detail.

Above the ozonosphere reactions (5.49) and (5.54) are very rapid. Recombination occurs through three-body collisions as mentioned earlier, but due
to decrease of air density with height three-body collisions become less
probable with height. The results of these various processes is that ozone
concentration becomes negligible above about 80 km, while atomic oxygen
concentration increases with height and exceeds diatomic oxygen concentration above about 110 km. Above 90 km O_2 dissociates more rapidly than
it recombines; the excess O diffuses and mixes downward to recombine
below 85 km. On the other hand, O_2 diffuses and mixes upward to replace

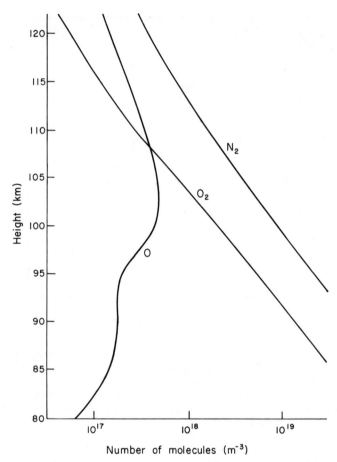

FIG. 5.21.   Number of molecules of atomic oxygen (O), diatomic oxygen (O_2), and diatomic nitrogen (N_2) per cubic meter as a function of height (adapted from *COSPAR International Reference Atmosphere*, CIRA 1972, p. 19, Akademie Verlag Berlin, 1972).

the shortage due to dissociation. Above about 150 km atomic oxygen is the dominant constituent of the atmosphere because of the combined effect of dissociation and diffusive separation. The vertical distribution of O_2, O, and N_2 from 80 to 120 km is shown in Fig. 5.21.

5.19   Effects of Trace Gases on Stratospheric Ozone

In addition to molecular species of natural origin, such as H, OH, NO, HO_2, and NO_2, there are also trace gases arising from human activities which may modify the natural distribution of ozone in the stratosphere. Two such trace gases in particular have aroused scientific and public interest because of the possibility that they might cause reductions in ozone concentrations which would have serious biological or climatological consequences.

Chlorine is released in the stratosphere when chlorofluoromethane molecules are photolytically dissociated in the stratosphere by absorption of solar radiation of wavelengths between 190 and 210 nm. The most common chlorofluoromethanes are the two "freons," $CFCl_3$, which has been widely used as a propellant in spray cans, and CF_2Cl_2, used as a refrigerant. Both are inert in the troposphere; they are released at the earth's surface and are carried upward to the stratosphere by organized air currents and turbulence. Their potential effectiveness in depleting the ozone layer results from the fact that chlorine acts catalytically to destroy ozone by absorption of a photon without change in its own concentration as described by the following sequence of reactions.

$$Cl + O_3 \rightarrow ClO + O_2$$

$$O_3 + hv \rightarrow O_2 + O$$

$$ClO + O \rightarrow Cl + O_2$$

The net result is $2O_3 + hv \rightarrow 3O_2$ with the Cl remaining to enter another cycle. The primary loss mechanism for Cl is believed to be formation of HCl by abstraction of a hydrogen atom from hydrocarbons and then downward turbulent diffusion of HCl to the troposphere where it can be washed out by rain or snow.

Nitric oxide (NO) is formed in the stratosphere by the reaction of nitrous oxide (N_2O) with monatomic oxygen. Nitrous oxide is formed by bacteria in the soil; it is inert in the troposphere and therefore is transported by air motions to the stratosphere without loss. Nitric oxide also is formed in smaller amounts by cosmic rays interacting with the atmosphere and by lightning. The catalytic cycle by which NO destroys ozone is the following.

$$NO + O_3 \rightarrow NO_2 + O_2$$

$$O_3 + hv \rightarrow O_2 + O$$

$$NO_2 + O \rightarrow NO + O_2$$

The net result is $2O_3 + hv \rightarrow 3O_2$ with NO remaining to enter another cycle. The primary removal process for NO is turbulent diffusion downward to the troposphere where it is washed out in precipitation. The chlorine and nitric oxide cycles also can interact as described by the following cycle.

$$ClO + NO \rightarrow Cl + NO_2$$
$$NO_2 + O \rightarrow NO + O_2$$
$$Cl + O_3 \rightarrow ClO + O_2$$

The net result is $O + O_3 \rightarrow 2O_2$.

Depletion of the ozone layer depends on the reaction rates of these chemical processes and on how rapidly gases are transferred vertically and horizontally in the stratosphere. Numerical models which represent stratospheric motions and transport to and from the troposphere, as well as the various chemical and photochemical processes, have been used to calculate the effect of chlorofluoromethanes and nitrogen oxides and other trace gases on the equilibrium distribution of ozone. Results have indicated that significant reductions in average ozone concentrations may be possible and that because the residence time of the depleting agents in the stratosphere is many years, even after the removal of the source of the depleting agents, many years would be required to recover the ozone loss. However, uncertainties in reaction rates, the possibility that important chemical reactions are still unknown, and approximations in the physics of the numerical models justify caution in accepting quantitative model results.

Observations of total ozone content in the atmosphere have so far not shown conclusive evidence of depletion of ozone by manmade trace gases. The natural variability of ozone in space and time, typically 10–20% within 2 days or over 1000 km in middle latitudes, is an order of magnitude larger than effects calculated on the basis of man's current activities. This variability makes it difficult to detect small but systematic effects.

Public concern over the effects of chlorofluoromethanes on the ozone shield has resulted in the U.S. Government requiring removal of these gases from all but essential uses. United States industry has been successful in developing other gases for use as propellants in spray cans, but other uses continue. Total worldwide release of chlorofluoromethanes continue to increase. Possible effects of nitrogen oxides on the ozone layer are of great concern because the increasing use of fertilizer around the world is resulting in increased release of nitrous oxide from the soil.

In addition to the biological consequences of reduced stratospheric ozone, climatic effects also may result from the lessened absorption of short-wave and long-wave radiation. The combined effect should be to cool the stratosphere and to warm the earth's surface, especially at high latitudes. Cooling of the stratosphere in the tropics where the tropopause reaches its lowest temperature could reduce the humidity of air entering the stratosphere, thereby producing a drier stratosphere over extensive regions of the earth. At the same time presence of trace gases such as the chlorofluoromethanes or nitrogen oxides may directly affect radiative warming or cooling. These climatic effects, although plausible, have not been proven.

5.20 Photoionization

Ionization of molecules and atoms by ultraviolet radiation can be treated relatively simply when it is assumed that absorption takes place in a narrow band ($\Delta\lambda$). Equation (5.51) may then be written in the form

$$E_\lambda = E_{\lambda 0} \exp\left(-k_\lambda \sec\theta \int_z^\infty \rho_c \, dz\right) \tag{5.64}$$

where ρ_c is the density of the absorbing gas. If again uniform temperature is assumed, this equation may be combined with Eq. (2.20), yielding

$$E_\lambda = E_{\lambda 0} \exp[-k_\lambda \rho_{c0} H \sec\theta \exp[-(z/H)]] \tag{5.65}$$

The production rate of ions is proportional to the amount of radiation absorbed per unit volume, that is

$$p = \beta_\lambda k_\lambda \rho_c E_\lambda \, \Delta\lambda$$

Substitution of Eqs. (2.20) and (5.65) into this equation yields

$$p = \beta_\lambda k_\lambda E_{\lambda 0} \rho_{c0} \, \Delta\lambda \exp[-(z/H) - k_\lambda \rho_{c0} H \sec\theta \exp[-(z/H)]] \tag{5.66}$$

The height of maximum ion production rate is found by differentiation to be

$$z_m = H \ln(H k_\lambda \rho_{c0} \sec\theta) \tag{5.67}$$

and the maximum production rate is

$$p_m = \frac{\beta_\lambda E_{\lambda 0} \, \Delta\lambda \cos\theta}{H} \exp(-1) \tag{5.68}$$

Figure 5.22 illustrates schematically the vertical distribution of ionizing radiation, density and ion production described by Eqs. (5.64), (2.20), and (5.51). Equations (5.67) and (5.68) show that for a particular reaction the height z_m and the production rate p_m depend only on the angle of incidence of the radiation.

The ion density resulting from this production rate can now be computed for photochemical equilibrium from Eq. (3.29). In the upper atmosphere

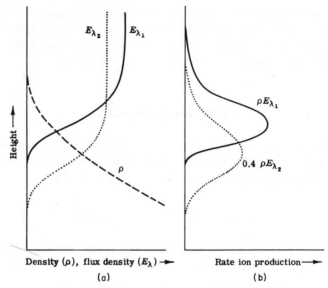

Density (ρ), flux density (E_λ) →   Rate ion production →
(a)   (b)

FIG. 5.22.   (a) Schematic representation of vertical distribution of air density (ρ) and of irradiance for two wavelengths $E_{\lambda 1}$ and $E_{\lambda 2}$. (b) Products $\rho E_{\lambda 1}$ and $\rho E_{\lambda 2}$ which are proportional to rate of ion production.

electron attachment to large ions may be neglected, and decay occurs only by recombination. Under these conditions

$$p = \alpha n^2 \tag{5.69}$$

where α is the recombination coefficient. Combination of Eq. (5.69) with (5.66) yields the vertical distribution of n in the form

$$n = \left(\frac{\beta_\lambda k_\lambda E_{\lambda 0} \rho_{c0} \, \Delta\lambda}{\alpha}\right)^{1/2} \exp\left[-\frac{z}{2H} - k_\lambda \rho_{c0} \frac{H}{2} \sec\theta \exp\left(-\frac{z}{H}\right)\right] \tag{5.70}$$

The maximum ion concentration is found from Eq. (5.70) to be

$$n_m = \left(\frac{\beta_\lambda E_{\lambda 0} \, \Delta\lambda \cos\theta}{\alpha H}\right)^{1/2} \exp(-\tfrac{1}{2})$$

A region where the ion distribution is in agreement with Eq. (5.70) is called a simple Chapman region in honor of the geophysicist who developed the preceding theory. The D layer at a height of 60 km most nearly fulfills the assumptions required by the Chapman theory. Extension of the theory to include continuous absorption and height variation of temperature and composition renders the equations rather complex and requires numerical or graphical techniques for solution.

List of Symbols

		First used in Section
a	Absorptance	5.2
A	Area, albedo	5.1, 5.8
b	Stefan–Boltzmann constant for radiance	5.4
c	Speed of light	5.1
c_p	Specific heat at constant pressure	5.16
d	Thickness	5.17
E	Irradiance	5.1
$\bar{E}_s$	Solar constant	5.7
E'	Spherical irradiance	5.8
g	Force of gravity per unit mass	5.18
h	Planck's constant, height	5.3, 5.8
H	Volume of cell in phase space, scale height	5.3
$H_n(x)$	Gold function of nth order	5.14
L	Radiance	5.1
k	Absorption coefficient, Boltzmann's constant	5.2, 5.3
K	Reaction coefficient for chemical reactions	5.18
m	Equivalent mass of photon, mass of molecule	5.3, 5.18
n	Number of cells per frequency interval per degree of freedom, number of molecules per unit volume	5.3, 5.18
N	Number of photons per frequency interval	5.3
N_i	Average occupation number per cell in phase space	5.3
N_0	Avogadro's number	5.18
p	Momentum of photon, pressure, production rate of ions	5.3, 5.14, 5.19
q	Specific humidity	5.14
q_0	Energy received per unit horizontal area per day at top of the atmosphere	5.9
Q	Radiant energy	5.1
r	Reflectance	5.2
R	Radius	5.9
R_m	Specific gas constant for gas with molecular mass M	5.19
s	Specific entropy, path length	5.3, 5.20
t	Time	5.1
T	Absolute temperature	5.3
u	Optical thickness, specific internal energy	5.2, 5.3
V	Volume	5.3
W	Thermodynamic probability	5.3
x, y, z	Cartesian coordinates	5.2, 5.16
X	Dimensionless optical thickness ($u/2\pi\alpha$)	5.5
Z	Density weighted height	5.14
α	Wien's displacement constant, half-width of spectral line, recombination coefficient	5.4, 5.5, 5.19

		First used in Section
β	Lagrange multiplier, fraction of molecules participating in photochemical reaction per unit absorbed energy	5.3, 5.18
ε	Emissivity	5.2
ε_f	Flux emissivity	5.14
θ	Zenith angle	5.1
λ	Wavelength, thermal conductivity	5.1, 5.17
ν	Frequency	5.1
ρ	Density	5.2
σ	Stefan–Boltzmann constant for irradiance	5.4
σ_λ	Extinction coefficient	5.7
τ	Transmittance	5.2
Φ	Geopotential	5.18
ψ	Half the plane angle subtended by the earth from a satellite	5.8
ω	Solid angle	5.1

Subscripts (other than those derived from defined symbols)

a	Air
b	Black, bottom
c	Constituent
e	Effective
i	Index
l	Pertaining to spectral line, long-wave
m	Maximum
n	Net
0	Pertaining to earth's surface or other base
r	Reflected, reference
s	Short-wave
t	Top
w	White

Problems

1. Find an expression for $L_\lambda(x_0)$ in Fig. 5.3 in terms of $L_{\lambda 0}$, x_0, k_λ, and r_λ, where r_λ represents both the external and internal reflectivity at $x = 0$ and $x = x_0$, and show that it is accurate to within 1% for $r_\lambda = 0.2$.

2. Derive Wien's displacement law as expressed in Eq. (5.20).

3. Show that $L^*_{\lambda\,\mathrm{max}} = \mathrm{const}\ T^5$.

4. If the average output of the sun is $6.2 \times 10^4\ \mathrm{kW\ m^{-2}}$, the radius of the sun is 0.71×10^6 km, the distance of the sun from the earth is 150×10^6 km, and the radius of the earth is 6.37×10^3 km, what is the total amount of energy intercepted by the earth in 1 s?

5. Using the information given in Problem 4 together with the Stefan–Boltzmann law, find the temperature of the solar surface.

6. The quantities E'_{sr} and E'_l introduced in Eq. (5.26) are not irradiances in the strict sense of the definition because the spherical sensors are equally sensitive for all directions. Find an expression for the actual irradiances E_{sr} and E_l passing through a horizontal plane at the position of the satellite in terms of E'_{sr} and ω or E'_l and ω.

7. Using the solar constant as $1370\ \mathrm{W\ m^{-2}}$, compute the average effective temperature of the earth for average albedos of 0.3 and 0.4.

8. Show that the total annual insolation at the top of the atmosphere is the same for corresponding latitudes in Northern and Southern Hemispheres.

9. If one assumes that the earth reflects 35% of the incident solar radiation and that the atmosphere, transparent for short-wave radiation, acts as a single isothermal layer which absorbs all long-wave radiation falling on it, find the global average radiative equilibrium temperatures of the atmosphere and the earth's surface.

10. If the average surface temperature of the earth is 288 K and the average albedo of the earth and atmosphere for solar radiation is 35%, find the "effective absorptance" of the atmosphere for long-wave radiation.

11. Suppose that a cloud layer whose temperature is $7°$ C moves over a snow surface whose temperature is $0°$ C. What is the maximum rate of melting of the snow by radiation from the cloud if the absorptance of the cloud is 0.9 and of the snow is 0.95?

12. Find the downward irradiance from water vapor in the atmosphere for the following vertical distribution of temperature and specific humidity taken at Quillayute, Washington, 8 June 1979, 12 Z.

Pressure (kPa)	Temp. ($°$ C)	Spec. hum. (‰)	Pressure (kPa)	Temp. ($°$ C)	Spec. hum. (‰)
101.9	4	4.80	58.0	−7	1.10
100.0	8	6.80	50.0	−16	1.00
97.2	10	4.30	40.0	−28	0.53
94.6	10	2.80	37.1	−31	0.60
85.0	7	3.30	33.8	−36	0.35
78.1	1	2.50	31.6	−39	0.35
76.2	2	1.10	30.0	−40	0.30
70.0	0	1.20	25.0	−50	0.05
65.7	−3	1.00	20.0	−58	0.04
62.9	−3	0.90	17.7	−63	0.02

13. For the data given in Problem 12, find the rate of radiational temperature change at a reference height corresponding to 700 mb.

14. Derive Eqs. (5.46) and (5.47).

15. An ideally designed radiometersonde is one in which absorptivity of the black surface is unity and reflectivity of the shield and conductivity of the insulation are negligible. If such an instrument is raised by a balloon at 5 m s^{-1} and the temperature difference between upper and lower plate changes from $+2.0$ to $+1.0°$ C in 1 min, find the mean radiational temperature change for the following conditions:

Pressure: 700 mb
Air Temperature: $-7.5°$ C constant with height
Conductivity of air: $1.8 \times 10^{-2} \text{ W m}^{-1} \, °\text{C}^{-1}$
Separation of plates from shield: 1.0 cm
Conductivity of space between plates: $1.8 \times 10^{-2} \text{ W m}^{-1} \, °\text{C}^{-1}$
Separation of plates: 1.0 cm

Solutions

1. The beam of radiation will experience a series of partial internal reflections at the surfaces $x = 0$, x_0. The radiance of the total beam emerging at the right of $x = x_0$ can be expressed from the following schematic diagram. The λ subscripts are omitted for convenience.

$x = 0$		$x = x_0$	
$\to L_0$	$L_0(1 - r) \to$	$\to L_1$	$L_1(1 - r) \to$
	$L_2 \leftarrow$	$\leftarrow L_1 r$	
	$L_2 r \to$	$\to L_3$	$L_3(1 - r) \to$
	$L_4 \leftarrow$	$\leftarrow L_3 r$	

Absorption between $x = 0$ and $x = x_0$ is expressed by Beer's law. Therefore, the successive components of the radiance are related as follows

$$L_1 = L_0(1 - r)e^{-k\rho x_0}$$

$$L_2 = L_0(1 - r)re^{-2k\rho x_0}$$

$$L_3 = L_0(1 - r)r^2 e^{-3k\rho x_0}$$

Therefore, the sum of the first two components on the right of $x = x_0$ can be expressed by

$$L(x_0) = L_0(1 - r)^2 e^{-k\rho x_0} + L_0(1 - r)^2 r^2 e^{-3k\rho x_0}$$
$$= L_0(1 - r)^2 e^{-k\rho x_0}[1 + r^2 e^{-2k\rho x_0}]$$

The next term will be $r^4 e^{-4k\rho x_0}$. For $r = 0.2$, this term will be less than 0.0016, so can be neglected.

This problem can be solved somewhat more simply using the method developed in Section 5.13.

2. In Eq. (5.17) let $x = hc/k\lambda T$. Differentiating with respect to λ and setting $dL_\lambda/d\lambda = 0$ gives

$$-5 + x(e^x - 1)^{-1}e^x = 0$$

or $x = 5 - 5e^{-x}$. The solution is close to $x = 5$. Therefore

$$\lambda_m T \approx \frac{hc}{5k} = 0.290 \times 10^{-2} \text{ m K}$$

3. Substitution of Eq. (5.20) in Eq. (5.18) gives

$$L^*_{\lambda m} = \text{const } T^5$$

4. 18×10^{16} J
5. 5733 K.
6. Recalling that $d\omega = \sin \theta \, d\theta \, d\phi$ Eqs. (5.1) and (5.25) may be written

$$E = \int_0^\theta L \sin \theta \cos \theta \, d\theta \int_0^{2\pi} d\phi$$

$$E' = \int_0^\theta L \sin \theta \, d\theta \int_0^{2\pi} d\phi$$

Upon averaging L over the solid angle and integrating

$$E = \pi \bar{L} \sin^2 \theta$$

$$\omega = 2\pi(1 - \cos \theta)$$

$$E' = 2\pi \bar{L}(1 - \cos \theta)$$

Therefore, $\sin^2\theta = (\omega/\pi)[1 - (\omega/4\pi)]$, and $E = E'[1 - (\omega/4\pi)]$
It follows that for small ω, $E \approx E'$, and as $\omega \to 2\pi$, $E \to \frac{1}{2}E'$.

7. 255 K ($\alpha = 0.3$), 245 K ($\alpha = 0.4$)

8. Draw two lines passing through the sun's center from one side of the earth's orbit to the other. The angle between the two lines is $\delta\theta$. Conservation of angular momentum of the earth about the sun requires that

$$R^2 \frac{\delta\theta}{\delta t} = \text{const}$$

where R represents the distance of the earth from the sun. Recognizing that in the two triangles $\delta\theta_1 = \delta\theta_2$, we have that

$$\frac{R_1^2}{\delta t_1} = \frac{R_2^2}{\delta t_2}$$

The solar irradiance reaching the earth varies as the inverse square of distance from the sun, so that

$$\frac{E_1}{E_2} = \frac{R_2^2}{R_1^2}$$

Therefore

$$E_1\,\delta t_1 = E_2\,\delta t_2$$

showing that for all periods of equal angular displacement, the solar energy falling on the earth in opposite seasons of the year is equal. Since the zenith angle of the sun at corresponding latitudes is the same in opposite seasons, it follows that total annual insolation is the same for corresponding latitudes in the two hemispheres.

9. $T_a = 250$ K, $T_e = 298$ K

10. $a = 0.86$

11. $dm/dt = 8.8 \times 10^{-5}$ kg m^{-2} s^{-1} (0.03 g cm^{-2} hr^{-1})

12. $E^\downarrow = 218$ W m^{-2}

13. $(\partial T/\partial t)_R = -1.7 \times 10^{-5}$ K s$^{-1} = -1.4$ K (24 hr)$^{-1}$

14. The energy balance at the top plate may be written in the form

$$E_+^\downarrow - E_+^\uparrow = L_t$$

where $E_+^\downarrow$ is the total downward irradiance that arrives at the top plate and $E_+^\uparrow$ is the total upward irradiance which leaves from the top plate. These irradiances may be written separately

$$E_+^\downarrow = (1 - r)E^\downarrow + rE_+^\uparrow$$

and

$$E_+^\uparrow = a\sigma T_t^4 + (1 - a)E_+^\downarrow$$

Upon substituting in the balance equation

$$E^\downarrow = \sigma T_t^4 + \frac{1 - r(1 - a)}{a(1 - r)} L_t$$

Equation (5.47) is derived similarly.

15. $(\partial T/\partial t)_R = -3.5 \times 10^{-5}$ K s$^{-1} = -0.12$ K hr^{-1}

General References

Chandrasekhar, *Radiative Transfer*, is a classical text providing a clear and comprehensive account of the problems associated with radiative transfer on a more advanced level.

Charney, "Radiation," in *Handbook of Meteorology* (ed. Berry *et al.*), gives a concise introduction to the principles of radiative transfer.

Craig, *The Upper Atmosphere*, discusses upper atmosphere processes from the point of view of one interested in the behavior of the whole atmosphere.

Elsasser with Culbertson, *Atmospheric Radiation Tables*, provides essential information concerning the absorbing constituents of the atmosphere.

Goody, *Atmospheric Radiation*, Vol. I. *Theoretical Basis*, gives an authoritative and comprehensive discussion of radiative transfer on an advanced level.

Haltiner and Martin, *Dynamical and Physical Meteorology*, complements Parts I to III of this chapter. The organization is similar and the treatment is at about the same level as presented here.

Kondratyev, *Radiation in the Atmosphere*, covers to a large extent the same material as Goody but is less theoretical and more experimental.

Kourganoff, *Basic Methods in Transfer Problems*, gives an introductory discussion similar to that of Chandrasekhar but with special emphasis on methods.

Möller, "Strahlung in der Unteren Atmosphäre," in *Handbuch der Physik*, gives a detailed account of radiative transfer in the lower atmosphere including scattering.

CHAPTER VI

Transfer Processes

*"The scientific method, so far as it is a method,
is nothing more than doing one's damndest with
one's mind, no holds barred."* P. W. BRIDGMAN

Of the radiant energy absorbed by the earth and its atmosphere, most is absorbed by the earth's surface; here it is transformed into internal energy with the result that large vertical and horizontal temperature gradients are developed. As a consequence, the immediate source of the energy which drives the atmosphere may be identified with the earth's surface and, specifically, with its internal energy distribution. A variety of processes may bring about subsequent energy transformations: evaporation, conduction into the earth, long-wave radiation, and upward conduction and convection of heat into the atmosphere, each depending on physical properties of the earth's surface and the atmosphere. These processes and some of their interactions are discussed in this chapter.

PART I: CONDUCTION AND TURBULENCE

6.1 Energy Transfer near the Earth's Surface

Because the interface between the earth's surface and the atmosphere has no heat capacity, conservation of energy requires that the fluxes into and out of the earth must balance. This statement is expressed by the equation

$$-E_n + E_g = E_e + E_h \tag{6.1}$$

where E_n represents the net irradiance at the surface, E_g the energy flux per unit area (subsequently, shortened to flux) by conduction upward from the earth, and E_e and E_h the energy fluxes upward from the earth to the atmosphere by evaporation and conduction of heat, respectively. Because evaporation and conduction of heat are particularly difficult to measure, Eq. (6.1) is often used to measure their sum when E_n and E_g are known.

The principle of conservation of energy also may be applied to the layer of air in contact with the earth's surface. Assuming horizontal uniformity conservation of energy for the layer is expressed by

$$E_h(0) = E_h(z) + \int_0^z \frac{\partial E_n}{\partial z}\, dz + \int_0^z \rho c_p \frac{\partial T}{\partial t}\, dz - \int_0^z L \frac{\partial \rho_d}{\partial t}\, dz \qquad (6.2)$$

where $E_h(z)$ represents rate of upward heat transfer across surface z, ρ density of air, c_p specific heat at constant pressure, L latent heat of phase change, ρ_d mass of water in the form of droplets per unit volume, and $E_n(0)$ and $E_n(z)$ the net irradiances at 0 and z, respectively. The three integrals may be determined by measurement or by calculation, but for layers only 10 m or so in thickness these terms are usually small compared to $E_h(0)$ and $E_h(z)$. In fact, it is useful to define a *surface layer* by the requirement that within this layer the rate of transfer of heat is essentially independent of height. Within this layer

$$E_h(0) = E_h(z_1) = E_h(z_2), \qquad \text{etc.} \qquad (6.3)$$

Whereas $E_h(0)$ represents heat flux by conduction, $E_h(z_1)$ and $E_h(z_2)$ represent transport (or *convection*) of heat by moving elements of fluid.

Although the terms in Eq. (6.2) which represent condensation and absorption or emission of radiation have been neglected in Eq. (6.3) because these terms are small compared to the individual terms representing heat flux, it does not follow that these processes are unimportant. These processes will be discussed later in this chapter in connection with fog formation.

The contributions to energy transfer near the ground surface which are made by photosynthesis and by oxidation have not been considered in the above discussion. In general, these contributions may be considered negligible, but there are conditions in which they become important or even dominant, e.g., the photosynthesis of a cornfield in the late afternoon in early summer or the release of chemical energy by a forest fire.

Equations (6.1) and (6.3) are important statements of energy conservation, but they tell very little about how energy transfer occurs. In the following sections this aspect will be explored in some detail.

6.2 Heat Conduction into the Earth

The rate of heat conduction into or out of the solid earth is an application of Eq. (2.70) and may be expressed by

$$E_g(0) = -\left(\lambda_g \frac{\partial T}{\partial z} \right)_{-0} \qquad (6.4)$$

where λ_g represents the thermal conductivity of the earth at the surface

where $E_g(0)$ is measured. The temperature gradient within the earth is measured as close as possible to the surface, and this is indicated in Eq. (6.4)
by the subscript -0. Thermal conductivity ranges from about 2 W m^{-1}
$^\circ\text{C}^{-1}$ for wet soil or ice to about 0.4 for dry sand and about 0.1 for new snow.
Consequently, the heat conduction may vary between rather wide limits.

The principle of conservation of energy may be applied to the layer of
earth into which heat penetrates in a certain interval of time, say a day or
a year. The heat conduction during this time must equal the total internal
energy change within the layer. Therefore, if integration is begun at a small
finite depth $(-z_1)$ to facilitate measurement

$$\int_0^t \left(\lambda_g \frac{\partial T}{\partial z} \right)_{-z_1} dt = -\int_{-z_1}^{-\infty} \rho_g c_g \, \Delta T \, dz \qquad (6.5)$$

where ρ_g represents the density of the ground, c_g the specific heat of the
ground, and ΔT the change of temperature during the time interval. If
ρ_g, c_g, and λ_g are known, temperature profiles suffice to calculate the heat
conduction. An example is given in Problem 1. Figure 6.1 clearly shows

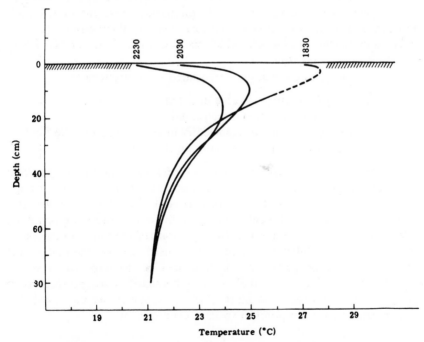

FIG. 6.1. Temperature profiles measured in the ground at O'Neill, Nebraska on 8 August
1953 (after H. Lettau and B. Davidson, *Exploring the Atmosphere's First Mile*, p. 398. Pergamon. New York, 1957).

the downward progress of the diurnal temperature wave. This wave continues to advance downward and at the same time decreases in amplitude until it can no longer be identified below a depth of about 80 cm. A similar annual temperature wave is identifiable to greater depths. The details of course depend strongly on the density and specific heat as well as on the rate of heat conduction at the upper surface.

The ideas developed above are applicable at least in a formal sense to water and air as well as to the solid earth. However, water and air motions result in much more efficient transfer of heat downward into the water and upward into the air, so that temperature changes are normally very small (tenths of degrees in water and degrees in air as compared to tens of degrees in the solid earth) and the layer of heat storage is likely to extend to greater depths in water and to still greater heights in air.

6.3   Turbulence

The motions which facilitate the transfer of heat or other properties have a random character and are usually so effective in transfer that conduction can be neglected except very near the boundaries where the motions are constrained. The random three dimensional velocity fluctuations are aptly described by the term *turbulence*, which stems from the Latin words *turbare* (to disturb) and *lentus* (full of).

The following characteristics are associated with turbulence:

(a)   the motions are unpredictable in detail,
(b)   the fluctuations are three dimensional,
(c)   properties of the fluid are efficiently mixed throughout the fluid,
(d)   energy must be supplied to maintain the turbulence

Not all random motions are turbulent, e.g., the sea surface may appear to be in three-dimensional random motion, but transfer of properties through the interface takes place only by molecular diffusion. Therefore, the air–sea interface is not in turbulent motion, except where waves are breaking.

Because an analytic solution of the equations of motion is not possible for turbulent flow, statistical treatments are used. We assume that all the concentrations of properties of the fluid can be decomposed in slowly varying average values and rapidly varying fluctuations around these averages. If s represents the concentration of the property per unit mass, the average value at a given point in space is expressed by

$$\bar{s} = \frac{1}{T} \int_0^T s \, dt \tag{6.6}$$

where T is the averaging time, usually 30 min–1 hr. The fluctuating part is given by

$$s' = s - \bar{s} \qquad (6.7)$$

This type of averaging, called *Reynolds averaging*, is only useful when the averaged quantities do not vary appreciably in time, so that for all practical purposes $\bar{s}' = 0$. When this condition is fulfilled, which is often the case in the atmospheric surface layer, the time series $s(t)$ is called *stationary*.

Now let us consider the vertical transport of property s. At any instant the rate of vertical transport at the point of observation, which may be a few meters above the surface, is expressed by $\rho w s$, where ρ is the density of the air and w is the vertical velocity component. Upward and downward currents are randomly distributed in space and time, and we assume that in sufficient time a representative average sample passes the observation point. We recognize that the net vertical transport of air must be zero in the average if the total mass below the observation point is to be constant in the average. Therefore

$$\overline{\rho w} = \overline{\bar{\rho} \bar{w}} + \overline{\rho' w'} = 0 \qquad (6.8)$$

On the other hand, $\rho w s$ may well in the average have a value different from zero, for the upward currents may systematically carry higher values of s than the downward currents. The net average rate of vertical transport or flux is thus expressed by

$$E_s = \overline{\rho w s} = \bar{\rho}\,\overline{w's'} + \bar{w}\,\overline{\rho's'} + \overline{w'\rho's'} \qquad (6.9)$$

where Eq. (6.8) was used to eliminate two terms of the expansion and the averages of all terms with a single prime are zero. We shall see in Section 6.4 that usually $\bar{\rho}\,\overline{w's'}$ is much larger than the other two terms and can be identified as the flux of s. Equation (6.9) shows that accurate continuous observations of ρ, w, and s are needed to determine the turbulent flux of s in the vertical. These observations can be made for temperature, specific humidity, specific momentum, and other properties.

Before discussing the individual fluxes in more detail we must discuss the equation of state in a turbulent medium. Upon expanding Eq. (2.10) in mean and fluctuating quantities

$$\bar{p} + p' = R_m(\bar{\rho} + \rho')(\bar{T} + T') \qquad (6.10)$$

Averaging yields

$$\bar{p} = R_m \bar{\rho} \bar{T}\left(1 + \frac{\overline{\rho' T'}}{\bar{\rho}\bar{T}}\right) \qquad (6.11)$$

Because in the atmosphere $|\rho'/\bar{\rho}| \lesssim 10^{-2}$ and $|T'/\bar{T}| \lesssim 10^{-2}$, the last term of Eq. (6.11) is negligible; thus

$$\bar{p} = R_m \bar{\rho} \bar{T} \tag{6.12}$$

Dividing Eq. (6.10) by this result yields

$$\frac{p'}{\bar{p}} = \frac{\rho'}{\bar{\rho}} + \frac{T'}{\bar{T}} \tag{6.13}$$

From observations it appears that $|p'/\bar{p}| \ll |\rho'/\bar{\rho}|$ or $|T'/\bar{T}|$. This can be understood when we realize that the pressure is related to the mass of a column of air from the surface to the top of the atmosphere. A fluctuation in temperature of a few degrees Celsius in the lowest 10 m is unlikely to generate a significant pressure fluctuation. Therefore, a good approximation is given by

$$\frac{\rho'}{\bar{\rho}} = -\frac{T'}{\bar{T}} \tag{6.14}$$

which means that the processes in the surface layer are essentially at constant pressure.

6.4   Fluxes of Heat, Water Vapor, and Momentum

Turbulent heat flux (E_h) requires special consideration. In order to express the transport of energy by turbulence past a reference plane at height

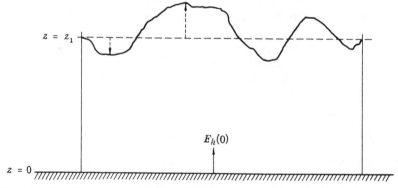

$z = z_1$

$E_h(0)$

$z = 0$

FIG. 6.2.   Heat flux at the earth's surface $[E_h(0)]$ into a layer of air of semi-infinite horizontal extent bounded at the top at time $t = 0$ by the dashed line and at the later time by the solid line.

z_1, consider the volume enclosed by the earth's surface, a horizontal reference plane, and vertical walls as shown in Fig. 6.2. It is assumed that the surface is the only heat source for this system and that the average properties of the system are constant in time. It follows from Eq. (6.2) that heat flux is independent of height. Because local mass transports take place through the upper boundary of the system, it is called an *open system*. Consider now a closed system that at time $t = 0$ coincides with the open system. The first law of thermodynamics requires for the closed system that the heat added equals the work done by the system plus the change in internal energy. The work is done at the upper surface and is expressed per unit area and unit time by pw. Because the internal energy of the open system does not change, the change in internal energy of the closed system must be equal to the net amount of internal energy passing through the reference level. Upon applying the first law to the closed system, the heat entering the system at the ground may be expressed by

$$\left(\int E_h(0)\, dA \right) \Delta t = \left(\int E_h(z_1)\, dA \right) \Delta t = \left\{ \int \rho wu\, dA + \int pw\, dA \right\}_{z_1} \Delta t$$

where A represents horizontal area and u specific internal energy. Upon taking the space average and omitting the height designation, the turbulent heat flux density at z_1 is expressed by

$$E_h = \overline{\rho w(u + p\alpha)} \tag{6.15}$$

and from Eqs. (2.10), (2.38), and (2.41)

$$E_h = \overline{c_p \rho w T} \tag{6.16}$$

Because $c_p T$ for an ideal gas is equivalent to the specific *enthalpy*, the turbulent flux of heat is properly referred to as the turbulent flux of enthalpy.[†]

We see that Eq. (6.16) is a form of Eq. (6.9) with s replaced by $c_p T$. If c_p is assumed for the present to be constant, expansion of the right side of Eq. (6.16) yields

$$c_p \overline{\rho w T} = c_p(\bar{\rho}\,\overline{w'T'} + \bar{w}\,\overline{\rho'T'} + \overline{\rho'w'T'}) \tag{6.17}$$

The last term on the right-hand side is small compared to the first term. This reflects the facts that $|\rho'| \ll \bar{\rho}$ and that the product of ρ' and T' is nearly always negative while w' fluctuates in sign. Experimental evidence indicates that third-order correlations of the form $\overline{\rho'w's'}$ are usually negligible in comparison with second-order correlations.

The second term on the right-hand side may be rewritten with the aid of

[†] The derivation of Eq. (6.16) is due to R. B. Montgomery, *J. Meteorol.* **5**, 265, 1948.

Eqs. (6.8) and (6.14) in the form $\overline{w'T'} \cdot \overline{\rho'^2}/\rho$, which can be recognized as 4 orders of magnitude smaller than the first term. Therefore, a very good approximation of the heat flux is

$$E_h = c_p \overline{\rho} \, \overline{w'T'} \tag{6.18}$$

The latent heat flux may be obtained by replacing s by Lq in Eq. (6.9). This yields

$$E_e = L(\overline{\rho} \, \overline{w'q'} + \overline{w} \, \overline{\rho'q'} + \overline{\rho'w'q'})$$

Again the last two terms on the right-hand side are negligible in comparison with the first term. Thus a good approximation is

$$E_e = L\overline{\rho} \, \overline{w'q'} \tag{6.19}$$

Similarly, we find for the horizontal momentum flux [replace s by u in Eq. (6.9) and neglect the last two terms]

$$E_m = \overline{\rho} \, \overline{w'u'} \tag{6.20}$$

Equations (6.18)–(6.20) express the fluxes in terms of a correlation between the fluctuations of the vertical velocity component and fluctuations of the quantity that is being transported. Although these correlations can be measured, it is cumbersome and expensive to do so. Therefore, a great deal of effort has been made to relate the fluxes to mean quantities, which are much easier to measure. This problem is addressed in Section 6.6.

Secondary Effects of Humidity Fluctuations

As we have seen in Section 2.19 the gas constant R_m and the specific heat at constant pressure c_p vary somewhat with humidity; consequently, the density fluctuations which are associated with humidity fluctuations may be expressed for constant p and T by combining Eqs. (2.95) and (6.14)

$$\rho'_q \simeq -0.61 \, \rho_a q' \tag{6.21}$$

where the subscript q refers to humidity and a refers to dry air. Therefore, humidity has an effect on the buoyancy and stratification of the air.

Furthermore, it has been argued that a sensible heat flux is induced by the vapor flux because the humidity fluctuations generate fluctuations in c_p. From Eq. (2.98) we may express this fluctuation by

$$c'_p = 0.84 c_{pa} q' \tag{6.22}$$

It follows that c_p should not be considered a constant in Eq. (6.17) but should be included in the averaging operation. Brook[†] carried out this averaging

† R. R. Brook, *Boundary Layer Meteorol.* **15**, 481, 1978.

operation and found that approximately 10% of the latent heat flux should be added to the sensible heat flux. Problem 2 requires this exercise. However, as has been pointed out in several recent papers, Brook did not apply the equation of continuity (2.28) correctly for steady state and horizontal uniformity. He assumed $\overline{\rho w} = 0$, whereas in this case

$$\overline{\rho w} = \bar{\rho}\,\overline{w'q'} \tag{6.23}$$

because the vapor mass that enters at the surface due to evaporation must also pass through any level in the surface layer. Equations (6.22) and (6.23) lead to a term equal in magnitude and opposite in sign to the correction introduced by Brook. Therefore, Eq. (6.18) correctly represents the heat flux.

6.5 The Mixing Length Concept and the Eddy Transfer Coefficients

In order to gain insight into the complex mechanism of turbulent transfer, simplified models of turbulent motions have been constructed, and these models have been useful in many developments in the theory of turbulence. A simple model of the elementary molecule was introduced in Chapter II, and this was used to derive a relation between temperature and kinetic energy of the molecules which agrees well with observations over an important range. However, one should be careful not to assume that agreement of theory and observation implies that the model provides a complete account of the phenomenon.

The *mixing length* model has been useful in describing properties of turbulent transfer. In this model we imagine a parcel of air, variously called an *eddy*, a *turbulon*, or *blob*, which may have any size and shape, which is homogeneous in the property s, and which moves as a coherent unit. We imagine also that this eddy passes through a horizontal reference surface after originating a distance l_s (the *mixing length*) above or below the surface. At the point of origin of the eddy the property s is assumed to have been equal to the average value of s at that height. It is recognized then that if s is conserved during vertical displacement

$$s' = -l_s \frac{\partial \bar{s}}{\partial z} \tag{6.24}$$

l_s is positive when the eddy comes from below and negative when the eddy comes from above. When we substitute this expression into Eq. (6.9) and neglect the small terms, we obtain

$$E_s = -\bar{\rho}\,\overline{w l_s}\,\frac{\partial \bar{s}}{\partial z} \tag{6.25}$$

A *turbulent transfer coefficient* may now be defined by

$$K_s \equiv \overline{wl_s} \qquad (6.26)$$

so that Eq. (6.25) becomes

$$E_s = -\rho K_s \frac{\partial \bar{s}}{\partial z} \qquad (6.27)$$

By replacing s successively by specific enthalpy, latent heat, and momentum, equations may be written representing the turbulent transfer of these quantities. Thus

$$E_h = -\rho c_p K_h \left(\frac{\partial \overline{T}}{\partial z} + \Gamma \right) \qquad (6.28)$$

$$E_e = -\rho L K_e \frac{\partial \bar{q}}{\partial z} \qquad (6.29)$$

$$E_m = -\rho K_m \frac{\partial \bar{u}}{\partial z} \qquad (6.30)$$

By analogy with the molecular transfer coefficients discussed in Section 2.16, the turbulent transfer coefficients K_h, K_e, and K_m are called, respectively, *eddy thermal diffusion coefficient, eddy diffusion coefficient*, and *eddy viscosity*. The second term on the right of Eq. (6.28) represents the rate of change of temperature with height following an individual parcel as it moves vertically (the adiabatic lapse rate). Verification of Eq. (6.28) is left for Problem 3. Subscripts have been added to the individual transfer coefficients in Eqs. (6.28)–(6.30) to indicate that they are not necessarily identical. Differences may arise from the fact that the behavior of the eddies may be biased in certain ways. For example, due to buoyancy warm eddies may tend to ascend, whereas cold eddies may tend to descend, suggesting that K_h may be larger than K_m when the atmosphere near the surface is unstable.

Another flux which will be used later in this chapter should be distinguished from the heat flux. The *buoyancy flux* expresses the correlation between the vertical velocity and virtual temperature (or density) and therefore the rate of work that is being done by the buoyancy force. Buoyancy flux may be written in the form

$$E_b = -\rho c_p K_b \left(\frac{\partial T_v}{\partial z} + \Gamma \right) \simeq E_h + 0.02 E_e \qquad (6.31)$$

where K_b is the buoyant flux transfer coefficient and T_v represents virtual temperature as defined in Section 2.19.

There is substantial experimental evidence over land that the profiles of potential temperature and specific humidity are similar; consequently, we have reason to set $K_h \simeq K_e \simeq K_b$. Over water there is still need for more experimental information, but here also we shall assume that these turbulent transfer coefficients are equal.

Introduction of eddy transfer coefficients does not solve the problem of turbulent transfer. It only expresses turbulent flux as proportional to a gradient and consolidates the complexities of turbulence within the transfer coefficients. The problem now is to find explicit relations for these coefficients in time and space. The next two sections represent an attempt to solve this problem by using the mixing length model again.

The Closure Problem

So far the discussion of turbulence may appear to have ignored the basic physical principles introduced in the previous chapters. It may seem that suddenly we are groping for a completely new set of rules. There is, however, a close connection with what has been developed in earlier chapters. The equations of motion, the equation of continuity, the equation of heat conduction, which is an expression of conservation of energy, and the equation of state, constitute a set of six equations with six independent variables, i.e., the three wind components, the density, the temperature, and the pressure. Since the equations are independent, the system should be fully determined or *closed*. However, the governing equations are nonlinear differential equations which have certain stable solutions and other unstable solutions. The unstable solutions apply to most geophysical situations and particularly to shear flow in the planetary boundary layer. The mathematical instabilities of the equations are reflected in the physics of turbulent motions. As we have seen, for example, in the treatment of the equation of state in this section, the basic equations may be split into equations for the mean and fluctuating parts of the variables. The set of equations for the mean flow are now no longer a closed set of equations. New unknowns such as $\overline{u'w'}$ and $\overline{w'T'}$ appear, which makes the number of unknowns larger than the number of equations. As we have seen these covariances may be replaced by the product of a turbulent transfer coefficient and a mean gradient, and the set of equations may be closed by making assumptions concerning the turbulent transfer coefficients. This *first-order closure* method will be followed in the following sections. It is also possible to use the basic equations for constructing new equations for the covariances, but in doing so, new unknowns in the form of triple correlations will appear. This procedure may be repeated, but at each step a higher-order correlation appears as an unknown. Thus it is fundamentally impossible to close the set of equations in a mathematical sense without assumptions. The problem thus encountered is referred to as the *closure problem* of turbulence.

To develop the second- and higher-order closure methods requires information concerning higher-order statistics of turbulence as a basis for closure assumptions. To carry this out is beyond the scope of this book; therefore, in the following sections turbulent transfer is based on first-order closure methods.

PART II: THE BOUNDARY LAYER

6.6   Transfer by Turbulence in the Adiabatic Atmosphere

The simplest form of turbulent transfer takes place in the absence of buoyant forces, that is, in a layer characterized by the adiabatic lapse rate of temperature (neutral static stability). Although transfer of heat may be absent or very small under these conditions, transfer of momentum and other properties may be quite important.

Consider the relation between the vertical flux of momentum and the turbulent motions which accompany it in the normal case of increasing wind speed with height. At a reference height, eddies with a deficit in momentum arrive from below, and eddies with an excess in momentum arrive from above. Under steady conditions the mean velocity at the reference surface does not change, so the deficits and excesses of momentum of the various eddies add up to zero in the average. Formally, the average specific momentum from below and from above which arrives per eddy at the reference surface is given by

$$\bar{u} - l_{\mathrm{m}} \frac{\partial \bar{u}}{\partial z}$$

A similar consideration for the kinetic energy of these eddies shows that they carry an excess of energy to the reference surface which is on the average

$$\overline{l_{\mathrm{m}}^2} \left(\frac{\partial \bar{u}}{\partial z} \right)^2$$

This is the average specific kinetic energy per eddy which is converted from the mean flow to turbulent flow. The reservoir of turbulent kinetic energy is maintained by a balance of energy input from the mean flow or other sources to turbulence and viscous dissipation of kinetic energy into heat.

We assume now that the reservoir of turbulent kinetic energy, as well as the individual components, is proportional to the input. Of special interest is the amount of kinetic energy associated with the vertical velocity component. Therefore

$$\overline{w^2} \propto \overline{l_m^2} \left(\frac{\partial \bar{u}}{\partial z}\right)^2 \tag{6.32}$$

Proportionality (6.32) may be combined with Eq. (6.26) and transformed to

$$K_m^2 \propto \overline{l_m^4} \left(\frac{\partial \bar{u}}{\partial z}\right)^2 \quad \text{or} \quad K_m \propto \overline{l_m^2} \frac{\partial \bar{u}}{\partial z}$$

The vertical momentum flux which is expressed by Eq. (6.30), therefore, may be written

$$\frac{E_m}{\rho} \propto -\overline{l_m^2} \left(\frac{\partial \bar{u}}{\partial z}\right)^2 \tag{6.33}$$

This downward vertical momentum flux (E_m) represents a force per unit area, or stress (τ), which acts in the positive x direction on the fluid below the reference level and is expressed by

$$\frac{\tau}{\rho} \propto \overline{l_m^2} \left(\frac{\partial \bar{u}}{\partial z}\right)^2 \tag{6.34}$$

At this point we need an assumption concerning l_m. It is clear that very close to the surface the eddies are restricted in their vertical motions by the surface. The higher above the surface the more freedom in the vertical exists. This argument led Prandtl, who originally introduced the mixing length concept, to assume that $l_m \propto z$. Introducing this assumption into proportionality (6.34), we find

$$\left(\frac{\tau}{\rho}\right)_{z=0}^{1/2} \equiv u_* = kz \frac{\partial \bar{u}}{\partial z} \tag{6.35}$$

where u_* represents the *friction velocity* and the constant of proportionality k is known as *von Karman's constant*. The friction velocity is introduced as a shorthand notation of $(\tau/\rho)_{z=0}^{1/2}$ and does not provide new physical insight, but it may be used as a convenient scaling velocity. Equation (6.35) does not hold very close to the surface because $\partial \bar{u}/\partial z$ remains finite as $z \to 0$. It may be integrated to give the wind profile equation for conditions of neutral stability in the form

$$\bar{u} = \frac{u_*}{k} \ln \frac{z}{z_0} \tag{6.36}$$

which specifies that $\bar{u}$ vanishes at a height z_0 (the *roughness length*) above the surface. This does not imply that the real wind must vanish at this height, for the assumptions made in the derivation of Eq. (6.35) may well fail very close to the boundary. Observations over many types of surfaces have shown that for $z \gg z_0$ the logarithmic profile appears with remarkable fidelity when the conditions of horizontal uniformity, steady state, and neutral static stability are fulfilled. Although this result strengthens faith in the validity of the assumptions on which (6.32) and (6.35) rest, it does not constitute proof, for other assumptions might also lead to Eq. (6.36) as is indeed shown in Section 6.8. An example of the adiabatic wind profile is shown in Fig. 6.3, and Problem 4 provides a practical application.

The model is incomplete because it fails to predict the values of the von Karman constant and the roughness length. These have to be determined by experiment.

The von Karman constant has been determined in wind tunnels and in the atmosphere. Most estimates are in the range from 0.35 to 0.40. Wind tunnel estimates are usually higher than atmospheric estimates suggesting

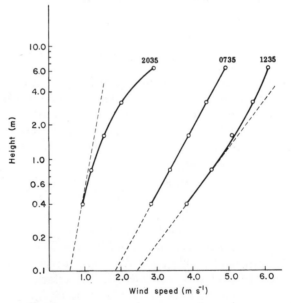

FIG. 6.3. Vertical distribution of average wind speed observed at O'Neill, Nebraska on 19 August 1953 under conditions of stability (2035 CST), neutral stability (0735 CST), and instability (1235 CST). The roughness length where the three straight lines converge is at a height of about 0.9 cm (after H. Lettau and B. Davidson, *Exploring the Atmosphere's First Mile*, p. 444. Pergamon, New York, 1957).

that k may be dependent on the Reynolds number† of the flow. Therefore it may depend on the roughness of the surface and on the wind speed.

The roughness length is a variable which depends on the character of the surface. It may vary from about 10^{-5} m for smooth surfaces such as mud flats, ice, or water before ripples form to 0.1 m for very rough terrain such as sagebrush. Larger roughness elements such as trees and houses are associated with even larger z_0 (up to 1 m), but in such cases it may be difficult to define a surface layer.

The transfer coefficient for momentum (the eddy viscosity) may be expressed by combining Eqs. (6.30) and (6.35) in the form

$$K_m = k^2 z^2 \frac{\partial \bar{u}}{\partial z} = kzu_* \tag{6.37}$$

which may be transformed to

$$K_m = \frac{k^2 z(\bar{u} - \bar{u}_1)}{\ln z/z_1} \tag{6.38}$$

where $\bar{u}_1$ represents the average wind speed at height z_1 and $\bar{u}$ at height z. Problem 5 requires calculation of the eddy viscosity from Eqs. (6.37) and (6.38).

6.7 Transfer by Turbulence in the Diabatic Atmosphere

Although the discussion of turbulent transfer in the adiabatic atmosphere has provided useful insight, the air near the earth's surface is seldom in a state of neutral equilibrium. More often, the air is either statically stable or statically unstable, and these states are referred to as *diabatic*. In the diabatic case the turbulent energy is derived both from the kinetic energy of the mean wind and from buoyant energy associated with local density fluctuations.

The discussion in Section 6.6 leading to proportionality (6.34) may now be extended for the diabatic case. The buoyant force per unit mass acting on an eddy whose density differs from its surroundings by ρ' is expressed in Section 2.14 by

$$-(g/\bar{\rho})\rho'$$

When the eddy has traveled over a vertical distance l_m and has arrived at

† The Reynolds number, defined by $\bar{u}D/\nu$, where D is a characteristic spatial dimension, represents the ratio of acceleration of flow to shear stress.

the reference height z, the buoyant force has performed specific work proportional to

$$-(g/\bar{\rho})\rho' l_m$$

Energy equivalent to this work enters directly into the vertical component of the turbulent velocity; on the other hand, energy produced by shear of the mean wind enters directly into one horizontal component but indirectly into the vertical component of turbulent velocity. It may now be recognized that proportionality (6.32) is a special form of a more general proportionality expressed by

$$\overline{w^2} \propto \overline{l_m^2}\left(\frac{\partial \bar{u}}{\partial z}\right)^2 - \alpha' \frac{g}{\bar{\rho}}\overline{l_m \rho'} \tag{6.39}$$

where α' is an empirical coefficient relating the work done by the buoyant force to the work done by vertical shear.

It follows from the definition of virtual temperature as given in Section 2.19 that neglecting pressure fluctuations, as before in Section 6.3

$$\rho'/\bar{\rho} \simeq -T_v'/\overline{T}_v \tag{6.40}$$

which says that buoyancy fluctuations may be expressed by fluctuations in the virtual temperature. By substituting Eq. (6.40) into (6.39), then multiplying by l_m^2 and applying Eqs. (6.24) and (6.26), the result is

$$K_m^2 \propto \overline{l_m^4}\left(\frac{\partial \bar{u}}{\partial z}\right)^2 - \alpha \frac{g}{\overline{T}_v}\overline{l_m^4}\left(\frac{\partial \overline{T}_v}{\partial z} + \Gamma\right)$$

which may be compared with the adiabatic case. Because of the statistical operation the coefficient α is probably not the same as α'. Upon eliminating K_m by use of Eq. (6.30), replacing the vertical momentum flux by the surface stress, and again assuming l_m is proportional to z, this proportionality can be written as

$$u_*^4 = \left(\frac{\tau}{\rho}\right)^2 = \left(kz\frac{\partial \bar{u}}{\partial z}\right)^4 - \alpha \frac{g}{\overline{T}_v}\left(\frac{\partial \overline{T}_v}{\partial z} + \Gamma\right)(kz)^4\left(\frac{\partial \bar{u}}{\partial z}\right)^2$$

where k again represents von Karman's constant. Equation (6.35) now may be generalized to

$$u_*^2 = k^2\left(\frac{\partial \bar{u}}{\partial(\ln z)}\right)^2\left(1 - \frac{\alpha(g/\overline{T}_v)(\partial \overline{T}_v/\partial z + \Gamma)}{(\partial \bar{u}/\partial z)^2}\right)^{1/2}$$

$$= k^2\left(\frac{\partial \bar{u}}{\partial(\ln z)}\right)^2(1 - \alpha\,\mathrm{Ri})^{1/2} \tag{6.41}$$

which shows clearly the effect of temperature lapse rate on the stress. The dimensionless combination

$$\mathrm{Ri} \equiv \frac{g(\partial \overline{T}_v/\partial z + \Gamma)}{\overline{T}_v(\partial \overline{u}/\partial z)^2} = \frac{g}{\overline{\theta}_v} \frac{\partial \overline{\theta}_v/\partial z}{(\partial \overline{u}/\partial z)^2} = \frac{N^2}{(\partial \overline{u}/\partial z)^2} \tag{6.42}$$

is the Richardson number, which was introduced in Section 4.13. θ_v is the virtual potential temperature which was introduced in Section 2.19, and N is the Brunt–Väisälä frequency as introduced in Section 4.12. Equation (6.41) may be transformed to a simpler form by introducing the dimensionless wind shear

$$\phi_m \equiv \frac{kz}{u_*} \frac{\partial \overline{u}}{\partial z} \tag{6.43}$$

Thus

$$\phi_m = (1 - \alpha \, \mathrm{Ri})^{-1/4} \tag{6.44}$$

When $\mathrm{Ri} = 0$, we have again the adiabatic case, and it is easy to see that $\phi_m = 1$ is equivalent to Eq. (6.35) in the new notation.

By measuring the profiles of wind and temperature Ri may be determined and ϕ_m calculated from Eq. (6.44). From experiments where both the profiles and the fluxes were measured, the empirical coefficient α has been determined. For unstable stratification, i.e., $\mathrm{Ri} < 0$, α is about 16. For stable stratification α is not quite constant and appears to be a function of Ri. It should follow in principle that it is possible to obtain u_* from the profiles alone.

A relation similar to (6.43) for a dimensionless gradient of virtual potential temperature is given by

$$\phi_h \equiv \frac{kz \overline{T}_v}{\theta_* \overline{\theta}_v} \frac{\partial \overline{\theta}_v}{\partial z} \tag{6.45}$$

where $\theta_* \equiv -E_b/c_p \rho u_*$, a scaling temperature, and E_b is the buoyancy flux given by Eq. (6.31). By taking the ratio of Eqs. (6.30) and (6.31) and solving for K_h/K_m using definitions (6.44) and (6.45) we find

$$\frac{K_h}{K_m} = \frac{\phi_m}{\phi_h} \tag{6.46}$$

Equation (6.46) expresses the fact that when $K_h = K_m$ the profiles of wind and virtual temperature must have similar shapes. For the stably stratified surface layer this appears to be a good approximation, but for the unstable case experiments indicate that $K_h \neq K_m$. Returning now to the derivation of Eq. (6.44) we observe that Ri expresses the ratio of the two turbulence production terms; the buoyant production over the shear production. The shear production term, which is always dominant near the surface, decreases more rapidly with height than the buoyancy term; consequently, there is usually a height where both terms are equal. Obukhov[†] was the first to recognize that this height might provide a useful scaling length for the surface layer. In order to estimate this height he assumed the profiles of wind and temperature to be similar and logarithmic. This means that $K_h = K_m = u_* kz$, $\partial \bar{u}/\partial z = u_*/kz$ and $\partial \bar{\theta}_v/\partial z = (\bar{\theta}_v/\bar{T}_v)(E_b/u_* kz c_p \rho)$. If we substitute this into the Richardson number, the result is

$$\text{Ri} = -\frac{g}{\bar{T}_v}\frac{E_b kz}{c_p \rho u_*^3} \equiv \frac{z}{L} \equiv \zeta \tag{6.47}$$

where

$$L \equiv -\frac{\bar{T}_v c_p \bar{\rho} u_*^3}{kgE_b}$$

L is called the *Obukhov length*. Because of the assumptions made in the derivation, L is actually somewhat larger than the height where the two production terms are equal. Nevertheless L as a useful scaling length and ζ as a dimensionless height are used widely. If Obukhov's assumptions are not made, Eqs. (6.28), (6.30), (6.42), (6.43), and (6.45) lead to

$$\zeta = (K_h/K_m)\,\text{Ri}\,\phi_m \tag{6.48}$$

Equations (6.44), (6.46), and (6.48) provide three relations between ϕ_m, ϕ_h, K_h/K_m, Ri, and ζ. We need one more relation to express the first four variables as functions of ζ alone. Such a relation is given in the case of similarity between the wind and temperature profiles by $\phi_m = \phi_h$, and consequently $K_h = K_m$. In Problem 6 the reader is invited to find the relation between ϕ_m and ζ for this case.

Unstable Stratification

Experimental evidence, however, indicates that $K_h \neq K_m$ in the unstably stratified surface layer and that another independent relation is needed between the five quantities mentioned above. A number of independent experiments have provided data which determines a relation between ζ and Ri. Figure 6.4 presents the empirical relation obtained from an experiment carried out by the Air Force Cambridge Research Laboratories' Boundary

---

[†] A. M. Obukhov, Turbulence in an atmosphere with nonuniform temperature. *Tr. Akad. Nauk. SSSR Inst. Teoret. Geofi.*, No. 1, 1946 (translation in *Boundary Layer Meteor* 2, 7–29, 1971).

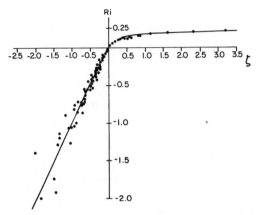

FIG. 6.4. The relation between the Richardson number and the dimensionless height ζ. For unstable stratification ($\zeta < 0$) the solid line represents $\mathrm{Ri} = \zeta$ and for stable stratification ($\zeta > 0$) the solid line represents $\mathrm{Ri} = \zeta(1 + 4.7\zeta)^{-1}$.

Layer Group in Kansas during the summers of 1967 and 1968. The surprising result is that for the unstable boundary layer a good approximation is

$$\zeta = \mathrm{Ri} \tag{6.49}$$

This equation combined with Eq. (6.44) gives

$$\phi_m = (1 - \alpha\zeta)^{-1/4} \tag{6.50}$$

Problem 7 requires application of this equation. By combining Eq. (6.49) with Eqs. (6.46), (6.48) and (6.50), we find

$$\frac{K_h}{K_m} = \phi_m^{-1} = (1 - \alpha\zeta)^{1/4} \tag{6.51}$$

and

$$\phi_h = \phi_m^2 = (1 - \alpha\zeta)^{-1/2} \tag{6.52}$$

Equations (6.50) and (6.52) specify the wind and virtual temperature profiles in the unstably stratified surface layer. These equations may be integrated to yield

$$\frac{\bar{u}}{u_*} = \frac{1}{k}\left(\ln\frac{z}{z_0} - \psi_1\right) \tag{6.53}$$

and

$$\frac{\bar{T}_v - T_{v0} + \Gamma z}{\theta_*} = \frac{1}{k}\left(\ln\frac{z}{z_0} - \psi_2\right) \tag{6.54}$$

where T_{v0} is the extrapolated virtual temperature for $z = z_0$ (this is not necessarily the actual surface virtual temperature) and

$$\psi_1 = 2\ln[(1 + \phi_m^{-1})/2] + \ln[(1 + \phi_m^{-2})/2] - 2\tan^{-1}\phi_m^{-1} + \pi/2$$

$$\psi_2 = 2\ln[(1 + \phi_m^{-2})/2]$$

Problem 8 provides an exercise with measured profiles.

Stable Stratification

From Fig. 6.4 it is clear that Eq. (6.49) is not valid for the stably stratified surface layer and that this equation may be replaced by

$$Ri = \frac{\zeta}{1 + \beta\zeta} \tag{6.55}$$

where β is an empirical constant. This equation suggests that for large ζ Ri reaches an asymptotic value, $Ri = 1/\beta \simeq 0.21$. Therefore, $\beta \simeq 4.7$.

Furthermore, experimental evidence indicates that $K_h = K_m$, and accord-

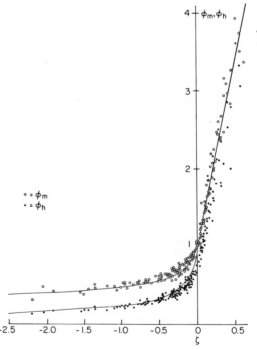

FIG. 6.5.   The dimensionless wind gradient ϕ_m and temperature gradient ϕ_h as function of the dimensionless height. For $\zeta < 0$ the solid lines are $\phi_m = (1 - \alpha\zeta)^{-1/4}$ and $\phi_h = (1 + \gamma\zeta)^{-1/2}$. For $\zeta > 0$ the solid line is $\phi_m = \phi_h = 1 + \beta\zeta$.

ing to Eq. (6.46) $\phi_m = \phi_h$. This combined with Eqs. (6.48) and (6.55) yields

$$\phi_m = \phi_h = 1 + \beta\zeta \tag{6.56}$$

which may be integrated to

$$\frac{\bar{u}}{u_*} = \frac{1}{k}\left(\ln\frac{z}{z_0} + \beta\zeta\right) \tag{6.57}$$

and

$$\frac{T_v - T_{v0} + \Gamma z}{\theta_*} = \frac{1}{k}\left(\ln\frac{z}{z_0} + \beta\zeta\right) \tag{6.58}$$

These are referred to as the log-linear profiles. Equations (6.57) and (6.58) have the same functional form as Eqs. (6.53) and (6.54). Thus for the stable case, we find $\psi_1 = \psi_2 = -\beta\zeta$. Figure 6.5 shows that observational data fit the preceding description reasonably closely for the range $-2.5 < \zeta < 1.0$ although there are minor systematic departures.

It may be noted that Eq. (6.56) can be reached by an alternate line of reasoning. By substituting Eq. (6.55) into Eq. (6.44) it follows that α is related to ζ by

$$\alpha = \beta + \zeta^{-1} - \zeta^{-1}(1 + \beta\zeta)^{-3} \tag{6.59}$$

This means that the simple mixing length model which assumed α to be constant and led to Eq. (6.44) is not useful for the stable case. Nevertheless, a mixing length argument can be given as follows. In the neutral case, Eq. (6.37) shows that $K_m = kzu_*$; this may also be expressed as $K_m = l_m u_*$. If this relation is interpreted as defining the mixing length in the stable case, Eqs. (6.30) and (6.43) yield the result

$$l_m = kz/\phi_m \tag{6.60}$$

and a similar expression can be written for l_h. Because in the stable case the buoyancy force is negative, and turbulent kinetic energy is converted into potential energy, the vertical displacement of eddies is more limited than the assumption $l_m = kz$ for the neutral case indicates. We expect that at some height the mixing length should be limited by the buoyancy and not by the distance to the surface, but that very close to the surface the neutral condition still prevails. Furthermore, it is reasonable to assume that the length scale to which the mixing length is limited can be related to the Obukhov length. With these assumptions the following expression accounts for the required limits

$$l_m = \frac{kz}{1 + \beta z/L} = \frac{kz}{1 + \beta\zeta} \tag{6.61}$$

This result combined with Eq. (6.60) yields Eq. (6.56) and following equations.

Latent Heat Flux

Because accurate measurements of humidity fluctuations are difficult, flux of latent heat has been less thoroughly studied than the fluxes of momentum and heat. Profile measurements suggest that there is similarity between the temperature and humidity profiles, which provides a strong argument to assume that $K_h = K_e$. It follows that

$$\phi_h = \phi_e \tag{6.62}$$

where

$$\phi_e \equiv \frac{kz}{q_*}\frac{\partial\bar{q}}{\partial z} \tag{6.63}$$

and

$$q_* = -\frac{E_e}{\rho L u_*}$$

Thus Eqs. (6.52) and (6.54), as well as (6.56) and (6.58), may be used for the humidity profiles by replacing ϕ_h with ϕ_e and $(\bar{T}_v - T_{v0} + \Gamma z)/\theta_*$ with $(\bar{q} - q_0)/q_*$.

Role of the Richardson Number

The Richardson number was introduced in Chapter IV, and it has appeared again earlier in this section. This dimensionless number represents the ratio of buoyant energy input into turbulence to the energy input by shear of the mean flow. Two extreme conditions in the diabatic surface layer may be considered.

In the limit that the buoyant force is positive and the only force contributing to turbulence, we speak of *free convection*. In this case the first term on the right side of Eq. (6.39) is omitted, and development similar to that given on p. 276 leads to

$$\phi_h \propto (-\zeta)^{-1/3}$$

However, because for free convection $-\zeta \gg 1$, Eq. (6.52) leads to

$$\phi_h \simeq (-\alpha\zeta)^{-1/2}$$

Surface layer observations are consistent with this relation, and not with the earlier $-1/3$ power relation. This result can be understood if we recognize that in free convection large eddies are associated with varying vertical shear. Consequently, there is shear production of turbulent kinetic energy near the surface even though the shear averaged over many eddies vanishes.

In the opposite case, when the buoyancy force is large and negative, Ri is large and positive, turbulence is fully suppressed, and vertical transfer is negligible. Thus there exists a positive critical Richardson number (Ri_c) above which turbulence is suppressed and below which turbulence is generated. Determination of an accurate value of Ri_c has proved to be difficult. Figure 6.4 suggests that for large positive values of ζ, the Richardson number reaches an asymptotic value in the neighborhood of 0.25, and this is usually taken as the best value for Ri_c. As noted in Section 4.13, Taylor has developed a theory which shows that initially laminar flow becomes turbulent when $Ri \leq 0.25$.

If there is an appreciable wind, Ri in the surface layer is nearly always less than 0.25 and the surface layer is consequently turbulent. If we anticipate results from the following section, the solution to Problem 10 shows that for a geostrophic wind of 1 m s^{-1} the surface layer would be turbulent unless $\partial \bar{\theta}_v / \partial z > 21°\text{C m}^{-1}$, a very large temperature gradient indeed, demonstrating that even in light winds Ri_c is rarely exceeded near the surface.

Throughout most of the cloudless atmosphere above the boundary layer turbulence is suppressed, and this is consistent with the fact that the Richardson number here usually is well above 0.25.

6.8 The Planetary Boundary Layer

The transition region between the turbulent surface layer and the normally nonturbulent "free" atmosphere is called the *planetary boundary layer*. In this layer vertical turbulent fluxes may vary with height, and other new properties are encountered. The concepts and equations developed earlier in this chapter for the surface layer and those developed in Chapter IV for the free atmosphere can be combined to describe some of the properties of the planetary boundary layer. Under the restrictive conditions of steady state and horizontal homogeneity Eqs. (4.16), (4.17), (4.19), and (4.20) yield the following equations:

$$f(v_g - \bar{v}) = \frac{1}{\rho} \frac{\partial \tau_x}{\partial z} \tag{6.64}$$

$$f(u_g - \bar{u}) = -\frac{1}{\rho} \frac{\partial \tau_y}{\partial z} \tag{6.65}$$

The surface boundary conditions are $\tau_{x0} = -\rho\overline{(u'w')}_0$, and $\tau_{y0} = \rho\overline{(v'w')}_0$. $\bar{v} = 0$ and $\bar{u} = 0$ and at the top of the boundary layer $\bar{u} = u_g$, $\bar{v} = v_g$, and $\tau_x = \tau_y = 0$.

Equations (6.64) and (6.65) may be solved exactly for the special case of laminar flow. In this case, $\tau_x = \nu \, \partial\bar{u}/\partial z$, and $\tau_y = \nu \, \partial\bar{v}/\partial z$, and Eqs. (6.64) and (6.65) may be reduced to one second-order ordinary linear differential equation in the complex variable $w \equiv u + iv$. The reader is invited in Problem 9 to show that the solution is

$$u = u_g(1 - e^{-az} \cos az)$$
$$v = u_g e^{-az} \sin az \tag{6.66}$$

where $a = (f/2v)^{1/2}$ and the coordinate system is chosen so that $v_g = 0$. This solution was presented by Ekman in 1905 and is known as the *Ekman spiral*. The coefficient a^{-1} represents a characteristic height for the spiral, which for $f = 10^{-4}\,\mathrm{s}^{-1}$ and $v = 1.8 \times 10^{-5}\,\mathrm{m}^2\,\mathrm{s}^{-1}$ is 0.6 m. The lowest height where u is in the direction of u_g occurs where $az = \pi$ or $z = 1.88$ m. The atmospheric boundary layer is typically 2 to 3 orders of magnitude deeper than this, reflecting the fact that transfer of momentum occurs through turbulence rather than through molecular viscosity. By replacing v with a constant eddy viscosity (K_m) which is 4 to 6 orders of magnitude larger, we can obtain boundary layer heights which are realistic. The dependence of K_m on height and stability can be specified, and in the case of the stably stratified boundary layer this may lead to realistic results. In the case of the unstable layer the flux occasionally may be opposite in direction to the gradient, implying negative eddy viscosities. This constitutes a serious difficulty in the use of eddy viscosity models of the planetary boundary layer.

However, another defect of this theory is more serious. It turns out that the Ekman spiral solutions are, in general, unstable to small perturbations and that these perturbations grow into secondary circulations. This subject is beyond the scope of this book.

Matching Solutions

The vertical distribution of mean velocity in the planetary boundary layer may be represented in another way by matching the *inner* and *outer* layers of the boundary layer. As is shown by Eq. (6.36) for the neutral surface layer, z_0 is the scaling length for the wind profile and $\bar{u}/u_* = g_1(z/z_0)$. For heights which are a sizable fraction of the boundary layer height, $z \gg z_0$, and a more appropriate scaling length becomes u_*/f; therefore, $\bar{u}/u_* = g_2(zf/u_*)$. There may also be a range where

$$z_0 \ll z \ll u_*/f$$

and in this range we expect that $g_1 = g_2$ and that the height derivatives should match also. Therefore

$$\frac{\partial \bar{u}}{\partial z} = \frac{u_*}{z_0} \frac{\partial g_1}{\partial(z/z_0)} = f \frac{\partial g_2}{\partial(zf/u_*)}$$

or after rearranging

$$\frac{z}{u_*} \frac{\partial \bar{u}}{\partial z} = \frac{z}{z_0} \frac{\partial g_1}{\partial(z/z_0)} = \frac{fz}{u_*} \frac{\partial g_2}{\partial(fz/u_*)} \tag{6.67}$$

The first equation is only a function of z/z_0 and the second equation is only a function of fz/u_*. This can only be correct over a range of z if the left-hand side is a constant. We know from Eq. (6.35) that this constant is $1/k$. Thus the governing equation for the surface layer is

$$\frac{z}{u_*} \frac{\partial \bar{u}}{\partial z} = \frac{1}{k} \tag{6.68}$$

which is the same result as we found by introducing the mixing length leading to the logarithmic profile (6.36). We recognize that the assumption of a constant stress layer is not needed to arrive at the logarithmic profile. This means that the logarithmic profile is not only valid in the surface layer, but extends into the matching layer.

The wind distribution in the upper part of the boundary layer can be obtained by introducing Eq. (6.68) into the second of Eqs. (6.67) and integrating. The result may be written

$$\frac{\bar{u} - u_g}{u_*} = \frac{1}{k}\left(\ln\frac{zf}{u_*} + A\right) \tag{6.69}$$

where A is a constant of integration. This relation may be combined with Eq. (6.36) to yield

$$\frac{u_g}{u_*} = \frac{1}{k}\left(\ln\frac{u_*}{fz_0} - A\right) \tag{6.70}$$

A similar analysis can be carried out for the v component of velocity. Because there is no appreciable shift in wind direction in the surface layer and matched layer, $\bar{v} = 0$, and $\partial\bar{v}/\partial z = 0$, and this leads to the simpler result

$$\frac{v_g}{u_*} = -\frac{B|f|}{kf} \tag{6.71}$$

where B is a constant of integration and $|f|/f$ is introduced to account for the fact that the surface layer wind is to the left of the geostrophic wind in the Northern Hemisphere and to the right in the Southern. Equations (6.70) and (6.71) may be combined to yield

$$\frac{G}{u_*} = \frac{1}{k}\left[\left(\ln\frac{u_*}{fz_0} - A\right)^2 + B^2\right]^{1/2} \tag{6.72}$$

and

$$\sin \alpha_0 = \frac{B}{k} \frac{u_* |f|}{Gf} \tag{6.73}$$

where G is the magnitude of the geostrophic wind $(u_g^2 + v_g^2)^{1/2}$ and α_0 is the angle between the surface stress and the geostrophic wind. This result determines the cross isobar flow in the boundary layer which, as pointed out in Chapter IV, is closely related to Ekman pumping and its dynamic consequences.

The integration constants A and B depend on the height h where the geostrophic wind is reached. From Eq. (6.69) we have

$$A = -\ln(hf/u_*)$$

The height h is often determined by an inversion which caps the boundary layer and provides an independent constraint to it. Consequently, A and B are functions of h as well as of stratification as we shall see below. Experimental data have been used to determine that $A \approx 1.5$ and $B \approx 2.5$, but considerable uncertainty in these values remains. It follows that $h \simeq 0.22 u_*/f$.

The matching of the inner and outer layer of the boundary layer may be extended to the diabatic case. From the diabatic profile [Eq. (6.53)] in the surface layer the effect of stratification is represented by $\psi_1(z/L)$. This function is already a matched function because L is usually of the order of the matching height. Equation (6.68) is now replaced by

$$\frac{z}{u_*} \frac{\partial \bar{u}}{\partial z} = \frac{\phi_m(z/L)}{k} \tag{6.74}$$

and after integration for the surface layer, we have for unstable stratification Eq. (6.53) and for stable stratification Eq. (6.57).

In the upper part of the boundary layer

$$\frac{\bar{u} - u_g}{u_*} = \frac{1}{k} \left[\ln \frac{zf}{u_*} - \psi_1 \left(\frac{z}{L} \right) + A \left(\frac{hf}{u_*}, \frac{h}{L} \right) \right] \tag{6.75}$$

Equation (6.71) remains unchanged except that now $B = B(hf/u_*, h/L)$.

Because A and B are functions of both hf/u_* and h/L and have to be determined from experiment, a large data set is required to establish these functions. In order to satisfy the theory conditions of quasi-steady-state, horizontal uniformity, and no baroclinity are required. It is not surprising that the existing data set shows a large scatter of points.

The equations developed in this section provide the mean velocity components throughout the planetary boundary layer. They are important because they provide for coupling of the surface layer with the free atmosphere. However, they do not account for organized secondary flows which may involve varying velocities comparable to the mean velocities. And the requirements of a quasi-steady-state and horizontal homogeneity are serious limitations.

PART III: APPLICATIONS

6.9 Practical Determination of the Fluxes near the Surface

The mental excursions in the previous sections have provided us with some insight into the flux–profile relations. In this and following sections practical applications are considered.

The objective is to obtain the fluxes using as simple and as few observations as possible. Observing stations usually measure the wind, temperature, and humidity at one height. If the roughness of the terrain is known, the momentum flux can be determined under neutral conditions from Eqs. (6.30) and (6.38). In order to determine the heat and vapor fluxes, ΔT and Δq are needed, and this requires observations at two heights. This requirement is frequently fulfilled over the sea when the sea surface temperature is known. In order to relate the fluxes to these single mean quantities, *bulk transfer coefficients* are introduced, which are defined by

$$u_*^2 = C_d(\bar{u} - u_s)^2 \tag{6.76}$$

$$E_h = -c_p \bar{\rho} C_h(\bar{u} - u_s)(\bar{\theta}_v - \theta_{vs})T_v/\theta_v \tag{6.77}$$

$$E_e = -\bar{\rho} L C_e(\bar{u} - u_s)(\bar{q} - q_s) \tag{6.78}$$

where C_d is the drag coefficient, C_h is the bulk transfer coefficient for sensible heat, C_e is the bulk transfer coefficient for evaporation, u_s is the average interface current, and θ_v is the virtual potential temperature. The bulk transfer coefficients are a function of height of measurement and stability. The height chosen is usually 10 m above sea surface. The virtual potential temperature is used for convenience, recognizing that near the surface $T_v/\theta_v \simeq 1$ and remembering that $E_h \simeq E_b$.

The information given in the preceding sections is sufficient to give a specific formulation of the bulk transfer coefficient. The drag coefficient for the neutral case can be expressed using Eqs. (6.36) and (6.76) by

$$C_{dn}(z) = k^2 [\ln(z/z_0)]^{-2} \tag{6.79}$$

This can be combined with Eqs. (6.53) and (6.57) to give the drag coefficients for the unstable and stable case, respectively, in the forms

$$C_d = C_{dn}\left\{1 - C_{dn}^{1/2}\frac{\psi_1}{k}\right\}^{-2} \quad \text{(unstable)}$$

and

$$C_d = C_{dn}\left\{1 + C_{dn}^{1/2}\frac{\beta\zeta}{k}\right\}^{-2} \quad \text{(stable)}$$

(6.80)

Similarly, combining Eqs. (6.77), (6.54), and (6.58) yields

$$C_h = C_d^{1/2}C_{dn}^{1/2}\left(1 - C_{dn}^{1/2}\frac{\psi_2}{k}\right)^{-1} \quad \text{(unstable)}$$

and

$$C_h = C_d^{1/2}C_{dn}^{1/2}\left(1 + C_{dn}^{1/2}\frac{\beta\zeta}{k}\right)^{-1} \quad \text{(stable)}$$

(6.81)

Observations over land indicate that $C_e = C_h$, but over water the relation is not well established, and in fact, there is some evidence that over water $C_e \approx C_d$. Under conditions of neutral stability the three bulk transfer coefficients have been found to be about equal.

Equations (6.80) and (6.81) express the bulk transfer coefficients as functions of ζ; however, the Obukhov length already requires a knowledge of the fluxes we would like to compute with Eqs. (6.76)–(6.78). Therefore, it is necessary to express ζ in terms of the mean quantities that are measured. Using Eq. (6.47) in combination with Eqs. (6.76)–(6.78), we find that

$$\zeta = \frac{kC_h}{C_d^{3/2}}\left[\frac{gz}{\bar{\theta}_v}\frac{\bar{\theta}_v - \theta_s}{(\bar{u} - u_s)^2}\right]$$

(6.82)

The term in brackets on the right-hand side is a bulk Richardson number defined by

$$\text{Ri}_{bs} \equiv \frac{gz}{\bar{\theta}_v}\frac{(\bar{\theta}_v - \theta_s)}{(\bar{u} - u_s)^2}$$

(6.83)

From Eqs. (6.80)–(6.83) it is seen that C_d and C_h can be described as functions of Ri_{bs} and C_{dn}. In Fig. 6.6 these functions are given.

The surface values of u, θ, and q that are used in Eqs. (6.76)–(6.78) are not as easily obtained as one might think. The surface wind drift current is of the order of u_* over the open ocean and may be neglected. The "surface" temperature is usually obtained from a water sample below the surface, the "bucket temperature." The interface temperature is often lower than the bucket temperature because evaporation from the surface tends to reduce

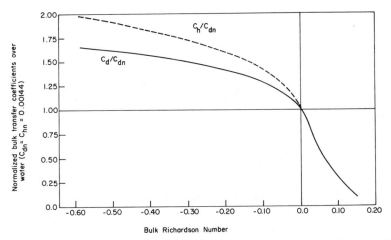

FIG. 6.6.   Ratios of bulk transfer coefficients over their neutral value as a function of the bulk Richardson number for a water surface.

its temperature. A relatively warm layer of 10 cm or so may be produced by absorption of solar radiation. For these reasons the difference between bucket temperature and interface temperature is often several tenths of a degree and may be significant when the heat flux is small and the evaporation relatively large, a situation quite common over the tropical oceans. The surface humidity is determined by the surface temperature and salinity.

Over land, Eqs. (6.76)–(6.78) are not as useful as over water because the surface values of θ and q are even more uncertain than over water. In this case it is necessary to have observations at two heights z_1 and z_2, and the bulk transfer coefficients must be modified to accommodate this as follows:

$$u_*^2 = C_{d1}(u_2 - u_1)^2 \tag{6.84}$$

and

$$\frac{E_h}{c_p\rho} = -u_*\theta_* = -C_{h1}(u_2 - u_1)(\theta_2 - \theta_1) \tag{6.85}$$

$$\frac{E_e}{L\rho} = -u_*q_* = -C_{e1}(u_2 - u_1)(q_2 - q_1) \tag{6.86}$$

From Eqs. (6.53) and (6.57) it is seen that

$$C_{d1} = k^2\left\{\ln\frac{z_2}{z_1} - [\psi_1(\zeta_2) - \psi_1(\zeta_1)]\right\}^{-2} \quad \text{(unstable)}$$

and
$$C_{d1} = k^2\left[\ln\frac{z_2}{z_1} + \beta(\zeta_2 - \zeta_1)\right]^{-2} \quad \text{(stable)}$$

$$\tag{6.87}$$

Similarly, from Eqs. (6.54) and (6.58)

$$C_{h1} = k^2 \left\{ \ln \frac{z_2}{z_1} - [\psi_1(\zeta_2) - \psi_1(\zeta_1)] \right\}^{-1}$$

$$\times \left\{ \ln \frac{z_2}{z_1} - [\psi_2(\zeta_2) - \psi_2(\zeta_1)] \right\}^{-1} \qquad \text{(unstable)}$$

and   (6.88)

$$C_{h1} = C_{d1} = k^2 \left[\ln \frac{z_2}{z_1} + \beta(\zeta_2 - \zeta_1) \right]^{-2} \qquad \text{(stable)}$$

Also, it is convenient to introduce a bulk Richardson number in the form

$$\text{Ri}_{b1} = \frac{g}{\theta} \frac{\theta_2 - \theta_1}{(u_2 - u_1)^2}(z_2 - z_1) \qquad (6.89)$$

From Eqs. (6.84)–(6.88), the following relations can be derived

$$\text{Ri}_{b1} = (\zeta_2 - \zeta_1) \frac{\{\ln \zeta_2/\zeta_1 - [\psi_2(\zeta_2) - \psi_2(\zeta_1)]\}}{\{\ln \zeta_2/\zeta_1 - [\psi_1(\zeta_2) - \psi_1(\zeta_1)]\}^2} \qquad \text{(unstable)}$$

and   (6.90)

$$\text{Ri}_{b1} = (\zeta_2 - \zeta_1)[\ln \zeta_2/\zeta_1 + \beta(\zeta_2 - \zeta_1)]^{-1} \qquad \text{(stable)}$$

It is assumed that for the unstable case $C_{e1} = C_{h1}$ and that for the stable case $C_{d1} = C_{h1} = C_{e1}$.

In Fig. 6.7 the relations between the transfer coefficients and Ri_{b1} are plotted for two values of $\zeta_2/\zeta_1 = z_2/z_1$. Because the preceding method uses exclusively mean characteristics of the profiles, it is called *the aerodynamic method*.

If the surface energy balance is known, the simpler *energy method* may be applied to obtain E_h and E_e. No specific use is made of the profiles, but it is assumed that temperature and moisture profiles are similar and that $K_h = K_e$. Equation (6.1) provides one relation connecting E_h and E_e; an independent relation is obtained by dividing Eq. (6.28) by Eq. (6.29). The result is

$$\frac{E_h}{E_e} \equiv \beta = \frac{c_p(\partial \bar{T}/\partial z + \Gamma)}{L \, \partial \bar{q}/\partial z} \simeq \frac{c_p}{L} \frac{\bar{\theta}_2 - \bar{\theta}_1}{\bar{q}_2 - \bar{q}_1}$$

The ratio β is called the *Bowen ratio*; it can be calculated from temperature and humidity observations at two heights. By combining this equation with Eq. (6.1), the two fluxes may be expressed by

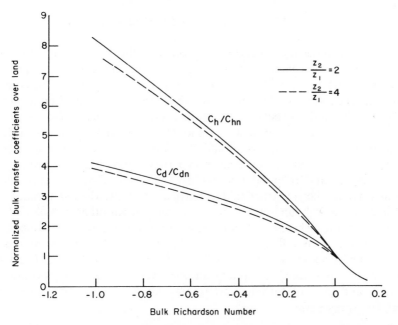

FIG. 6.7. Ratios of bulk transfer coefficients over their neutral value as a function of the bulk Richardson number over land for the two height ratios of $\zeta_2/\zeta_1 = 2$ and $\zeta_2/\zeta_1 = 4$.

$$E_h = \frac{\beta}{1 + \beta}(E_g - E_n) \tag{6.91}$$

and

$$E_e = \frac{1}{1 + \beta}(E_g - E_n) \tag{6.92}$$

The accuracy of the energy method depends strongly on accurate determination of β. The energy balance is easier to determine over land than over water, so this method is used more frequently over land than over the ocean. Over the ocean the value of β is often small (approximately 0.2) and about half the sensible heat flux arises through the effect of humidity on the specific heat at constant pressure. When $E_e \ll E_h$, β is large, and E_h can be determined accurately even if the assumption that $K_h = K_e$ is not quite correct. When $E_h \ll E_e$, β is small, and E_e can be determined accurately. When β is of order one, as occurs frequently over land, the energy method has some of the same limitations as the aerodynamic method.

Although subject to obvious deficiencies, the aerodynamic and energy methods are in common use. The energy method is simpler, and therefore is usually preferable as long as E_g can be determined. When this is not possible or where $E_g \simeq E_n$, the aerodynamic method is preferable.

6.10 Vertical Aerosol Distribution

Turbulent eddies transport dust, smoke, and other aerosol particles from regions of high concentration to regions of low concentration by the process discussed in Section 6.4. Here we are interested in the case of drifting aerosol particles, such as sand, dust, or snow, which have been picked up from the surface. The initial process that lifts particles from the surface is called *saltation*. When the wind is strong enough, some particles at the surface gain momentum from the wind and start rolling over the surface. A particle which has attained sufficient momentum may collide with other surface particles and rebound into the air. This will allow for even more momentum to be imparted to the particle, so that when it reaches the ground, it may rebound even higher and may loosen other particles which in turn become airborne. When the saltation process is well developed and the particles reach sufficient height (5–10 cm), turbulent eddies may pick up the particles and carry them higher up in the boundary layer. During a sustained strong wind, an equilibrium may develop between gravitational settling and upward turbulent transport. Above the saltation level, this may be expressed by

$$w_s s = -K_s \frac{ds}{dz} \tag{6.93}$$

where w_s represents the settling speed and s the concentration of particles. This equation is valid in this simple form only for particles having the same settling rate. Because strong winds are needed for this process, the stability near the ground will be close to neutral. It is tempting to assume that $K_s = K_m = ku_* z$ [Eq. (6.37)]. This is a reasonable assumption as long as the settling speed w_s is small in comparison to the variance of the vertical velocity component; i.e., $w_s \ll (\overline{w'^2})^{1/2} \equiv \sigma_w$. From Eqs. (6.32) and (6.35) we see that $\sigma_w \simeq u_*$; therefore, the above condition may be expressed by $w_s \ll u_*$. In this case Eq. (6.93) may be integrated to

$$s = s_1 \left(\frac{z}{z_1}\right)^{-w_s/ku_*} \tag{6.94}$$

where z_1 is a reference level near the ground but above the saltation height.

In case w_s is of the same order as u_*, $K_s \neq K_m$ because only those eddies that have a vertical velocity larger than w_s will be able to carry the particles upward. The consequence is that $K_s > K_m$, but K_s will remain proportional to u_* and z. Equation (6.94) will still be valid, with the von Karman constant replaced by a larger constant which is a function of w_s/u_*.

Particles suspended in the atmosphere near the earth's surface limit the visual range.† If the relation between particle size and number and extinction coefficient can be determined, measurements of the extinction coefficient with height may specify the vertical particle distribution. The particle distribution also can be measured using remote sensing instruments as described in Part IV of Chapter VII. Equation (6.93) may then be used to calculate the coefficient of eddy transfer. To design and carry out experiments for this purpose provides a stimulating challenge.

6.11 Nocturnal Cooling

In the absence of solar radiation, the earth and atmosphere radiate to space, and the ground surface or the tops of clouds cool rapidly. The resulting nocturnal cooling may be represented as the sum of several readily understood phenomena: essentially black-body radiation of the earth's surface, radiation from the atmosphere, radiation from clouds and from hills, trees, and other obstacles, evaporation or condensation at the surface, heat conduction within the earth, and turbulent transfer within the atmosphere. These processes are expressed formally by Eq. (6.1). Determination of the surface temperature as a function of time is a particularly complex mathematical problem because important interactions between terms occur. For example, as the surface cools a temperature inversion develops in the air just above the surface, which limits the turbulent transfer severely. Also, as surface temperature falls the radiation emitted by the surface falls more rapidly than the radiation received from the atmosphere. Because interactions of this kind are extensive and complicated, only a very simplified analysis is carried out here.

An hour before sunset on 8 August 1953 at O'Neill, Nebraska, the vertical distribution of temperature in the neighborhood of the ground surface was as shown in Fig. 6.8. The surface radiated strongly to the atmosphere and to space at the rate of 450 W m^{-2}. The atmosphere supplied only 390 W m^{-2} so that the surface cooled. Four hours later the surface had cooled by almost 10° C, and the direction of heat conduction within the earth and the atmosphere had reversed from its earlier direction. Subsequent cooling proceeded only very slowly, and by midnight the total flux toward the surface had become equal to the total flux away from the surface. At this time, assuming negligible evaporation, Eq. (6.4) combined with Eqs. (5.38) and (6.28) may be written

$$\int_0^{\varepsilon_{f\infty}} \sigma T^4 \, d\varepsilon_f - \left[\lambda_g\left(\frac{\partial T}{\partial z}\right) \right]_{-0} = \sigma T_g^4 - \left[\rho c_p K_h\left(\frac{\partial \overline{T}}{\partial z} + \Gamma\right) \right]_{+0} \quad (6.95)$$

† See Section 7.11 for quantitative discussion of visual range.

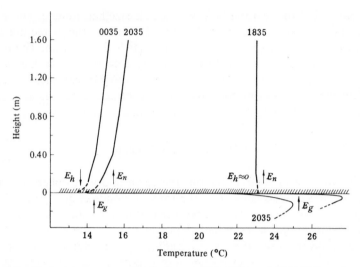

Fɪɢ. 6.8.   Vertical temperature distribution at 1835, 2035, and 0035 CST on the night of 8–9 August 1953 at O'Neill, Nebraska. Directions of vertical fluxes are indicated by arrows (after H. Lettau and B. Davidson, *Exploring the Atmosphere's First Mile*, p. 402. Pergamon, New York, 1957).

where the earth's surface has been assumed to be black and the subscripts -0 and $+0$ refer, respectively, to values just below and just above the surface. In order to clarify the major physical processes a series of approximations is useful. Under the inversion conditions which prevail at the time of minimum temperature turbulent motions are severely inhibited, so that K_h is small and the turbulent transfer of enthalpy may be neglected. Heat conduction within the upper layer of the earth may be represented by the linear function

$$E_g(0) = -\lambda_g \frac{T_g - T(-z_1)}{z_1} \tag{6.96}$$

where z_1 represents the thickness of the surface layer of assumed constant conductivity. Where the surface layer is thin and of relatively small conductivity the temperature approaches a linear function of height after only a few hours of surface cooling. Such conditions are found with a thin snow cover, in ice overlying water, and with less accuracy, in dry cultivated soil overlying compact moist soil.

Atmospheric irradiance may be represented by integrating the first term in Eq. (6.95). The result may be expressed by either side of the identity

$$\sigma \overline{T_a^4} \varepsilon_f \equiv \sigma T_{ef}^4 \tag{6.97}$$

where $\overline{T}_a$ represents the mean air temperature, ε_f the flux emissivity of the atmosphere, and T_{ef} the effective temperature of the atmosphere.

Equation (6.95) may now be solved for T_g as a function of $T(-z_1)$ and λ_g. An approximate solution is given by expressing T_g by $T_{ef} + \Delta T$ and expanding in a binomial series. This procedure yields the minimum temperature at the ground surface in the form

$$T_{gm} = \left[\frac{4 + \kappa T(-z_1)/T_{ef}}{4 + \kappa} \right] T_{ef} \tag{6.98}$$

where $\kappa \equiv \lambda_g / \sigma T_{ef}^3 |z_1|$. Upon introducing measured soil conductivity and atmospheric data obtained by radiosonde into Eq. (6.98), the minimum surface temperature can be calculated as is required in Problem 11. For small conductivity or great thickness of the surface insulating layer (small κ), T_{gm} approaches the effective temperature. If a value of 0.70 is used for ε_f, Eqs. (6.97) and (6.98) yield for no heat conduction from air or earth, $T_{gm} \approx 0.92\overline{T}_a$, which means that for a uniform air temperature of $0°$ C, the lowest possible temperature of the surface is about $-22°$ C. The effects of conduction from within the earth, or from the air, and of condensation on the surface are to raise the lowest possible minimum. If a term representing solar irradiance is introduced into Eq. (6.95), the maximum surface temperature during the day can be expressed by an equation similar to Eq. (6.98).

Although we have assumed here that the air temperature remains constant as the surface cools, this cannot be true very close to the surface. Here, the large positive vertical temperature gradient results in transfer toward the surface. The vertical temperature profile assumes a form qualitatively similar to that of the wind profile. We have seen in Section 6.7 that the temperature and wind profiles are similar in a stably stratified surface layer which implies that the transfer coefficients for enthalpy and for momentum are the same.

6.12 Fog Formation

In Chapter III it was emphasized that condensation almost always occurs when the dew point even slightly exceeds the temperature. Because the daytime dew point depression near the ground often amounts to only a few degrees, it appears that evaporation from the surface or cooling of the air at night by contact with the ground should lead to the formation of fog. Indeed, fog does form under certain conditions, but not nearly so readily as might be expected from the discussion so far. It is necessary to look in more detail at the processes which bring the temperature and the dew point together.

Two processes act to prevent supersaturation and fog formation. First, in the case of a cold surface, the turbulent eddies which transfer heat downward also may transfer water vapor downward; the situation may be understood most easily by inspection of Fig. 6.9. Initial isothermal temperature and dew point distributions are modified by transfer to the colder surface, so that they assume the forms shown by the T_a and T_d profiles in Fig. 6.9. Both heat and water vapor are transferred downward to the cold surface by the same turbulent eddies and with approximately equal effectiveness, so that the dew point may remain below the air temperature. In the case of a warm wet surface, heat and water vapor are transferred upward, and in this case also the dew point may be less than the temperature. Second, the hygroscopic nature of much of the earth's surface (for example, sea water, dry soil, or dry vegetation) results in depression of the dew point below the temperature even at the surface, as indicated in Fig. 6.9. This depression amounts to 0.2° C at the sea surface.

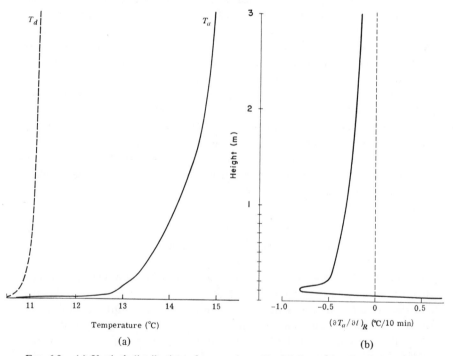

FIG. 6.9. (a) Vertical distribution of temperature T_a and dew point T_d observed on 26 August 1954 above a water surface at Friday Harbor, Washington. (b) Vertical distribution of radiative temperature change calculated from Eq. (5.45) for the curves shown on the left.

In spite of these inhibiting factors, the following processes bring about supersaturation and fog formation under favorable conditions:

(a) Mixing of parcels of saturated, or nearly saturated, air at different temperatures and vapor densities may produce a supersaturated mixture.

(b) Divergence of net irradiance may result in cooling air below its dew point.

(c) Adiabatic cooling accompanying upward motion or falling pressure may result in supersaturation and condensation. "Upslope" fog is formed in this way.

(d) Condensation may occur on giant hygroscopic nuclei at relative humidities below 100%. This is especially effective in industrial regions, and the product is sometimes known as "smog."

(e) The molecular diffusion coefficient for water vapor in air exceeds the thermal diffusivity so that water vapor may evaporate so rapidly from warm surfaces that supersaturation occurs within the laminar layer adjacent to the surface.

Processes (a) and (b) deserve extended discussion.

Mixing of two masses of different temperatures and vapor densities results in a mixture in which these properties are, respectively, linear averages of the initial properties, appropriately weighted by the masses. The dew point of the mixture is higher than the corresponding linear average of the individual dew points because of the exponential form of the Clausius–Clapeyron equation (Section 2.18). The dew point increment, the difference between the linear average of the dew points and the dew point corresponding to the average specific humidity, may then be expressed by

$$\delta T_{\rm d} \approx \frac{d^2 T_{\rm d}}{d q_{\rm s}^2} \frac{\Delta q_{\rm s}^2}{2}$$

where $\Delta q_{\rm s}$ represents the difference between the initial and the average specific humidity. Upon recalling that $dq_{\rm s}/q_{\rm s} \approx de_{\rm s}/e_{\rm s}$, the Clausius–Clapeyron equation [Eq. (2.88)] may be substituted into the above equation with the result

$$\delta T_{\rm d} \approx -(R_{\rm w} T_{\rm d}^2 / 2L q_{\rm s}^2)\, \Delta q_{\rm s}^2$$

For a mixture of equal masses $\Delta q_{\rm s}$ is half the difference between the individual specific humidities. For the range of atmospheric conditions in which fog is common the coefficient is slowly varying, and the right-hand side of the above equation has the approximate value $7(\Delta q_{\rm s}/q_{\rm s})^2$ °C. For the case in which $\Delta q_{\rm s}/q_{\rm s}$ is 0.1, the dew point increment produced by mixing of equal masses is about 0.07° C. If the dew point depression of the two masses were initially within this increment, mixing may result in saturation and fog formation. The conditions favorable for fog formation by this process are high relative humidity, large temperature gradient, and the presence of turbulent eddies.

Fog formation by divergence of net irradiance will be discussed by reference to radiative processes in air above a cold surface. An air parcel at temperature T_g in contact with the surface also of temperature T_g exchanges radiation with the surface with no gain or loss of energy. It exchanges radiation with the atmosphere above with gain of energy because T_g is below the air temperature. An air parcel at a height of, say, 10 cm is warmer than the surface and so exchanges radiation with the surface with net loss. It exchanges with the warmer atmosphere with net gain, so that there may exist a certain height where the total net exchange vanishes. At a height of, say, 1 m the air is much warmer than the surface, and so loses energy by exchange with the surface while it gains only slightly by exchange with the atmosphere above. These calculations may be made using Eq. (5.45). The results are summarized by Fig. 6.9b. The effect of the radiational temperature change shown is to prevent saturation just above the surface and to drive the temperature curve strongly toward the dew point curve in the neighborhood of 10–20 cm. The modified temperature profile results in a compensating heat flux by turbulence so that net cooling and warming is likely to be very small. In the case shown in Fig. 6.9, the dew point was too low to be reached by radiational cooling, but it may be inferred that under more nearly saturated conditions fog should be expected to form first at a height of a few decimeters above a cold moist surface. The effect of radiational cooling is detectable at a height of 10–20 cm in the T_a curve of Fig. 6.9 and appears to amount to about 0.1° C. The conditions favorable for fog formation by this process are high relative humidity and large temperature gradient. Turbulent eddies act to prevent fog formation by radiation.

Above a warm moist surface the temperature and humidity decreases with height; the results are modified to the extent that the sign of the radiational temperature change is reversed. Thus there is strong cooling just above the surface with warming above 10 cm or so, and fog forms in contact with the warm surface. In this case, processes (a)–(c) above may all be effective.

After fog has formed, the physical processes are altered greatly. In the warm surface case, the air is statically unstable, so buoyant eddies carry the fog upward, the depth of the layer increases and the top is uneven and turbulent. Fog which has formed in this way sometimes moves from the warm surface to a cold surface, e.g., warm water to cold land. The surface

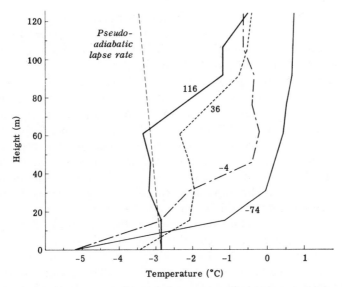

FIG. 6.10. Vertical temperature distribution at Hanford, Washington, on the night of 9 December 1947. The numbers indicate the number of minutes before (−) or after (+) fog was first observed (after R. G. Fleagle, W. H. Parrott, and M. L. Barad, *J. Meteorol.*, **9**, 53–60, 1952).

energy source is cut off, and the character of the upper surface of the fog is altered. A sharp temperature inversion may develop at the top of the fog as shown in Fig. 6.10 as a result of upward radiation from the fog. This stable stratification leads to a smooth, uniform top to the fog layer, and the unstable stratification within the fog leads to a well mixed homogeneous layer. The combined effect is to confine smoke or other pollutants within the fog layer and to thoroughly mix them within this layer. This is why the air pollution disasters in the Meuse Valley of Belgium in December 1930, at Donora, Pennsylvania, in October 1948, and at London, England, in December 1952 each occurred within a persistent blanket of fog.

6.13 Air Modification

In previous sections the various forms of energy transfer that occur in the atmosphere have been described. In Sections 6.11 and 6.12 we sought to provide insight into two special examples of energy transfer, nocturnal cooling, and fog formation. Here the more general problem of air modification by energy transfer from the earth's surface will be considered. The objective is to gain an understanding of the factors which determine the vertical fluxes and of the resulting vertical distribution of temperature and water vapor. The vertical fluxes continually modify the lower atmosphere on a worldwide scale and thereby cause the development of storms and winds of all descriptions, in short, of the weather.

In Section 6.7 it was pointed out that the fluxes near the ground are much smaller in stably stratified air than in unstably stratified air. The result is that upward fluxes of heat and water vapor are much more important than the downward fluxes and that the earth's surface functions as an energy source for the atmosphere. An example of very large upward fluxes and consequent strong air modification occurs with cold air flowing over the much warmer Great Lakes; the famous snowstorms of Buffalo are the direct result. The opposite effect of a strong downward heat flux accompanied by considerable cooling of the air is rare but may occur locally in chinook or foehn winds; however, the widespread and vigorous upward fluxes are much more effective in air modification.

In this section we shall discuss two simple examples of air modification due to heating of the boundary layer from below. The first one deals with a uniform land surface heated by the sun and the second deals with cold air from a continent flowing over a warm ocean. In both cases the boundary layer is well-mixed and capped by an inversion. In order to keep the discussion simple we will assume that the buoyancy flux may be replaced by the sensible heat flux and that the contribution of the vapor flux to the sensible heat flux may be neglected. The errors introduced by this assumption are small when we only calculate the change in potential temperature and the change in specific humidity. The heat content of the boundary layer may then be calculated separately.

In each of the two cases air heated by the warm surface rises and penetrates somewhat into the stable air above the boundary layer. As a result, some of the warmer air above the boundary layer is entrained into it and a sharp inversion is formed. Although the details of this process are complicated, a practical result can be reached by assuming that the boundary layer is so well mixed that its potential temperature is uniform and that the downward heat flux due to entrainment at the top of the boundary layer $[E_h(h)]$ is a constant fraction of the surface heat flux. This is expressed by

$$E_h(h) = -aE_h(0) \qquad (6.99)$$

where a is a constant equal to about 0.2 and h is the height of the boundary layer.

The boundary layer is heated by fluxes both at the bottom and at the top;

thus conservation of energy may be expressed by

$$E_h(0) - E_h(h) \simeq 1.2E_h(0) = \int_0^h c_p\rho \frac{T}{\theta} \frac{d\theta}{dt} dz \qquad (6.100)$$

Case of Horizontally Uniform Heating

In the case of a uniform land surface advection may be negligible, so that $d\theta/dt = \partial\theta/\partial t$. Since θ is uniform throughout the boundary layer, $\partial\theta/\partial t$ must be uniform also. Equation (6.100) now may be integrated, assuming $c_p\rho(T/\theta)$ to be approximately constant. The height of the boundary layer is found to be

$$h = 1.2 \frac{\theta}{T} \frac{E_h(0)}{c_p\rho} \left(\frac{\partial\theta}{\partial t}\right)^{-1} \qquad (6.101)$$

Because $\partial\theta/\partial t$ is independent of height, the heat flux decreases with height linearly according to

$$E_h(z) = E_h(0)(1 - 1.2z/h) \qquad (6.102)$$

In Fig. 6.11 the distributions of heat flux and potential temperature with height are sketched. The height of the boundary layer as a function of time can be calculated easily when there is no downward heat flux at the top

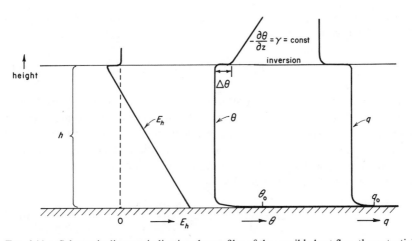

FIG. 6.11.   Schematic diagram indicating the profiles of the sensible heat flux, the potential temperature and the specific humidity with height in the boundary layer including the inversion at the top.

and when $\Delta\theta = 0$ (see Fig. 6.11). In this case the coefficient 1.2 in Eq. (6.101) is replaced by 1.0 and

$$\frac{\partial\theta}{\partial t} = -\gamma\frac{dh}{dt} \tag{6.103}$$

where γ is the lapse rate of potential temperature above the boundary layer $[\gamma \equiv -\partial\theta/\partial z]$. Equation (6.103) may be substituted in Eq. (6.101) and integrated for constant $E_h(0)$ with the result

$$h = \left\{h_0^2 - 2\frac{\theta}{T}\frac{E_h(0)}{c_p\rho\gamma}t\right\}^{1/2} \tag{6.104}$$

where h_0 is the height of the boundary layer at $t = 0$. When $E_h(0)$ is a function of time, a simple numerical procedure may be used to obtain the height. The potential temperature of the bulk of the boundary layer θ is simply proportional to h, according to Eq. (6.103).

In the more general case of a downward heat flux at the top of the boundary layer and a capping inversion ($\Delta\theta$), the calculation of h is more complex. In this case Eq. (6.103) is replaced by

$$\frac{d\,\Delta\theta}{dt} = -\left(\frac{\partial\theta}{\partial t} + \gamma\frac{dh}{dt}\right) \tag{6.105}$$

as can be seen from inspecting Fig. 6.11. Furthermore, the downward heat flux at the top of the boundary layer is given by

$$E_h(h) = -c_p\rho\,\Delta\theta\frac{dh}{dt}\frac{T}{\theta} \tag{6.106}$$

Equations (6.99), (6.101), (6.105) and (6.106) form a closed set and can be solved for h, θ, and $\Delta\theta$. However, the resulting relationships are very complex, obscuring the rather simple physical insight that is aimed for here. If the assumption $E_h(h) \simeq -0.2E_h(0)$ is somewhat relaxed and Eq. (6.104) is used as a model for the solution, a reasonable approximation for the boundary layer height is given by

$$h = \left\{h_0^2 - 2.4\frac{\theta}{T}\frac{E_h(0)}{c_p\rho\gamma}t\right\}^{1/2} \tag{6.107}$$

In Problem 12 some of the consequence of this assumption are explored.

Case of Steady State

When cold air flows from a continent or ice covered surface over a warm ocean, the cold air is modified on a large scale. Assuming steady-state conditions, Eq. (6.100) may be written

$$E_h(0) - E_h(h) = \int_0^h c_p \rho \frac{T}{\theta} u \frac{\partial \theta}{\partial x} dz \qquad (6.108)$$

In this case, the heat flux is a function of the temperature difference between the air and the ocean surface, and we can use the bulk aerodynamic method for calculating surface heat flux as discussed in Section 6.9. Above a certain height z_1, θ is assumed to be independent of height. If entrainment at the top of the boundary layer is neglected, Eq. (6.108) may be written

$$C_h \rho_1 u_1 (\theta_0 - \theta_1) = \frac{d\theta_1}{dx} \int_{z_1}^h \frac{T}{\theta} \rho u \, dz \simeq \frac{T}{\theta} \frac{d\theta_1}{dx} \overline{\rho u} h \qquad (6.109)$$

where $\overline{\rho u}$ is the vertically averaged momentum. The height h depends on the temperature distribution of the original cold air before it reaches the ocean and on the change in potential temperature which occurs during passage over the sea. This last difference is expressed by $\theta_1(x) - \theta_1(0)$, which must be distinguished from $\theta_0 - \theta_1$, measured in the vertical. Because the potential temperature is continuous at h, this height may be expressed by

$$h = \frac{\theta_1(x) - \theta_1(0)}{-\gamma} \qquad (6.110)$$

Upon substituting Eq. (6.110) into (6.109), for simplicity assuming that $\overline{\rho u}$ is independent of x, introducing a normalized potential temperature (θ^*), and integrating

$$\theta^* + \ln(1 - \theta^*) = -\xi \qquad (6.111)$$

where

$$\theta^* \equiv \frac{\theta_1 - \theta_1(0)}{\theta_0 - \theta_1(0)} \quad \text{and} \quad \xi \equiv \frac{C_h(-\gamma)\rho_1 u_1 x}{[\theta_0 - \theta_1(0)]\overline{\rho u}(T/\theta)}$$

Equation (6.111) is plotted in Fig. 6.12. Observations have been used to establish the value of C_h of about 2×10^{-3} for strong winds and large air–sea temperature differences.

Because water vapor and heat are transferred by the same turbulent mechanism, modification of humidity may be treated by a method essentially similar to that used for modification of temperature. A difference arises because the height of the layer of turbulent mixing is determined by the potential temperature distribution according to Eq. (6.110), whereas the humidity distribution has no appreciable effect on this height. Consequently, turbulent mixing may produce a discontinuity in humidity at the top of the layer. Following a development analogous to that introduced for temperature modification and introducing a discontinuity in specific humidity represented by $q_1 - q_h(0)$, the humidity modification may be expressed by

$$\frac{\rho_1 u_1}{\overline{\rho u}} C_e(q_0 - q_1) = \frac{dq_1}{dx} h + [q_1 - q_h(0)] \frac{dh}{dx} \qquad (6.112)$$

When the height h increases, dry undisturbed air with humidity $q_h(0)$ enters the mixing layer and quite suddenly attains by turbulent mixing the humidity $q_1(x)$.

Equation (6.112) is solved by eliminating h with Eq. (6.110) and x with Eq. (6.111). This yields an equation relating the humidity and the temperature; upon assuming that $q_h(0)$ is constant with height, this equation takes the form

$$\frac{d\eta}{d\theta^*} = \frac{\theta^* - \eta}{\theta^*(1 - \theta^*)}$$

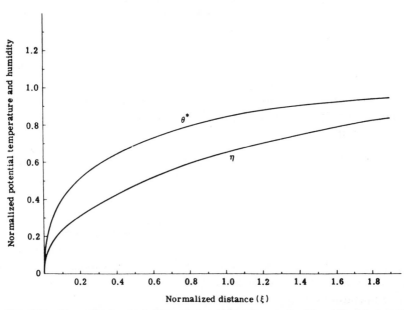

FIG. 6.12. Normalized potential temperature (θ^*) and normalized humidity (η) plotted as a function of normalized distance traveled over a warm ocean (ξ) as calculated from Eqs. (6.111) and (6.113) (after J. A. Businger, *Med. Verh. Kon. Ned. Met. Inst.*, **61**, 1954).

where η represents the normalized specific humidity $[q_1 - q_1(0)][q_0 - q_1(0)]^{-1}$. The boundary condition is $\eta = 0$ when $\theta^* = 0$. Integration of this equation yields

$$\eta = \frac{1 - \theta^*}{\theta^*} \ln(1 - \theta^*) + 1 \qquad (6.113)$$

In Fig. 6.12, η is shown as a function of ξ. Problem 13 provides an exercise in calculating the fluxes of sensible and latent heat at various distances from a coast line under steady state conditions.

The examples of air modification treated here serve to illustrate the process in simple but important cases. The analysis can be extended to include the effect of thermal wind, variable surface temperatures, and many other conditions. We have glimpsed the gate to the jungle.

6.14 Global Summary of Energy Transfer

The physical processes of energy transfer by radiation, evaporation, and turbulent conduction vary greatly from place to place. Among the factors responsible for this variation the most important are the effect of latitude on solar flux density and the different surface properties of water and land. Each of us is familiar with geographical variations in energy transfer, and we make use of this familiarity in planning travel to the beach in mid-July, in buying tire chains in preparation for a winter trip to Minneapolis, or in deciding to plant corn in Iowa.

The insolation absorbed at the earth's surface was described in Chapter V; this is the first step in the series of energy transformations which ultimately are responsible for winds and ocean currents as well as for all biological processes. In this chapter the mechanisms responsible for the intervening energy transformations have been discussed. The equations developed in the course of this discussion may be used to calculate evaporation and heat transfer at many points over the surface of the earth and at many times during the year. The distribution of the mean annual heat transfer with latitude is shown in Fig. 6.13 for land areas and for ocean areas; readers who are properly skeptical hardly need to be told that results of this sort reflect great perseverance and great courage, but accuracy is likely to be uncertain.

Figure 6.13 reveals that on the average over both land and sea heat is transferred upward from the earth to the atmosphere. The fact that this occurs in spite of the normal increase of potential temperature with height reflects the large values of eddy transfer coefficients which accompany superadiabatic lapse rates. Figure 6.13 also shows that between 45°N and

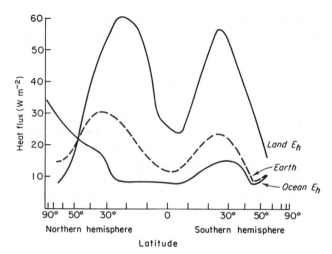

F<small>IG</small>. 6.13.   Mean annual vertical heat flux as a function of latitude for ocean areas and for land areas (after M. I. Budyko, "The heat balance of the earth," 85–113, in *Climatic Change,* John Gribbin, ed., Cambridge University Press, 1978).

45°S upward transfer over land is three or four times as large as it is over the oceans.

Of more interest is the latitudinal distribution of evaporation over land and over the oceans. By comparing Figs. 6.13 and 6.14 the energy utilized in evaporation over the oceans may be recognized as amounting to roughly ten times the vertical heat flux; over land, the two energy transfers are more nearly equal.

Evaporation from land areas reaches a strong maximum in the equatorial region, but at all latitudes precipitation exceeds evaporation. This implies because water vapor stored in the atmosphere is constant over long periods, that there is a net transport of water by the atmosphere from ocean to land areas. Evaporation from ocean areas exhibits maxima amounting to about 2 m year^{-1} in the subtropical regions with minima over the poles and the equator. Consequently, water must be transported by the wind systems from the subtropical oceans to middle latitudes (40° to 60°) and to the equatorial regions where the major precipitation systems of the world are concentrated. In studying Fig. 6.14 one should remember that land occupies only 30% of the surface area of the earth, so that the distributions of evaporation and precipitation for the whole earth are close to those shown for the oceans.

Aside from the slow accumulations of thermal energy called changes in climate, the annual insolation absorbed by the earth and its atmosphere must equal that given up in the form of radiation to space. Similarly, the

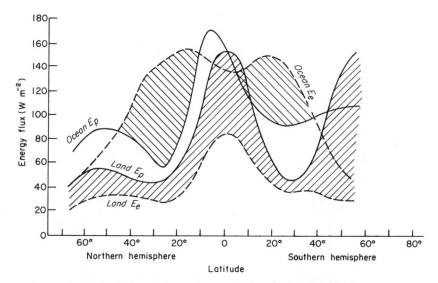

FIG. 6.14.   Energy fluxes corresponding to the mean annual evaporation E_e and precipitation E_p for land areas and ocean areas (after M. I. Budyko, "The heat balance of the earth," 85–113, in *Climatic Change*, John Gribbin, ed., Cambridge University Press, 1978).

total energy of all forms absorbed by earth and atmosphere separately must equal, respectively, that given up by the earth and atmosphere. Errors in calculation of individual terms are far larger than any possible small residual, but no calculations are needed to establish these propositions. However, the net radiation absorbed or emitted at a point or in particular regions may be quite large, as is shown in Fig. 6.15. The net radiation absorbed by the oceans, the earth, and atmosphere in the region between about 30°S and 30°N (the large positive area in Fig. 6.15) must be transported to higher latitudes by ocean currents and air currents. Recent estimates of the net energy absorbed by the oceans are also described in Fig. 6.15 as a function of latitude. Other independent estimates of energy absorbed by the oceans differ markedly from these estimates. The long term climatic average value of heat absorbed by the ocean is not known to better than about a factor of two, and we are completely in the dark as to possible interannual variability. Close to the equator the oceans transport most of the heat poleward, but at middle latitudes atmospheric winds carry more heat to the poles than the ocean currents.

The ocean and the atmosphere are two coupled thermodynamic engines which operate somewhat differently. For the atmospheric engine, the primary heat source may be regarded as the tropical land and ocean sur-

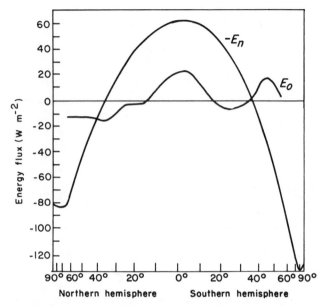

FIG. 6.15.   Net radiation E_n absorbed by the earth and atmosphere and the energy absorbed by the ocean E_o as a function of latitude (after M. I. Budyko, "The heat balance of the earth," 85–113, in *Climatic Change*, John Gribbin, ed., Cambridge University Press, 1978).

faces, and the primary heat sink is the upper part of the water vapor atmosphere. In transporting thermal energy from the source to the sink, the potential energy of the system tends to decrease with corresponding increase in kinetic energy. For the ocean engine, on the other hand, the primary source is the tropical ocean surface, and the primary sink is the polar ocean surface. Because the source and sink are at nearly the same geopotential, only weak fluid motion is to be expected. Wind stress acting on the ocean surface also is important in driving ocean currents.

6.15   Effects of Increasing Concentration of CO_2

Consumption of fossil fuels since the middle of the nineteenth century has released about 150×10^9 tons of carbon to the atmosphere. More than half of this release has occurred since 1958, and global consumption in 1979 is increasing at a rate of about 4% per year. Observations of CO_2 concentration in the atmosphere from 1958 to 1976 are shown in Fig. 6.16. Clearly shown here are the annual cycle and the secular trend. The annual cycle reflects the seasonal consumption and release of CO_2 by the Northern Hemisphere biosphere. The secular trend, which is the same for stations in

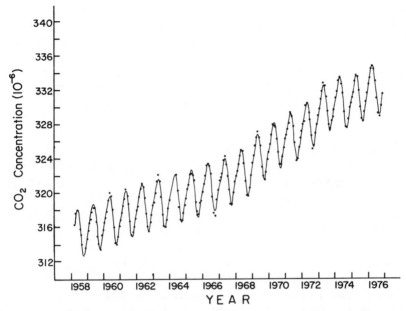

FIG. 6.16. Monthly mean values of atmospheric CO_2 observed at Mauna Loa, Hawaii (after C. D. Keeling *et al.*, *Tellus*, **28**, 538, 1976).

the Southern Hemisphere, reflects the increase in global atmospheric CO_2 concentration due to burning of fossil fuels. It is easy to calculate from the data given here that about half of the CO_2 added to the atmosphere since 1958 has remained in the atmosphere; the other half is assumed to have been absorbed in the oceans and biosphere. Estimates of the effect of the biosphere vary widely, but it appears likely that over the past century there has been a net release of CO_2 from the biosphere to the atmosphere because of large scale deforestation. Ocean storage also is uncertain; it appears that the upper mixed layer (100 m or less), which communicates readily with the atmosphere, has absorbed only a small fraction of the CO_2 released by burning. However, CO_2 may be carried downward to deeper layers of the ocean by the process of Ekman pumping (see Section 4.11) driven by the large anticyclonic wind systems which are situated over the middle latitude oceans. This mechanism could greatly increase the rate of storage of the ocean over that resulting from small scale mixing alone.

Projections of the future consumption of fossil fuels have indicated that if the present rate of increase in consumption continues atmospheric CO_2 concentration is expected to double by about the year 2030. Such an increase would be likely to have effects on global climate. As pointed out in Section 5.14, CO_2 absorbs infrared radiation and, together with H_2O

accounts for the blanketing effect of the atmosphere. Assuming that other atmospheric properties remain constant, the net rate of additional heating of the troposphere and the earth's surface resulting from doubling the CO_2 concentration has been calculated to be about 4 W m^{-2}.[†] This is equivalent to the negative net change in radiative flux at the tropopause. If it is assumed, further, that the earth acts as a black body, the Stefan–Boltzmann law [Eq. (5.19)] yields for the corresponding temperature change

$$\Delta T = \Delta E / 4\sigma T^3$$

where ΔE represents the incremental heating due to increased CO_2. Using $\Delta E = 4$ W m^{-2} this yields for the global average value, $\Delta T \approx 1$ K. Model calculations indicate that heating varies with latitude and season. The average temperature increase may be amplified or decreased by a variety of feedback effects. Increased surface temperature should increase the rate of evaporation from the earth's surface and increase atmospheric H_2O concentration. This in turn should result in greater absorption of infrared radiation from the earth and shortwave radiation from the sun. This positive feedback has been calculated using numerical models to about double the heating due to increase in CO_2 alone. A second positive feedback effect should result from the reduced albedo resulting from melting of ice and snow; this has been estimated as amounting to 0.3 W m^{-2} K^{-1}.[‡]

The change in clouds which result from global warming may have positive or negative feedback effects. Increased cloud cover increases the global albedo and reduces insolation at the earth's surface, but also traps more of the infrared radiation from the surface. The net effects may be quite variable and depend on how other atmospheric processes affect the types and distributions of clouds. The increase in CO_2 will be slowed somewhat by the net effect of more rapid growth and increased respiration of vegetation resulting from the increased atmospheric CO_2. Despite these and other uncertainties, it appears very likely that the average global surface temperature changes resulting from increasing rate of consumption of fossil fuels may be significant and positive in the next 50–100 years.

List of Symbols

		First used in Section
a	Scaling height, $(f/2v)^{1/2}$, entrainment constant	6.8, 6.13
A, B	Boundary layer parameters	6.8
c_g	Specific heat of ground	6.1
c_p	Specific heat at constant pressure	6.1

† V. Ramanathan et al., J. Geophys. Res. **84**, 4949, 1979.
‡ M. S. Lian, and R. D. Cess, J. Atmos. Sci. **34**, 1058, 1977.

		First used in Section
C_d	Bulk transfer coefficient for momentum (drag coefficient)	6.9
C_e	Bulk transfer coefficient for evaporation	6.9
C_h	Bulk transfer coefficient for sensible heat	6.9
e	Vapor pressure	6.9
E_b	Buoyancy flux per unit area	6.7
E_e	Latent heat flux per unit area	6.1
E_g	Heat flux per unit area into the ground	6.1
E_h	Enthalpy (sensible heat) flux per unit area	6.1
E_m	Momentum flux per unit area	6.4
E_n	Net irradiance	6.1
E_s	Flux per unit area of property s	6.3
f	Coriolis parameter	6.8
g	Force of gravity per unit mass	6.5
G	Magnitude of geostrophic wind	6.8
h	Height of boundary layer	6.8
k	von Karman constant	6.4
K	Coefficient of turbulent transfer	6.4
l	Mixing length	6.4
L	Latent heat of vaporization, Obukhov length	6.1, 6.7
N	Brunt-Väisälä frequency	6.7
p	Pressure	6.3
q	Specific humidity	6.3
q_*	Scaling humidity	6.7
R_m	Specific gas constant for air	6.3
R_w	Specific gas constant for water vapor	6.12
Ri	Richardson number	6.7
s	Any property to be transferred by turbulence	6.3
t	Time	6.1
T	Absolute temperature	6.1
T_v	Virtual temperature	6.7
u	Specific internal energy, horizontal velocity in x direction	6.4
u_*	Friction velocity, $(\tau/\rho)^{1/2}$	6.6
v	Horizontal velocity in y direction	6.8
w	Vertical velocity	6.3
x, y, z	Cartesian coordinates	6.1, 6.8
z_0	Roughness length	6.6
α	Specific volume	6.4
α, α'	Constants of proportionality	6.7
β	Constant of proportionality, Bowen ratio	6.7, 6.9
γ	Lapse rate of potential temperature	6.13
Γ	Adiabatic lapse rate	6.4
ε_f	Flux emissivity	6.11
ζ	Dimensionless height, z/L	6.7
η	Normalized specific humidity	6.13
θ	Potential temperature	6.9
θ_v	Virtual potential temperature	6.7
$\Delta\theta$	Jump in potential temperature at inversion	6.13
θ^*	Normalized potential temperature	6.13
θ_*	Scaling temperature	6.7

Subscripts

a	Air, dry air
d	Droplets, dew point, drag
ef	Effective
e	Evaporation
g	Ground, geostrophic
h	Heat transfer
m	Momentum transfer, minimum
n	Neutral
s	Saturated, any property to be transferred by turbulence, surface
v	Virtual

Problems

1. Calculate by two methods the rate of heat conduction into the soil for each 2-hr period shown in Fig. 6.1 for λ equal to 0.31 W m^{-1} °C^{-1} and the following values of $c_g \rho_g$.

Depth (cm)	$c_g \rho_g \times 10^{-6}$ (J m^{-3} °C^{-1})
0.5	1.2
1.5	1.2
2.5	1.2
3.5	1.2
4.5	1.2
5.5	1.2
10	1.6
20	1.5
30	1.2
40	1.1
80	1.2

Which method is likely to be more accurate?

2. Show that expansion of $\overline{c_p \rho w T}$ leads to $c_{pa} \bar{\rho} \, \overline{w'T'} + 0.1 E_e$, using $\rho w = 0$.

3. Starting from Eq. (6.23) show that Eq. (6.28) is the proper form of the transfer for heat.

4. Find the stress exerted by the wind on the ground surface and the roughness for 0630 9 August 1953 at O'Neill, Nebraska when the temperature near the ground was nearly independent of height and the following wind speeds were observed.

Height (m)	Speed (m s^{-1})
6.4	7.91
3.2	7.18
1.6	6.34
0.8	5.35
0.4	4.46

5. By using the wind profile given in Problem 4, calculate the eddy viscosity with Eqs. (6.37) and (6.38) for the height of 3.2 m. Which method is more accurate and why?

6. Assuming complete similarity between the wind and temperature profiles derive the relation for ϕ_m in the form

$$\phi_m^4 - \alpha\zeta\phi_m^3 = 1$$

which has been called the KEYPS† profile.

7. Although Obukhov's intention was to find the height where the buoyancy production of turbulent kinetic energy is equal to the shear production, he found a length (the Obukhov length L) that is greater than this height because he assumed the neutral profile for the shear production. Find the actual height where the two terms are equal using $K_m(\partial\bar{u}/\partial z)^2$ as the rate of shear production.

8. The following wind and temperature profiles were observed at Kansas on 26 July 1968.

Height (m)	Windspeed (m s^{-1})	Temperature (°C)
2.00	4.49	33.64
4.00	5.13	32.94
5.66	5.38	
8.00	5.64	32.52
11.31	5.86	
16.00	6.08	32.14
22.63	6.21	31.94
32.00	6.45	31.83

Derive u_* and E_h from these profiles by using Eqs. (6.49), (6.50), and (6.52). Use $\alpha = 16$, $\rho = 1.05$ kg m^{-3}, $c_p = 10^3$ J kg^{-1} K^{-1}.

9. Using Eqs. (6.64), (6.65) and the boundary conditions following these equations, derive Eq. (6.66).

10. Show that, for the boundary layer to be laminar, i.e., Ri > Ri$_{cr}$ everywhere, $\partial\bar{\theta}_v/\partial z > 21$ K m^{-1} at the surface when $u_g = 1$ m s^{-1}, $f = 10^{-4}$ s^{-1}, and $\nu = 1.8 \times 10^{-5}$ m^2 s^{-1}.

---

† KEYPS stands for the initials of Kazansky, Ellison, Yamamoto, Panofsky, and Sellers who independently derived this equation. It turns out that Obukhov was the first to derive it so O'KEYPS would have been the proper acronym.

11. Calculate from Eq. (6.98) the minimum possible surface temperature for an ice layer of 10 cm thickness having a conductivity of 2.1 W m^{-1} K^{-1}, when the vertical sounding of temperature and specific humidity is as follows:

Pressure (mb)	Specific humidity (g/kg)	Temperature (°C)
360	0.19	−38
400	0.40	−32
460	0.58	−23
500	1.00	−20
540	1.45	−18
600	1.40	−12
690	1.30	− 6
700	0.95	− 5
820	1.20	+ 1
850	1.90	+ 3
900	2.50	+ 4
930	3.70	+ 3
970	3.20	+ 1
990	2.30	− 2

12. (a) Show that Eq. (6.107) can be obtained by ignoring Eq. (6.106), assuming $\Delta\theta$ is const in Eq. (6.105) and $a = 0.2$ in Eq. (6.99).

(b) From the set of Eqs. (6.99), (6.101), (6.105), and (6.106) develop equations showing how $\Delta\theta$ changes with time if a = const, and how a changes with time when $\Delta\theta$ is constant.

13. (a) Find the fluxes of sensible and latent heat at the ocean surface at distances of 0, 50, 100, 200, 300, 500, 1000 km from the coast line for conditions specified below. Neglect the effect of vapor flux in calculation of the heat flux.

$$\theta_1(0) - \theta_0 = -10 \text{ K} \qquad v_1 = 10 \text{ m s}^{-1}$$
$$q_0 = 8 \text{ g (kg)}^{-1} \qquad C_h = C_e = 2 \times 10^{-3} \text{(representative value)}$$
$$q_1(0) = 2 \text{ g (kg)}^{-1} \qquad -\gamma = g/c_p = 10^{-2} \text{ K m}^{-1}$$

(b) Find the height of the boundary layer and the Bowen ratio at the designated points.

(c) Plot the results as a function of distance from the coast.

(d) In the expression for ξ (p. 303) an average of C_h of 2×10^{-3} has been used in parts (a)–(c). However, Fig. 6.6 shows that C_h is a function of stability. Discuss qualitatively how the results under (a)–(c) would change if this effect were taken into account.

Solutions

1. The two methods are based on Eqs. (6.4) and (6.5). From Fig. 6.1 the average temperature gradients just below the surface for the periods 1830–2030 and 2030–2230 are, respectively, −1.00 and −0.85 K cm^{-1}. Therefore, from Eq. (6.4)

$$E_g(0) = 31 \text{ W m}^{-2} : \text{first period}$$

$$E_g(0) = 26 \text{ W m}^{-2} : \text{second period}$$

The second method results from substituting Eq. (6.4) into (6.5) with the result

$$E_g(0) = \frac{1}{\Delta t} \int_{-\infty}^{0} \rho_g c_g \, \Delta T \, dz$$

The contribution of each layer can be tabulated and added for each of the two periods with the results

$$E_g(0) = 50 \text{ W m}^{-2} : \text{first period}$$

$$E_g(0) = 42 \text{ W m}^{-2} : \text{second period}$$

The second method is likely to be more accurate because it is an integral rather than a differential method, and therefore errors of measurement tend to be compensated. From a physical viewpoint the second method is more accurate also because changes in heat conduction at the surface during the period and changes in λ_g with depth very close to the surface may introduce significant errors in the first method, but not in the second.

2.   Equation (2.98) may be substituted in (6.16) and the result expanded as follows

$$E_h = \overline{c_p \rho w T} = \overline{c_{pa}(1 + 0.84\bar{q} + 0.84q')(\bar{\rho} + \rho')(\bar{w} + w')(\overline{T} + T')}$$

Because $0.84\bar{q} \ll 1$ this term may be neglected. Furthermore, all third- and fourth-order terms in the primed quantities are small compared to the second order terms. We have already seen that $\overline{T} \, \overline{\rho' w'} = -\bar{\rho} \, \bar{w} \, \overline{T}$ and that $\bar{w} \, \overline{\rho' T'} \ll \bar{\rho} \, \overline{w' T'}$, therefore, the terms left to be considered are

$$E_h = c_{pa}(\bar{\rho} \, \overline{w' T'} + 0.84\bar{\rho} \, \overline{T} \, \overline{w' q'} + 0.84\bar{w} \, \overline{T} \, \overline{\rho' q'} + 0.84\bar{\rho} \, \bar{w} \, \overline{q' T'})$$

The last two terms cancel each other approximately because $\rho'/\bar{\rho} = -T'/\overline{T}$. Therefore

$$E_h = c_{pa}\bar{\rho} \, \overline{w' T'} + 0.84c_{pa}\bar{\rho} \, \overline{T} \, \overline{w' q'}$$

Eliminating $\overline{w' q'}$ by Eq. (6.19) leads to

$$E_h + c_{pa}\bar{\rho} \, \overline{w' T'} + \frac{0.84c_{pa}\overline{T}}{L} E_e = \simeq c_{pa}\bar{\rho} \, \overline{w' T'} + 0.1E_e$$

3.   From Eq. (6.24)

$$\theta' = -l_\theta \frac{\partial \bar{\theta}}{\partial z}$$

where potential temperature (θ) is conserved during vertical displacement. The relation between potential temperature and temperature is discussed in Sections 2.13 and 2.14. From this we recognize that

$$\theta' = \frac{\theta}{T} T' \quad \text{and} \quad \frac{\partial \bar{\theta}}{\partial z} = \frac{\theta}{T}\left(\frac{\partial \overline{T}}{\partial z} + \Gamma\right)$$

where $\Gamma = g/c_p$, the adiabatic lapse rate. From Eq. (6.26) $K_h = \overline{wl_\theta}$, so that Eq. (6.28) may be written

$$E_h = -c_{pa}\bar{\rho} \, \overline{wl_\theta}\left(\frac{\partial \overline{T}}{\partial z} + \Gamma\right)$$

$$= -c_p\bar{\rho} K_h\left(\frac{\partial \overline{T}}{\partial z} + \Gamma\right)$$

4. By applying Eq. (6.36) to two heights z_1 and z_2, the difference in velocities is given by

$$\bar{u}_2 - \bar{u}_1 = \frac{1}{k}\left(\frac{\tau}{\rho}\right)^{1/2} \ln\frac{z_2}{z_1}$$

Therefore

$$\tau = \frac{k^2 \rho (\bar{u}_2 - \bar{u}_1)^2}{(\ln z_2/z_1)^2}$$

Data from the lowest layer (0.4–0.8 m) should now be used in order to minimize any effects of stability. Substitution leads to

$$\tau = 0.33 \text{ N m}^{-2}$$

From Eq. (6.36)

$$z_0 = z_2 \exp\left(-\frac{k\bar{u}_2}{(\tau/\rho)^{1/2}}\right) = 1.24 \times 10^{-2} \text{ m}$$

5. Using Eq. (6.35), u_* may be evaluated from data for the lowest layer. This yields $u_* = 0.514$ m s^{-1}. Substitution in Eq. (6.37) gives for the eddy viscosity $K_m = 0.66$ m^2 s^{-1}. Using Eq. (6.38), data for the layer from 1.6 m to 6.4 m can be substituted with the result $K_m = 0.58$ m^2 s^{-1}. The result using Eq. (6.37) is likely to be more accurate because u_* is determined from the lowest layer, thus minimizing effects of curvature of the profile, whereas Eq. (6.38) utilizes data up to a height of 6.4 m.

6. Introduction of Eqs. (6.48) and (6.43) into (6.41) with $K_m = K_h$ results in

$$1 = \phi_m^2\left(1 - \alpha\frac{\zeta}{\phi_m}\right)^{1/2}$$

Therefore

$$\phi_m^4 - \alpha\zeta\phi_m^3 = 1$$

7. Upon equating the shear production and the buoyant production of turbulent energy

$$K_m\left(\frac{\partial \bar{u}}{\partial z}\right)^2 = \frac{g}{\theta_v} K_b \frac{\partial \bar{\theta}_v}{\partial z}$$

Substituting from Eqs. (6.46), (6.43), and (6.45)

$$\frac{\phi_m u_*^2}{kz} = \frac{g\theta_*}{\theta_v}$$

where $\theta_* = -E_b/c_p \rho u_*$. This yields

$$z = -\frac{\phi_m \theta_v}{gk}\frac{c_p \rho u_*^3}{E_b} = -\phi_m L$$

Then substituting from Eq. (6.50)

$$-\frac{z}{L} = \frac{1}{(1 - \alpha z/L)^{1/4}}$$

The solution to this transcendental equation can be found easily by successive approximations or by plotting the two sides, to be $z = -0.574L$.

One may also try to solve by expanding the 1/4 root in a binomial series and retaining only two terms. The result in this case is $z = -0.39L$, showing that two terms are insufficient for an accurate result. This result might be used as the first approximation.

8. Upon combining Eqs. (6.43), (6.49), and (6.50)

$$u_* = kz(\partial \bar{u}/\partial z)(1 - \alpha \, \mathrm{Ri})^{1/4} = k\frac{\partial \bar{u}}{\partial \ln z}\left[1 - \alpha \frac{gz}{\bar{\theta}_v} \frac{\partial \bar{\theta}_v/\partial \ln z}{(\partial \bar{u}/\partial \ln z)^2}\right]^{1/4}$$

where z is the logarithmically averaged height $[z = dz/d(\ln z)]$. For the layer from 2 to 4 m, $z = 2.89$ m. Substitution of the numerical values gives

$$u_* = 0.476 \text{ m s}^{-1}$$

From Eqs. (6.45) and (6.52), upon equating T_v and θ_v

$$\phi_h = -\frac{kzc_p\rho u_*}{E_b}\frac{\partial \bar{\theta}_v}{\partial z} = (1 - \alpha \, \mathrm{Ri})^{-1/2}$$

And, upon introducing Eq. (6.43) and rearranging

$$E_b = -\frac{c_p\rho u_*^3}{k}\frac{\partial \bar{\theta}_v/\partial \ln z}{(\partial \bar{u}/\partial \ln z)^2}$$

Since no information is given regarding humidity, we assume $\theta_v = \theta$. Upon substituting numerical values for the layer from 2 to 4 m

$$E_b = 355 \text{ W m}^{-2} = E_h$$

9. Equations (6.64) and (6.65) may be combined to yield

$$\frac{\partial^2 w}{\partial z^2} - \frac{if}{v}(w - u_g) = 0$$

where $w \equiv u + iv$, and v_g has been set equal to zero by orienting the x axis in the direction of the geostrophic wind. If u_g and v are independent of height, the general solution is

$$w = u_g + A \exp[z(if/v)^{1/2}] + B \exp[-z(if/v)^{1/2}]$$

Applying the boundary conditions $u = v = 0$ at $z = 0$ and $u \to u_g$, $v \to 0$ as $z \to \infty$ we find that $A = 0$ and $B = -u_g$. Therefore

$$u + iv = u_g\{1 - \exp[-z(if/v)^{1/2}]\}$$

To separate real and imaginary parts we recall that $(i)^{1/2} = (1 + i)/(2)^{1/2}$ and therefore

$$u + iv = u_g\left\{1 - \exp\left[(1 + i)\left(\frac{f}{2v}\right)^{1/2}z\right]\right\}$$

And upon recalling that $e^{-ix} = \cos x - i \sin x$ and separating real and imaginary parts

$$u = u_g(1 - e^{-az}\cos az)$$

$$v = u_g e^{-az}\sin az$$

where

$$a \equiv (f/2v)^{1/2}$$

10. For flow to be laminar

$$\text{Ri} \equiv \frac{g}{\bar{\theta}_v} \frac{\partial \bar{\theta}_v / \partial z}{(\partial \bar{u} / \partial z)^2} > \frac{1}{4}$$

If the flow is laminar, the Ekman solution can be applied. Therefore, on differentiating Eq. (6.66)

$$\frac{\partial \bar{u}}{\partial z} = u_g a e^{-az} (\cos az + \sin az)$$

Applying this to the Richardson criterion at $z = 0$, where it is largest, gives

$$\frac{\partial \bar{\theta}_v}{\partial z} > \frac{\bar{\theta}_v u_g^2 a^2}{4g} \approx 21 \text{ K m}^{-1}$$

11. Upon solving for $E_a^{\downarrow}$ by the tabular method of Chapter V, we find that $E_a^{\downarrow} = 185 \text{ W m}^{-2}$. Because $E_a^{\downarrow} = \sigma T_{ef}^4$ we find $T_{ef} = 239$ K. This temperature and the given constants lead to a value for κ, as defined in Section 6.11, of 27.13. The temperature at the bottom of the ice layer is 273 K. With the above information Eq. (6.98) yields

$$T_{gm} = 268.6 \text{ K.}$$

12. (a) Assuming $\Delta \theta$ constant in Eq. (6.105), we write (6.101) in the form

$$h = -1.2 \frac{\theta}{T} \frac{E_h(0)}{c_p \rho \gamma} \frac{dt}{dh}$$

Integrating from h_0 to h and $t = 0$ to t gives

$$h = \left[h_0^2 - 2.4 \frac{\theta}{T} \frac{E_h(0)}{c_p \rho \gamma} t \right]^{1/2}$$

(b) When a is constant but $\Delta \theta$ changes with time, Eq. (6.101) becomes

$$h \left[\gamma \frac{dh}{dt} + \frac{d}{dt} \Delta \theta \right] = -(1 + a) \frac{\theta}{T} \frac{E_h(0)}{c_p \rho}$$

Substituting from Eq. (6.106) for dh/dt leads to

$$\frac{d}{dt} (\Delta \theta) = -\frac{\theta}{T} \frac{E_h(0)}{c_p \rho} \left(\frac{1 + a}{h} + \frac{\gamma a}{\Delta \theta} \right)$$

Since $\gamma < 0$ for an inversion, the sign of $d(\Delta \theta)/dt$ may be positive or negative. When $\Delta \theta$ is constant, we may combine Eqs. (6.99) and (6.106) with the result

$$a = \frac{T}{\theta} \frac{c_p \rho}{E_h(0)} \Delta \theta \frac{dh}{dt}$$

The coefficients in front of dh/dt are constant and dh/dt tends to decrease with time, therefore a tends to decrease with time when $\Delta \theta$ is constant.

13. (a)–(c). Numerical values of θ^* and η can be read from the curves of Fig. 6.12. The calculated results are as follows:

x (km)	ζ	h (m)	$E_h(0)$ (W m^{-2})	$E_e(0)$ (W m^{-2})	Bowen ratio
0	0	0	250	375	0.67
50	0.1	400	150	289	0.52
100	0.2	500	125	263	0.48
200	0.4	620	95	226	0.42
300	0.6	720	70	194	0.36
500	1.0	840	40	150	0.27
1000	2.0	960	10	56	0.18

(d) The effect of stability is proportional to $\theta_1 - \theta_0$, which decreases with distance from the coast. Therefore, the height of the boundary layer initially will grow faster and later slower than the calculated height. Similarly, the sensible and latent heat fluxes will be initially larger and later smaller than the results shown in the table above.

General References

Brunt, *Physical and Dynamical Meteorology*, relates the energy transfer processes to the large scale atmospheric structure and to the dominant systems of atmospheric motions.

Budyko, *The Heat Balance of the Earth's Surface*, gives the most complete description of the global distribution of vertical energy transfer. More up-to-date information is given in Gribbin (see below).

Geiger, *The Climate near the Ground*, provides a great quantity of data and discussion of the details of the temperature and humidity distributions.

Gribbin (ed.), *Climatic Change*, contains a collection of recent papers concerning climate including the heat balance of the earth.

Haugen (ed.), *Workshop on Micrometeorology*, contains a useful collection of papers on turbulent transfer in the surface layer and on the structure of the boundary layer.

Hinze, *Turbulence*, 2nd ed., is a complete up-to-date monograph on turbulence with emphasis on engineering applications.

Lettau and Davidson, *Exploring the Atmosphere's First Mile*, provides the most complete set of data relating to ground surface energy transfer. The tables are a gold mine, but the discussion is uneven and uncritical.

Lumley and Panofsky, *The Structure of Atmospheric Turbulence*, provides a good introduction to the basic equations and statistical characteristics of turbulence. It also provides a description of the structure of atmospheric turbulence based on observations before 1964.

Pasquill, *Atmospheric Diffusion*, relates diffusion to the fundamentals of atmospheric turbulence, and extends substantially the theory presented in this chapter.

Priestley, *Turbulent Transfer in the Lower Atmosphere*, distinguishes clearly between mechanical turbulence and free convection, and discusses the evidence concerning the dependence of the eddy transfer coefficients on static stability.

Sellers, *Physical Climatology*, provides a comprehensive account of the energy balances and transfer processes.

Tennekes and Lumley, *A First Cource in Turbulence*, is an up-to-date account of the structure of turbulence and turbulent transport. It is somewhat more advanced than the material presented in this chapter.

CHAPTER VII

Atmospheric Signal Phenomena

"To observations which ourselves we make,
We grow more partial for th' observer's sake." ALEXANDER POPE

In earlier chapters the properties of atmospheric gases and aerosols and the vertical distribution of density, pressure, temperature, and humidity have been set forth. In this chapter the effects of these atmospheric properties on the transmission of electromagnetic and acoustical waves will be considered. These effects we shall call *signal phenomena*. Signal phenomena are sometimes deliberately sought for in the observational study of certain properties of the atmosphere; in many cases they afford unique measurements; in other cases they are nuisances disturbing communication or introducing errors, for example, in the work of the surveyor. They appear, also, as curious, or remarkable, or beautiful sights whose wonder is adequate reason for trying to understand them. Everyone is familiar with the most common and spectacular phenomena such as the rainbow and the halo, but to the keen observer there is a wealth of more subtle but no less fascinating phenomena including supernumerary bows, focusing of sound, and the green flash. The observer may be a naturalist or a bird watcher, who makes careful notes of what is observed; he or she may be a poet inspired by the striking beauty of a visual phenomenon, or a scientist who is challenged to achieve physical understanding.

The scientist's interest is to interpret specific phenomena as applications of as few general principles as possible. Only a few of the many signal phenomena will be treated in this chapter, but these examples should establish a basis in understanding such that readers may provide their individual interpretations of many phenomena which must go unmentioned here.

PART I: GENERAL PROPERTIES OF WAVES

7.1 Nature of Waves

Most solid substances are so constituted that when a particle is displaced from its equilibrium position by an external force, the displaced particle exerts forces on adjacent particles. This property is called *elasticity*; it is defined as the ratio of the applied force to the displacement, or the ratio of stress to strain. Elasticity varies widely from one substance to another. For example, steel has a very high elasticity because when a particle which forms part of a steel object is displaced from its equilibrium position, relatively large forces act on adjacent particles which in turn result in their displacement. The reaction force results in the return of the first particle to its equilibrium position; a periodic oscillation then occurs as a result of the inertia of the moving particle. On the other hand, putty or rubber has a low elasticity since the forces exerted by the displaced particle are relatively small. As a result of elasticity, individual particles perform an oscillatory motion and the oscillation is transmitted from particle to particle. The elastic material is said to perform *wave motion*, and it is easy to recognize that transmission of energy is associated with wave motion.

In solid bodies the forces which are transmitted by wave motion from one particle to an adjacent particle are intermolecular forces. For small displacements, these forces and the reactive forces are large, so that it may be concluded intuitively that the resulting wave motion is rapid. Other forces also result in wave motion. For example, the tension in a stretched string produces a restoring force in a portion of the string displaced from the equilibrium position. The reaction to this force acting on adjacent portions of the string results in their displacement, and the displacement moves as a wave along the string. The force of gravity acting on water displaced from its equilibrium position results in the familiar motion known as surface water waves, which have been discussed in Chapter IV. In this last example the wave motion is not dependent on the elasticity of the medium, but on the geopotential force field acting on water displaced from its equilibrium position. Electromagnetic forces acting on the ionized plasma of the magnetosphere induce *hydromagnetic* waves in an analogous manner. And finally, as has been shown in Chapter IV, rotation of the spherical earth accounts for large scale atmospheric waves which are closely related to large scale pressure systems and their attendant weather.

Waves are commonly classified as *longitudinal* if the oscillation of the particles of the medium is predominantly parallel to the direction of wave propagation and as *transverse* if the oscillation of the particles is predominantly perpendicular to the direction of wave propagation. Sound waves and tidal waves in both ocean and atmosphere are of the longitudinal type, whereas waves on a stretched string and the Rossby–Haurwitz waves associated with rotation on the spherical earth are of the transverse type.

For simple harmonic oscillation along the z axis the displacement may be expressed as a function of x and t by

$$z = A \cos(2\pi/\lambda)(x - ct) \tag{7.1}$$

where A represents the amplitude of the wave, λ the wavelength, and c the wave speed (more rigorously, the *phase* speed) in the x direction. The quantity $(2\pi/\lambda)(x - ct)$ is called the phase of the wave.

More complicated forms of waves may be represented by the sum of a series of sine or cosine terms with different values of λ and c.

7.2 Phase Speed

The speed of waves in many cases depends upon the characteristics of the transmitting medium. This dependence will be illustrated by developing the speed of several simple types of waves.

Wave on a Stretched String

Consider a transverse wave traveling along a stretched string with a velocity c. If the coordinate system is permitted to move at the speed c in the direction of the wave, the principles of ordinary mechanics must hold in the moving system as well as in the stationary system. In the moving system the wave is stationary and the string moves with velocity $-c$. The forces acting on a short portion of the deformed string are the centrifugal force resulting from motion of the string in a curved path and the centripetal force resulting from tension in the curved portion of the string. The two forces must be equal since the wave is stationary. The centrifugal force of a length element Δs may be expressed by $m\,\Delta s\,c^2/r$, where m represents the mass per unit length and r the radius of curvature of the segment Δs. The centripetal force F may be developed from Fig. 7.1. Upon isolating the segment Δs, it may be recognized from the force diagram that in the limit as $\Delta s \rightarrow 0$

$$F/T = \Delta s/r$$

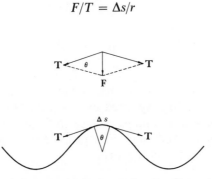

FIG. 7.1. Waves on a stretched string under tension T and corresponding force diagram for segment Δs.

where T is the tension in the string. Upon equating centrifugal and centripetal forces

$$T(\Delta s/r) = m\,\Delta s\,(c^2/r)$$

and

$$c = (T/m)^{1/2}$$

The speed of the wave depends only on the tension and the mass per unit length of the string. It should be noted that it has been assumed implicitly that the amplitude of the waves is sufficiently small so that the tension may be considered constant.

The Acoustic Wave

Consider a longitudinal wave traveling with velocity c through a compressible fluid enclosed in a cylinder. If the coordinate system is permitted to move at the velocity c, the wave is motionless with respect to the moving coordinate system. The wave system is composed of alternate stationary regions of compression and rarefaction, and the fluid moves through these regions with a velocity $-c$.

In the half-wavelength between adjacent planes of maximum and minimum pressure the fluid particles must undergo a change in speed dc; they slow down in approaching a plane of compression and speed up in approaching a plane of rarefaction. The difference in pressure between these two planes may be computed from Newton's second law. Thus for unit cross-sectional area

$$dp = -\rho\,ds\frac{dc}{dt} = -\rho c\,dc$$

and therefore

$$\frac{1}{c}\frac{dp}{dc} = -\rho \tag{7.2}$$

In order to eliminate dc, the volume of fluid flowing through unit cross section in time dt may be expressed by $V = c\,dt$. Now if the change in c is given by dc, the corresponding change in V may be expressed by the relation $dV/V = dc/c$. And since V is proportional to specific volume α, Eq. (7.2) may be rewritten in the form

$$\frac{dp}{d\alpha} = -\frac{c^2}{\alpha^2} \tag{7.3}$$

From Chapter II one should expect that the ratio of pressure change to volume change depends upon the flux of heat between adjacent regions of compression and rarefaction. Consider the following two extreme cases: (1) heat flux is sufficiently rapid so that no temperature gradient exists within the fluid; (2) heat flux is negligible. Express the ratio of pressure change to volume change in the first case by writing the equation of state (2.10) for constant temperature. Thus

$$p \, d\alpha + \alpha \, dp = 0$$

Substitution of this equation in (7.3) gives

$$c^2 = p\alpha$$

and by again substituting the equation of state

$$c = (R_m T)^{1/2}$$

This is the velocity of sound for isothermal expansion, often called the "Newtonian" velocity of sound.

In the second case adiabatic conditions prevail. From Eq. (2.46c) the ratio of pressure change to volume change may be written

$$\frac{dp}{d\alpha} = -\frac{c_p}{c_v} \frac{p}{\alpha}$$

Substitution of this equation in (7.3) gives

$$c = [(c_p/c_v)R_m T]^{1/2} \tag{7.4}$$

This is the velocity of sound for adiabatic conditions. The fact that observations of the velocity of sound agree almost exactly with Eq. (7.4) indicates that the compressions and expansions which occur in acoustic waves are very nearly adiabatic. For very long acoustic waves gravity plays a significant role, and as is shown in Section 4.14, the speed is greater than that given by Eq. (7.4).

The Surface Water Wave

Imagine the cylinder of the preceding discussion to be lying in a horizontal position and to be partly filled with water. As surface waves move through the cylinder, pressure changes occur within the water. At the bottom of the cylinder the particles must move longitudinally, back and forth, as the surface waves pass. These particles therefore behave just as corresponding particles in the case of sound waves. In the surface wave case, however, the fluid is incompressible, and dV/V is expressible by dh/h or dc/c. Also, dp may be expressed by $\rho g\, dh$, where h represents water depth. Equation (7.2) now yields for the *shallow water wave speed*

$$c = (gh)^{1/2}$$

In deep water waves, the water particles move very nearly in circles; consequently, the pressure at any reference depth depends both on the depth and on the vertical acceleration of the water particles above the reference. Vertical acceleration depends on wavelength, so the wave speed in deep water turns out to depend on wavelength as shown in Eq. (4.89). Similar dependence of wave speed on wavelength is found in electromagnetic waves in air. Media in which wave speed depends on wavelength are called *dispersive media*.

7.3   Electromagnetic Waves

Electromagnetic waves exhibit many of the features which are characteristic of the waves already mentioned, but they are different in that they may move through vacuum. No motion and no medium are needed for their transmission although the medium does influence the phase speed. The theory of electromagnetic waves rests on Maxwell's equations which in turn can be developed from the two fundamental experimental laws of Ampère and Faraday.

Ampère's law may be written in the two forms

$$\oint \frac{\mathbf{B}}{\mu} \cdot d\mathbf{l} = \int\int \mathbf{J} \cdot d\mathbf{A} = I \tag{7.5}$$

where μ represents the *permeability* of the medium, $\mathbf{B}$ is the *magnetic induction* (or *magnetic flux density*), $\mathbf{J}$ is the current density (rate of flow of charge per unit area), $d\mathbf{l}$ is the differential of length along a closed contour, $d\mathbf{A}$ is the differential of surface area enclosed by the contour, and I is the current flow through the area enclosed by the contour. The magnetic induction may be defined by the equation

$$\mathbf{F_M} = q\mathbf{v}_q \times \mathbf{B} \tag{7.6}$$

Equation (7.6) is based on experiment and states that a magnetic force $\mathbf{F_M}$ acts on a charge q moving at velocity $\mathbf{v}_q$ in the field $\mathbf{B}$.

Faraday's law may be written in the form

$$\oint \mathbf{E} \cdot d\mathbf{l} = -\int\int \frac{\partial \mathbf{B}}{\partial t} \cdot d\mathbf{A} \tag{7.7}$$

where $\mathbf{E}$ represents the *induced electric field*. The electric field is defined by Eq. (3.23), which states that a charge experiences a force in the direction of the electric field. The induced electric field is distinct from the electrostatic field discussed in Chapter III, and in general the electric field is the sum of the induced and electrostatic fields.

Maxwell considered a circuit containing a capacitor, and he defined the scalar *displacement* by the identity

$$D \equiv Q/A \tag{7.8}$$

where Q represents the charge on one plate of the capacitor and A the area of the plate. More generally, Eq. (7.8) applies to the charge on any surface A, say the surface of a sphere. For the sphere we may easily calculate from Eqs. (3.24) and (7.8) that the vector displacement is related to the electric field by

$$\mathbf{D} = \varepsilon\mathbf{E} \tag{7.9}$$

and this may be considered the definition of the vector displacement. Further, the *displacement current* per unit area may be defined by

$$\mathbf{J_D} \equiv \frac{\partial \mathbf{D}}{\partial t} \tag{7.10}$$

and if ε is independent of time

$$\mathbf{J_D} = \varepsilon\frac{\partial \mathbf{E}}{\partial t}$$

Now recall that within a dielectric positive and negative charges are impelled to move in opposite directions, and in consequence, electrical dipoles are created within the dielectric. The product of charge and separation of charges, the dipole moment per unit volume, is called the polarization $\mathbf{P}$. Evidently $\mathbf{P}$ and $\mathbf{D}$ have identical units, and the displacement within a dielectric may be represented by the sum of the displacement in vacuum and the polarization. Thus

$$\mathbf{D} = \varepsilon\mathbf{E} = \varepsilon_0\mathbf{E} + \mathbf{P}$$

Upon scalar multiplication of this vector equation with $\mathbf{E}$, the permittivity may be expressed by

$$\varepsilon = \varepsilon_0\left(1 + \frac{\mathbf{P}\cdot\mathbf{E}}{\varepsilon_0 E^2}\right) = \varepsilon_0(1 + \chi) \tag{7.11}$$

where χ represents the property of the dielectric known as the *susceptibility*. Susceptibility may be positive, negative, or zero and, indeed, may be complex. The latter case arises later in this chapter in discussion of the index of refraction.

It is convenient to define the *magnetic field intensity* by the identity

$$\mathbf{H} \equiv \mathbf{B}/\mu \tag{7.12}$$

The relation between $\mathbf{H}$ and $\mathbf{B}$ is analogous to the relation between $\mathbf{D}$ and $\mathbf{E}$. $\mathbf{H}$ and $\mathbf{D}$ are dependent only on the source of the respective fields, whereas $\mathbf{B}$ and $\mathbf{E}$ depend also on the local properties of the medium.

Now Ampère's law in the form of Eq. (7.5) should hold for a dielectric if the displacement current is added to the current I. Thus with the aid of Eqs. (7.10) and (7.12), Eq. (7.5) becomes

$$\oint \mathbf{H} \cdot d\mathbf{l} = \int\int (\mathbf{J_D} + \mathbf{J}) \cdot d\mathbf{A}$$

This equation applies to closed curves of finite and arbitrary form, whereas we should like to have relations between the field vectors at a point. This can be achieved by applying Stokes' theorem with the result

$$\boxed{\nabla \times \mathbf{H} = \frac{\partial \mathbf{D}}{\partial t} + \mathbf{J}} \tag{7.13}$$

The following additional equation results from applying Stokes' theorem to Eq. (7.7)

$$\boxed{\nabla \times \mathbf{E} = -\frac{\partial \mathbf{B}}{\partial t}} \tag{7.14}$$

Equations (7.13) and (7.14) are two of the four fundamental equations of electromagnetic theory known as Maxwell's equations. For the important case of no current and for μ and ε independent of time they may be written

$$\nabla \times \mathbf{H} = \varepsilon \frac{\partial \mathbf{E}}{\partial t} \tag{7.15}$$

$$\nabla \times \mathbf{E} = -\mu \frac{\partial \mathbf{H}}{\partial t} \tag{7.16}$$

These equations reveal the symmetry in the roles played by $\mathbf{E}$ and $\mathbf{H}$ in electromagnetic waves.

To develop an equation for the speed of electromagnetic waves we consider variation in one space dimension, as is illustrated in Fig. 7.2. The vector

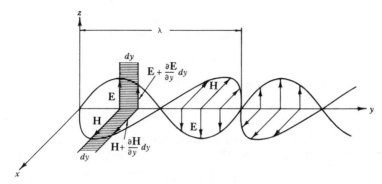

FIG. 7.2. Electric field vector **E** and magnetic intensity vector **H** associated with an electromagnetic wave of length λ traveling in the y direction.

E is assumed to oscillate in the z direction and the vector **H** to oscillate in the x direction. For this case, Eqs. (7.15) and (7.16) reduce to

$$\frac{\partial H}{\partial y} = -\varepsilon \frac{\partial E}{\partial t}$$

$$\frac{\partial E}{\partial y} = -\mu \frac{\partial H}{\partial t}$$

These equations may be combined by differentiating the first with respect to y and the second with respect to t with the result

$$\frac{\partial^2 H}{\partial y^2} = \mu\varepsilon \frac{\partial^2 H}{\partial t^2}$$

Cross differentiation in the opposite sense results in eliminating H and yields

$$\frac{\partial^2 E}{\partial y^2} = \mu\varepsilon \frac{\partial^2 E}{\partial t^2}$$

These equations are classic forms of the linear wave equation in one dimension. The reader is asked to show in Problem 1 that waves on the stretched string are governed by a similar differential equation. Solutions may be written in the form

$$H = H_m \sin(2\pi/\lambda)(y - ct)$$

$$E = E_m \sin(2\pi/\lambda)(y - ct)$$

Upon substituting the solutions into the wave equations, each equation yields

$$c^2 = 1/\mu\varepsilon$$

showing that the magnetic and electric fields maintain their phase relation; together they constitute an electromagnetic wave. The speed is given by

$$c = (\mu\varepsilon)^{-1/2} \qquad (7.17)$$

For vacuum the speed of electromagnetic waves is

$$c_0 = (\mu_0\varepsilon_0)^{-1/2} = 2.99793 \times 10^8 \text{ m s}^{-1}$$

For air, μ and ε differ only slightly from their respective values in vacuum, but the small differences play the central role in some of the applications which follow.

7.4 Dispersion and Group Velocity

If we look carefully at a series of water waves, we may observe that an "individual" wave experiences changes in shape and amplitude as it is overtaken by faster (longer) waves or as it overtakes slower (shorter) waves. In fact, a group of waves may move into undisturbed water as a coherent group at a speed less than the speed or phase velocity of the individual waves. Waves may be readily observed overtaking the group and increasing in amplitude, then moving ahead of the group and decreasing in amplitude. It is important to distinguish between the *phase velocity* and the *group velocity*.

Imagine two waves of slightly different wavelength moving through a medium in which phase speed increases with wavelength. The two waves alternately reinforce and interfere with each other with the result shown in Fig. 7.3. The displacement of the surface from its equilibrium position may be expressed by

$$z = A \sin(kx - \omega t) + A \sin(k'x - \omega' t)$$

where k represents wave number and ω angular frequency. This may be transformed to

$$z = 2A \cos\left(\frac{k - k'}{2}x - \frac{\omega - \omega'}{2}t\right) \sin\left(\frac{k + k'}{2}x - \frac{\omega + \omega'}{2}t\right)$$

Because, as already specified, the waves are of nearly equal length, k and ω are nearly equal to k' and ω', respectively. Therefore

$$z = 2A \cos\left(\frac{x}{2}\delta k - \frac{t}{2}\delta\omega\right) \sin(kx - \omega t) \qquad (7.18)$$

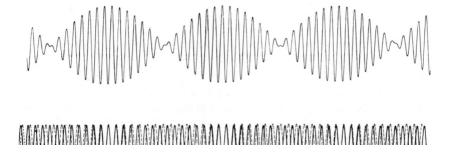

FIG. 7.3. Superposition of two sine waves of equal amplitudes and of slightly different wavelength.

Equation (7.18) describes an envelope with amplitude $2A$ within which there are frequent oscillations as illustrated in Fig. 7.3. To find the separation between successive maxima (separate groups), hold t constant and let $(x/2)\delta k$ increase from 0 to π. The corresponding increment in x is given by $\Delta x = 2\pi/\delta k$.

Similarly, if x is held constant and $(t/2)\delta\omega$ increases from 0 to π, the corresponding increment in t is given by $\Delta t = 2\pi/\delta\omega$. These increments represent, respectively, the wavelength and the period of the group; the group velocity is therefore defined as

$$c_g \equiv \frac{\Delta x}{\Delta t} \equiv \frac{\delta\omega}{\delta k} \approx \frac{d\omega}{dk}$$

This equation may be expressed in terms of wavelength λ and phase speed c by substituting $c = \omega/k$. Hence the group velocity becomes

$$c_g = c - \lambda\frac{dc}{d\lambda} \qquad (7.19)$$

PART II: SCATTERING OF RADIATION

7.5 The Physical Concept

Waves interact with matter on which they are incident in complex and wonderful ways. In trying to understand these processes it will be helpful to imagine a simple and familiar example. Consider a surface water wave incident on a floating block of wood. The block is set into vertical oscillation by the passage of the wave, and the vertical oscillation produces a circular wave which travels outward from the block in concentric circles. We say that some of the incident energy has been *scattered* in all directions by the block. Now if the block is very small or the wave very long, the block rises and falls with the surface of the water, and little energy is scattered. On the other hand, if the block is very large or the waves very short, the block is nearly motionless, and we say that the wave is *reflected*. However, reflection is clearly a special case of the general phenomenon of scattering. For a particular block, the energy scattered may be studied as a function of incident wavelength. One might expect to find that the proportion of energy scattered in a particular direction varies with wavelength, and that the total energy scattered in all directions reaches a maximum for a particular "resonant" wavelength.

Electromagnetic waves are scattered in an analogous manner. Passage of the wave induces oscillation of the atomic electrons with the result that waves are emitted from the atom. The proportion scattered in a particular direction depends on the incident wavelength and on the size and the *permittivity* of the scattering particle.

The scattered energy may be calculated by imagining that at a particular instant a plane polarized electromagnetic wave exerts forces on the electrical charges of an atom such that the positive charge is displaced in one direction (the direction of the **E** vector) and the negative charge is displaced in the opposite direction. The pair of opposite charges is called an induced electric *dipole*, and the product of the positive charge and the maximum separation of the charges is defined as the *dipole moment*. As the wave passes, the phase reverses and the charges are displaced toward each other; hence the electromagnetic wave produces oscillation of charges at the frequency of the wave. This form of polarization is called *dielectric* polarization. There exists a second form, *parelectric* polarization, which arises from the fact that individual molecules may possess an electric moment even when an electric field is absent. Although the moments are randomly oriented in the absence of an electric field, the moments may be aligned by an impressed field. In general, molecular orientation responds only rather slowly to an electric field, so that parelectric polarization is most important for low-frequency waves. Polarization plays a crucial role in scattering of radar waves by raindrops and in many other applications.

Development of the theory of scattering by electromagnetic waves requires consideration of the complete range of wavelengths and a range of size of scattering particles from electrons to raindrops or hailstones. A theory of this generality has been developed by Mie,[†] who integrated Max-

† G. Mie, *Ann. Physik* **25,** 377, 1908.

well's equations and expressed the energy scattered in a particular direction by a sphere by an infinite series of terms representing products of associated Legendre polynomials and spherical Bessel functions. The results reveal that the phenomena known as scattering, reflection, absorption, diffraction, and refraction are all contained in the same solution, the distinctions being determined by the nature of the scattering elements and the various scale parameters. For spherical particles small in diameter compared to the wavelength of the incident radiation, Mie scattering theory reduces to the simpler Rayleigh theory. In the quest for clarity and simplicity, we shall use Rayleigh theory; results of Mie theory will be introduced qualitatively, where necessary.

7.6 Complex Index of Refraction

In air or water the permeability μ is nearly equal to μ_0 and the directions of the electric field and the polarization are the same. Under these conditions substitution of Eq. (7.11) into (7.17) gives for the *index of refraction*

$$n' \equiv c_0/c = [1 + (\mathbf{P} \cdot \mathbf{E}/\varepsilon_0 E^2)]^{1/2} = (1 + \chi)^{1/2} \qquad (7.20)$$

The susceptibility χ depends strongly on frequency, so that the index of refraction varies with frequency or wavelength. For this reason electrostatic measurements of susceptibility may not be used to calculate index of refraction for light or other high-frequency waves.

Because, as mentioned in Section 7.4, the susceptibility may be complex, the index of refraction also may be complex. In this case

$$n' \equiv n + i\kappa \qquad (7.21)$$

where n represents the real index of refraction and κ the *absorption index*. The relation of κ to absorption of radiation can be recognized readily if the solution to the wave equation is written in the form

$$\mathbf{E} = \mathbf{E}_m \exp[i(2\pi/\lambda)(y - ct)]$$

For $t = 0$ this becomes, after combining with Eqs. (7.20) and (7.21)

$$\mathbf{E} = \mathbf{E}_m \exp(i2\pi\nu n y/c_0) \exp(-2\pi\nu\kappa y/c_0) \qquad (7.22)$$

where $\nu = c/\lambda$. The first exponent describes an oscillation, whereas the second exponent describes an exponential decay of the amplitude with y, which must be related to Beer's law [Eq. (5.11)]. Later in this section the energy radiated will be shown to be proportional to the square of the amplitude of the electromagnetic wave; therefore, the monochromatic absorption coefficient is expressed by

$$k_v = 4\pi v\kappa/\rho c_0 \tag{7.23}$$

In order to investigate the dependence of index of refraction on frequency, consider the polarization of a dielectric subjected to an oscillating electric field. For a single dipole the dipole moment may be expressed as proportional to electric field, and therefore the polarization may be expressed by

$$\mathbf{P} = N\alpha\mathbf{E} \tag{7.24}$$

where N represents the number of dipoles (atoms or molecules) per unit volume and α is called the polarizability.

The polarization of a dipole has been defined in Section 7.3 as the product of the charge e of the dipole and the displacement $\mathbf{s}$; therefore, the polarization of the dielectric is also expressed by

$$\mathbf{P} = N e\mathbf{s}$$

and upon combining this with Eq. (7.24)

$$\alpha = es/E \tag{7.25}$$

The polarizability may be calculated by considering a model of a dipole consisting of an electron of charge e and mass m which is displaced a distance z from its equilibrium position by a periodic force eE and is subject to a restoring force mkz and a damping force $m\gamma(dz/dt)$. For this system Newton's second law may be written

$$\frac{d^2z}{dt^2} + \gamma\frac{dz}{dt} + kz = \frac{eE}{m} \tag{7.26}$$

The particular solution of this nonhomogeneous differential equation may be written $z = z_0 \exp(-2\pi i v t)$, and this requires that

$$(-4\pi^2 v^2 + k - 2\pi i v\gamma)z = eE/m \tag{7.27}$$

This equation shows that the displacement z is proportional to the electric field, and therefore the work done on the dipole is proportional to E^2. Therefore, the energy passing unit area per unit time per unit solid angle, the radiance, is also proportional to E_m^2.

It is convenient to define the natural angular frequency v_0 by neglecting γ and setting the bracketed terms in Eq. (7.27) equal to zero. This results in $v_0 \equiv (1/2\pi)k^{1/2}$. Then Eq. (7.25) can be written in the form

$$\alpha = \frac{e^2}{m}\frac{1}{4\pi^2(v_0^2 - v^2) - 2\pi i\gamma v}$$

or upon rationalizing

$$\alpha = \frac{e^2}{m}\left[\frac{(v_0^2 - v^2)}{4\pi^2(v_0^2 - v^2)^2 + v^2\gamma^2} + \frac{i}{2\pi}\frac{v\gamma}{4\pi^2(v_0^2 - v^2)^2 + v^2\gamma^2}\right]$$

Substitution of this result into Eq. (7.24) and then into Eq. (7.20) yields

$$n'^2 = 1 + \frac{Ne^2}{\varepsilon_0 m}\left[\frac{(v_0^2 - v^2)}{4\pi^2(v_0^2 - v^2)^2 + v^2\gamma^2} + \frac{i}{2\pi}\frac{v\gamma}{4\pi^2(v_0^2 - v^2)^2 + v^2\gamma^2}\right]$$

and because, according to Eq. (7.21)

$$n'^2 = n^2 - \kappa^2 + 2in\kappa$$

it follows that

$$n^2 - \kappa^2 = 1 + \frac{Ne^2}{\varepsilon_0 m}\frac{(v_0^2 - v^2)}{4\pi^2(v_0^2 - v^2)^2 + v^2\gamma^2} \tag{7.28}$$

and

$$2n\kappa = \frac{Ne^2}{2\pi\varepsilon_0 m}\frac{v\gamma}{4\pi^2(v_0^2 - v^2)^2 + v^2\gamma^2} \tag{7.29}$$

For the clear atmosphere, $n \approx 1$ and $\kappa^2 \ll n^2$. For frequencies in the neighborhood of the resonant frequency, $v^2 - v_0^2 \simeq 2v_0\,\Delta v$, where Δv represents $v - v_0$. Under these conditions the index of refraction and the absorption coefficient may be expressed from Eqs. (7.28), (7.29), and (7.23) in the forms

$$n \approx 1 - \frac{Ne^2}{\varepsilon_0 m v_0}\frac{\Delta v}{16\pi^2(\Delta v)^2 + \gamma^2} \tag{7.30}$$

and

$$k_v = \frac{4\pi\kappa v}{\rho c_0} \approx \frac{Ne^2}{\rho c_0 m\varepsilon_0}\frac{\gamma}{16\pi^2(\Delta v)^2 + \gamma^2} \tag{7.31}$$

Equations (7.30) and (7.31) describe, respectively, the frequency dependence of the index of refraction and of the absorption coefficient in the neighborhood of the resonant frequency of an oscillating dipole. The functions are illustrated in Fig. 7.4. The frequency dependence of the index of refraction determines the mode of *dispersion* and, according to Eq. (7.19), the group velocity. For frequencies such that $v_0 - (1/4\pi)\gamma > v$, the index of refraction is greater than unity and the wave speed is less than c_0. As frequency increases, the index of refraction increases. This mode is referred to as *normal dispersion,* and under these conditions, the separate waves of light

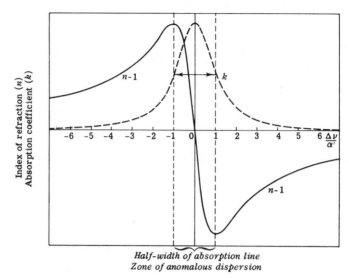

FIG. 7.4. The real and imaginary parts of the complex index of refraction computed from Eqs. (7.30) and (7.31).

are *dispersed* by a prism into the component colors with short wavelengths (blue) refracted most and long wavelengths (red) refracted least. For frequencies in the range defined by

$$v_0 + (1/4\pi)\gamma > v > v_0 - (1/4\pi)\gamma$$

the index of refraction decreases with increase in frequency; this is referred to as *anomalous dispersion*. Within this region the absorption coefficient k_v reaches a maximum as is shown in Fig. 7.4. For frequencies in the range $v > v_0 + (1/4\pi)\gamma$ normal dispersion again occurs, but the index of refraction is smaller than unity, and the phase speed is greater than c_0. A particular case of importance in this range occurs when electromagnetic waves are incident on free electrons or ions. In this case the individual charges are not bound to a nucleus and so do not experience the restoring force expressed in Eq. (7.26) by kz. Consequently v_0 vanishes, and when at the same time the damping constant is negligible, Eq. (7.28) may be written in the form

$$n = \left(1 - \frac{Ne^2}{4\pi^2 m \varepsilon_0 v^2}\right)^{1/2} \tag{7.32}$$

This equation is useful in explaining refraction of radio waves by the ionosphere.

Equation (7.31) is identical to Eq. (5.21) if

$$k_1 = \frac{Ne^2}{4\rho c_0 m \varepsilon_0} \quad \text{and} \quad \alpha' = \frac{1}{4\pi}\gamma$$

This may be interpreted to mean that damping generates an absorption line centered at the resonant frequency; the characteristic line shape prescribed by Eq. (7.31) is called the Lorentz line shape.

Damping of the oscillating dipole may occur by the following mechanisms. Electromagnetic waves may be radiated by the dipole, and the loss of energy associated with the emission of radiation results in reduced amplitude of the dipole. A second mechanism of damping results from collisions of molecules. If it is assumed that an oscillating dipole loses its energy during a collision, then the decay time is equal to the average time between two collisions. The damping constant is inversely proportional to the decay time as may be recognized from the solution for the damped linear oscillator [Eq. (7.26) with $E = 0$]. For this case Eq. (7.27) becomes

$$-4\pi^2(v^2 - v_0^2) - 2\pi i v \gamma = 0$$

or because $v \approx v_0$

$$2\pi v = 2\pi v_0 - i\gamma/2$$

and because $z = z_0 \exp(-2\pi i v t)$, the solution of the differential equation takes the form

$$z = z_0 \exp(-\tfrac{1}{2}\gamma t)\exp(-2\pi i v_0 t)$$

This shows that the decay time is equal to $2/\gamma$. The time between collisions is inversely proportional to the density and the velocity of the molecules. Therefore, for constant temperature, the damping constant is proportional to the pressure, and the width of absorption lines increases with pressure even though the total absorption is independent of pressure.

7.7   Radiation Emitted by an Oscillating Dipole

Oscillation of an electric dipole in a fixed direction produces a plane polarized electromagnetic wave whose amplitude at a particular point depends on its position with respect to the dipole, on the frequency of the incident electromagnetic wave, on the size, and on the electrical properties of the scattering particles. These parameters will be shown in the following discussion to be related to the irradiance. Conservation of energy requires that the irradiance be inversely proportional to r^2. Because the electric vector oscillates in a fixed plane, the amplitude of the radiated electric vector is a maximum at right angles to the dipole axis ($\theta = \pi/2$) and falls to zero along the axis ($\theta = 0, \pi$) as shown in Fig. 7.5. Therefore, the amplitude is proportional to $\sin \theta$, and the irradiance is proportional to $\sin^2 \theta$. Because electromagnetic waves arise from the acceleration of charges, the amplitude is proportional to the acceleration of the charge and to the magnitude of the charge. The displacement of charge in the dipole is proportional to $\sin(2\pi\nu)(t - r/c)$, so that the acceleration is proportional to $4\pi^2\nu^2$ or to $1/\lambda^2$. The amplitude also must be proportional to the number of dipoles per unit volume or (assuming constant density) to the volume of the scattering element. Therefore, upon combining all these considerations, the irradiance may be represented as proportional to

$$a^6 \sin^2 \theta / \lambda^4 r^2$$

where a represents the radius of the scattering element.

In addition to the factors mentioned above, the amplitude must be proportional to the polarizability of the dielectric of which the scattering particles are composed. For a material in which the separation of dipoles is large, interaction between dipoles is negligible, and the polarizability is expressed by Eq. (7.24). But for liquids or solids, interactions between di-

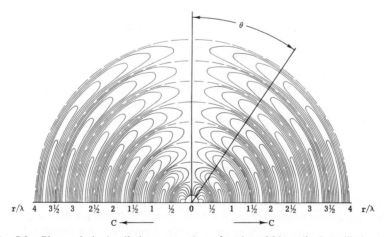

FIG. 7.5.   Plane polarized radiation scattered as a function of θ by a dipole oscillating along the vertical axis. The lines represent the distribution of absolute magnitude of the electric field at an instant (after G. Joos, *Theoretical Physics*, Hafner, New York, 1934, p. 327).

poles make the electric field dependent on the dipoles, and a further step is necessary. In this case a dipole embedded in a dielectric and surrounded by identical dipoles experiences electric field strength which may be represented by the sum of the external field $\mathbf{E}$ and the field $\mathbf{E}'$ arising from the surrounding electric charges. Its polarization $\mathbf{P}$ therefore is given by the right side of Eq. (7.24) plus an additional term which must be proportional to $\mathbf{P}$.

Imagine a minute spherical cavity in the material which does not affect polarization. The interior surface of the cavity has a certain charge density, which can be expressed by $|\mathbf{P}| \cos \theta$, where θ is the angle between the direction of $\mathbf{P}$ and the normal to the spherical surface. From Coulomb's law the increment of field strength contributed by a differential element of surface area $d\sigma$ is

$$|\mathbf{P}| \cos \theta \, d\sigma / 4\pi\varepsilon_0 a^2$$

where a represents the radius of the cavity and ε_0 is used because the cavity is considered to be empty. From symmetry, it may be recognized that the components of these incremental field strengths normal to $\mathbf{P}$ add to zero. Then upon recognizing that $d\sigma = a^2 \sin \theta \, d\theta \, d\phi$, the component of the increment of field in the direction of $\mathbf{P}$ is expressed by

$$d\mathbf{E}' = (\mathbf{P}/4\pi\varepsilon_0) \cos^2 \theta \sin \theta \, d\theta \, d\phi$$

Integrating over ϕ from 0 to 2π and over θ from 0 to π gives

$$\mathbf{E}' = \mathbf{P}/3\varepsilon_0$$

Instead of Eq. (7.24), the polarization now may be expressed by

$$\mathbf{P} = N\alpha[\mathbf{E} + (\mathbf{P}/3\varepsilon_0)]$$

Substitution of this equation into (7.20) leads to

$$\frac{n'^2 - 1}{n'^2 + 2} = \frac{N\alpha}{3\varepsilon_0}$$

This shows that the polarizability is proportional to $(n'^2 - 1)/(n'^2 + 2)$. Therefore, upon combining these several factors, the irradiance is proportional to

$$\frac{a^6}{r^2 \lambda^4} \left(\frac{n'^2 - 1}{n'^2 + 2} \right)^2 \sin^2 \theta \tag{7.33}$$

This is the result predicted by Rayleigh scattering theory.

7.8   Size Dependence

For particles which are small in diameter compared to the wavelength of the incident radiation, the results of Section 7.7 can be interpreted intuitively. Consider a wave which sets the electrical charges of a dipole into oscillation. The oscillating charges themselves produce an electromagnetic field which propagates outward away from the dipole with the speed of light. At a distance r from the dipole the phase lags behind the dipole phase by $2\pi r/\lambda$. Each time the oscillating charges reverse direction the electromagnetic field in the neighborhood of the dipole acts on them in such a way as to oppose the acceleration of the charges, and the energy of the field tends to return to the dipole. However, return of the energy also occurs at the speed of light, so that some energy will not have time to return to the dipole before the phase changes once again. If the frequency is high, much of the energy of the electromagnetic wave may in this way fail to be conserved by the dipole; it is radiated away.

A hydraulic analogy may help to clarify the phenomenon of radiation. Imagine a vertical cylinder which may be alternately pushed downward into and pulled upward out of a body of water. If the frequency is low, the work done in depressing the cylinder is regained in raising it. But if frequency is high, waves radiate from the cylinder, and mechanical energy must be continuously supplied at the cylinder. Proportionality (7.33) shows that the distribution of radiated energy is proportional to $\sin^2 \theta$, so that energy is scattered in the forward and backward directions with equal irradiance and drops to zero along the dipole axis as shown in Fig. 7.5. For a particular size of scattering particle, proportionality (7.33) shows that irradiance is inversely proportional to the fourth power of the wavelength of the incident radiation. A similar qualitative result must be expected in the case of randomly oriented dipoles; maximum radiation is emitted in the forward and backward directions, but radiance in the normal directions falls by only one-half due to contributions from obliquely oriented dipoles. The dependence of scattered radiation on wavelength may be used to explain the blueness of the sky and the redness of the sunset and to explain scattering of radar waves by water drops and ice crystals. These applications will be discussed in Parts III and IV of this chapter.

Scattering by particles whose diameter is comparable to the wavelength of incident radiation cannot be represented adequately by Rayleigh theory. The more complete Mie theory shows that the ratio of scattered energy to the incident energy intercepted by the geometrical cross section of the particles is given by

$$2\left(\frac{\lambda}{2\pi a}\right)^2 \sum_{j=1}^{\infty} (2j + 1)\{|a_j|^2 + |b_j|^2\}$$

where a_j and b_j are functions of spherical Bessel functions and Hankel functions of the second kind with complex arguments. The first three terms of the series representing the energy received at distance r and angle $\theta = \pi/2$ (forward or backward scatter) can be expressed as proportional to

$$\frac{a^6}{r^2\lambda^4}\left(\frac{n'^2 - 1}{n'^2 + 2}\right)^2\left\{1 + \frac{6}{5}\left(\frac{n'^2 - 2}{n'^2 + 2}\right)\left(\frac{2\pi a}{\lambda}\right)^2 + \frac{9}{25}\left(\frac{n'^2 - 2}{n'^2 + 2}\right)^2\left(\frac{2\pi a}{\lambda}\right)^4 + \cdots\right\}$$

(7.34)

The first term of this series represents the Rayleigh theory. The second term becomes significant for a/λ greater than about 0.1, so that Mie theory is necessary for particles of this size or larger. Mie scattering calculations for water drops and ice particles have been made for ratio of particle size to wavelength (a/λ) up to about 5. For a/λ greater than about 3, scattered energy is not strongly dependent on λ and the methods of geometrical optics are used. These methods are discussed in the following Sections 7.9 and 7.10. The ranges identified here are, of course, not precise and depend on n', as well as on a/λ.

7.9 Diffraction

For particles much larger in size than the incident wavelength the general but complex Mie treatment of scattering reduces to the simpler subjects of diffraction and refraction as formulated under geometrical optics. Two intuitively attractive "principles" may be used to explain diffraction. In each case the empirical statement long antedates the scattering theory which now provides its theoretical basis. *Huygen's principle* states that the shape of a wave may be predicted by considering that every point of equal phase simultaneously acts as a source of a spherical wave which diverges from each of these points. For a wave surface of infinite extent, all phases propagated obliquely to the wave front are cancelled by interference from phases from other sources. Only the energy propagated in directions normal to the wave surface remains; the wave propagated in the backward direction must be eliminated by a separate specification. Scattering theory shows that for particles comparable in size to the wavelength the energy contributed by each scattering element has a pronounced maximum in the forward direction.

Now apply Huygen's principle to a plane wave incident on an obstructing edge as shown in Fig. 7.6. Construct lines BP, CP, etc., such that each line is $\lambda/2$ longer than the preceding line. All waves originating at O, C, etc., arrive at P in phase and therefore reinforce each other, whereas waves originating at B, D, etc., are out of phase with those from O and C. If the

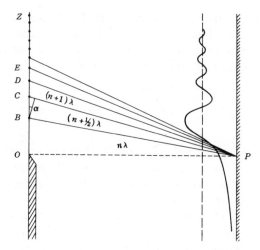

FIG. 7.6. Plane wave $O - Z$ incident on an obstructing edge. The point P represents the geometrical shadow, and the curve represents the one-dimensional diffracted wave amplitude.

amplitude which is observed at P due to the sector between O and B is called A_1, the amplitude observed at P due to the sector between B and C, A_2, etc., the total amplitude due to all sectors may be written

$$A_0 = A_1 - A_2 + A_3 - A_4 + \cdots = \sum_{i=1}^{n} -A_i(-1)^i$$

But

$$A_2 \approx \frac{A_1 + A_3}{2} \qquad A_i \approx \frac{A_{i-1} + A_{i+1}}{2}$$

and therefore

$$A_0 = \frac{A_1}{2} + \frac{A_n}{2} \approx \frac{A_1}{2} \qquad \text{for large } n$$

If the obstruction is removed, a like contribution from the lower portion of the wave surface results in an amplitude at P equal to A_1. It follows also that a greatly increased energy could be produced at P if alternate sectors were obstructed all along the wave; in this way Fresnel zones may be used to focus light much as a lens does.

Now consider the energy at a point below P. It is more convenient to imagine the obstruction moved upward, say to B. Then

$$A_B = A_2 - A_3 + A_4 - A_5 + \cdots$$

By the process demonstrated above $A_B \approx \frac{1}{2}A_2$. By moving the obstruction up to C, $A_C \approx \frac{1}{2}A_3$. Since the energy reaching P due to the individual sectors decreases gradually with increasing distance from P, the energy falls off gradually within the geometrical shadow.

The energy above the geometrical shadow may be investigated by moving the obstruction downward in steps. First consider it moved to $-B$. Then

$$A_{-B} \approx A_1 + \tfrac{1}{2}A_1 = \tfrac{3}{2}A_1$$

At $-C$

$$A_{-C} \approx A_1 - A_2 + \tfrac{1}{2}A_1 = \tfrac{3}{2}A_1 - A_2$$

Thus, as shown in Fig. 7.6, above the geometrical shadow there are alternate maxima and minima (the first minimum in the geometrical shadow of point C), and below the geometrical shadow the amplitude falls off monotonically with distance.

Now if an obstruction is introduced above point B, the diffraction pattern for the slit OB can be determined by analogous steps. The result is shown in Fig. 7.7. In a similar manner, an opaque obstacle produces the diffraction pattern shown in Fig. 7.8.

The second of the empirical principles to be used in this discussion is called *Babinet's principle*; it states that the disturbances produced by *complementary* screens are opposite in phase and identical in irradiance. Imagine a transparent screen containing one or more obstructing areas to be placed in the path of a series of wave fronts. Diffraction from the obstructions produces a disturbance in the wave amplitude reaching a particular point.

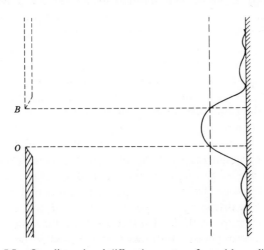

FIG. 7.7. One-dimensional diffraction pattern formed by a slit OB.

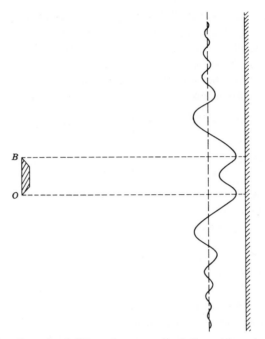

Fɪɢ. 7.8.   One-dimensional diffracted wave amplitude formed by an isolated obstacle.

Now if the original screen is replaced by a complementary screen in which transmission occurs only in the areas which in the original screen were opaque, the diffracted wave amplitude which reaches a particular point must be disturbed in just the opposite way from that produced by the original screen. The sum of these two amplitude disturbances must be zero as would be the case with a completely transparent screen, and because irradiance is proportional to the square of the amplitude, the two distributions of irradiance must be identical. Babinet's principle is useful in understanding the diffraction produced by small particles illuminated by the sun. The observer sees the same diffraction pattern which would be produced by a screen with corresponding small openings.

Figure 7.6 may be used to develop a provocative relation between size of obstacle and angular radius of the diffraction fringes. When the obstructing screen was moved through one interval (from $-B$ to $-C$), the resulting irradiance of P changed from a maximum to a minimum. Upon constructing a right triangle by dropping a perpendicular from B to CP, $\sin \alpha = \lambda/2BC$, and the relevant property of the distance BC is that it represents the dimension of an obstructing body. It follows that the smaller the size, the larger α must be in order that the lines BP and CP differ by half a wavelength.

Therefore, α represents the angular separation of minima and maxima, and the angular radius of the first minimum is given approximately by

$$\sin 2\alpha = \sin \theta = \lambda/2a$$

where a represents the radius of the obstructing body. The series of minima called the diffraction pattern may be represented approximately by

$$\sin \theta = n(\lambda/2a) \qquad n = 1, 2, 3, \ldots \tag{7.35}$$

The analysis which has been carried out for one dimensional diffraction may be extended to diffraction in two dimensions (by circular cross section, for example). The problem requires the use of Bessel functions, and the results differ from those developed here only in that in Eq. (7.35) n is replaced by $(n + 0.22)$.

Equation (7.35) shows that diffraction effects may occur for particles whose diameter is larger than the wavelength. For smaller particles, no diffraction fringes can develop, and for larger particles, the angular radius of the diffraction fringes (for small n) is so small that they are very difficult to observe. In this case also, the theoretical analysis becomes inaccurate. In order that diffraction fringes are well developed, it is also necessary that the particle size be fairly uniform because otherwise there are no fixed phase relations among the waves emitted from the various particles.

Equation (7.35) provides an immediate and simple explanation of the *corona*, the series of colored rings of light which sometimes appear to an observer to surround the sun or moon. In these cases there is between the observer and the sun or moon a thin cloud consisting of small droplets of uniform size. Diffraction fringes of characteristic radius are formed by each of the component colors of white light with blue on the inside and red on the outside. Upon measuring the angular diameter of the corona of a particular wavelength, it is a simple matter to compute the mean diameter of the diffracting particles, as is required in Problem 2.

7.10   Refraction

In a fundamental sense refraction depends upon the fact that the individual scattering elements within certain media like air, ice, water, and glass scatter with an effective phase change. In media which contain bound charges (air, water, glass, etc.) the scattered wave and the incident wave are superimposed to form a wave slightly retarded over the original incident wave. The phase speed in these media is less than in vacuum. On the other hand, free electrical charges scatter with an advance in phase, with the result that the phase speed is greater than the phase speed in vacuum.

Let us look at the problem of phase change in more detail. Consider an electron in a conductor which oscillates with the incident electromagnetic wave. If it is not bound to a positive atomic nucleus, it experiences no restoring force and therefore oscillates in phase with the incident electric vector as shown in Fig. 7.9. The electric vector radiated by the oscillating electron, however, is opposite in direction to the electric vector of the incident wave, for there must be no net electric field in the surface of the conductor. There is therefore a change of phase of 180° between incident and scattered wave; this occurs both in reflection by conductors and in scattering by free charges. The effect of this phase reversal will be considered presently. First, however, it must be recognized that bound electrons may scatter without this phase change. Such electrons experience a restoring force, and as shown in Section 7.6, they exhibit a natural frequency which increases as restoring force (or "stiffness" of the bond) increases. One may imagine an incident wave of constant frequency incident on an electron with adjustable natural frequency. As the natural frequency increases and approaches the incident frequency, the phase difference between the incident electric force and the acceleration of the electron decreases, the amplitude increases, and at the resonant frequency, the amplitude becomes so large that absorption occurs. For still greater natural frequencies, the phase relations are reversed and the direction of displacement of the electron may be opposed to the direction of the incident electric vector. A similar result occurs in a mechanical oscillator driven by a periodic force.

Now consider the effects of these two phase relations on the phase speed. Imagine a large number of scattering particles randomly distributed, each one of which scatters incident radiation. The electric vector at any point is

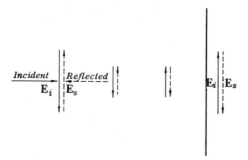

FIG. 7.9. Electric vectors ($\mathbf{E}_i$) of an electromagnetic wave incident on a surface containing free electrons and the resulting electric vectors of the reflected wave ($\mathbf{E}_s$).

the resultant of the electric vector of the incident wave and the electric vector of the scattered waves. The effect of the two scattering particles is illustrated in Fig. 7.10. If scattering occurs in phase with the incident wave, the incident wave reaches the two scattering particles P_1 and P_2 simultaneously and passes on to the right with slightly diminished amplitude. Scattered waves S_1 and S_2 originate in phase with the incident wave; S_1 reaches 0 in phase with I, but S_2 is slightly retarded due to the greater distance from the point of origin. Superposition of the scattered waves and the incident wave yields the heavy curve which is retarded in phase over the curve representing the incident wave. The decrease in phase speed may be expected to be directly related to the number of scattering particles per unit volume, but there is not necessarily any decrease in amplitude of the wave.

Next, consider the case of scattering 180° out of phase as illustrated in

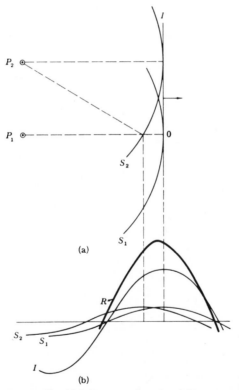

FIG. 7.10 (a) Plane wave (I) which has moved to the right past two scattering particles (P_1 and P_2) and scattered waves (S_1 and S_2) in the neighborhood of the point 0 for scattering in phase with incident wave. (b) Phase and amplitude relations of the incident and scattered waves and the resultant wave (R) in the neighborhood of 0.

Fig. 7.11. The incident wave reaches point 0 with slightly reduced amplitude due to out of phase scattering from P_1. At the moment that the incident wave reaches 0 with phase as shown, the scattered wave from P_2 is slightly retarded. Superposition of this scattered wave and the incident wave yields the heavy curve which is advanced in phase over the incident wave. Phase reversal also decreases the amplitude of the resultant wave as the scattered wave from P_1 illustrates, so that an electromagnetic wave can penetrate into a region of free electrical charges only a certain critical distance. At the same time each scattering event contributes radiation in the reverse direction; this is called *reflection*. In metals, charge density is so great that penetration is negligible, and we say that reflection occurs at the metallic surface.

Advance in phase by scattering results in an increase in the phase speed over the speed in vacuum. The phase speed, however, is a geometrical ab-

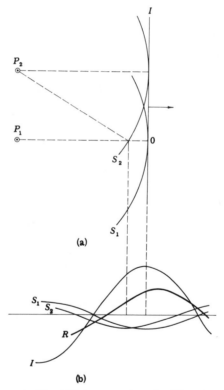

FIG. 7.11.   (a) Plane wave (*I*) which has moved to the right past two scattering particles (P_1 and P_2) and scattered waves (S_1 and S_2) in the neighborhood of the point 0 for scattering out of phase with incident wave. (b) Phase and amplitude relations of the incident and scattered waves and the resultant wave (*R*) in the neighborhood of 0.

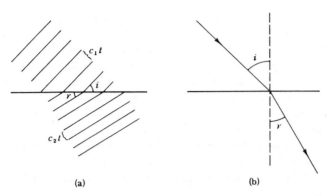

(a)   (b)

FIG. 7.12. (a) Positions of a single wave phase at successive time intervals separated by t. The phase speed in the upper mediums is c_1, in the lower medium is c_2. (b) The geometry of refraction of a ray drawn normal to successive positions of the wave phase.

straction, and this result does not imply that energy can be transmitted at speeds greater than the speed in vacuum.

The difference in phase speed of a wave when passing from one medium to another leads to a geometrical relation known as Snell's law. The series of lines in Fig. 7.12 represent the successive positions of a single wave phase at a series of equal time intervals. If the speed of the wave phase changes from c_1 to c_2 in passing the interface between the two media, then it is obvious from the diagram that

$$c_1/c_2 = \sin i/\sin r \qquad (7.36)$$

where i represents the angle of incidence and r the angle of refraction. The bending of the wave train or "beam" which occurs in passing from one medium to another in which phase speed is different is called *refraction*. The apparent bending of a stick thrust into water at an oblique angle is a familiar example of refraction.

PART III: NATURAL SIGNAL PHENOMENA

Parts I and II of this chapter provide the background from which many natural phenomena may be understood. The blueness of clear sky and the redness of sunsets have already been mentioned as illustrations of effects of scattering by air molecules and small particles. In these cases the fact that scattered energy is inversely proportional to the fourth power of the incident wavelength as shown by proportionality (7.33) results in scattering of more blue light than red from the solar beam with the result that if one looks away from the direction of the sun, predominantly blue light is seen, whereas if one looks toward the sun through a long scattering path as at sunset, one sees light which is deficient in blue wavelengths and therefore predominantly red. It can be shown easily that the ratio of blue light ($\lambda \sim 0.43$ μm) to red light ($\lambda \sim 0.65$ μm) is about 5. This is an example of the many natural phenomena which are easily observed. In the following sections only a few of these are discussed. Many questions are left to the reader to answer or to wonder about. For example, why is the blue of the sky deepest in directions at right angles to the sun? Why are distant snow-covered mountains somewhat orange, whereas forest-covered mountains appear blue?

The phenomena discussed in Part III of this chapter are limited to those detected by our human senses. Our senses are particularly good at detecting changes and in recognizing complex patterns. They are not very good in quantitative measurement, especially when stability over long periods is required. On the other hand, instruments can be used to make quantitative measurements, and this subject is discussed in Part IV of this chapter.

7.11   Visibility

The distance an object can be seen is called the *visual range*; the horizontal visual range near the surface of the earth is commonly called the *visibility*. Discussion of these atmospheric properties involves the characteristics of the human eye as well as the optical properties of the atmosphere.

The human eye is sensitive to light, the electromagnetic radiation between 0.4 and 0.7 μm. Sensitivity is strongly dependent on wavelength, and Fig. 7.13 indicates that green light of a certain radiance appears much brighter

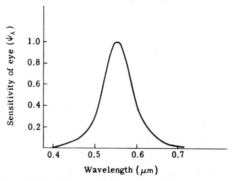

FIG. 7.13.   The sensitivity curve of the human eye.

to the eye than does violet or red light of the same radiance. The sensitivity (ψ_λ) of the eye at wavelength λ is defined by the ratio of radiance at 0.555 μm, where the eye has maximum sensitivity, to radiance at wavelength λ required to yield equal impressions of brightness. The *monochromatic brightness* is defined by

$$B_\lambda = \psi_\lambda L_\lambda$$

where L_λ represents the monochromatic radiance emitted or reflected by the object. The *brightness* is defined by $B \equiv \int_0^\infty B_\lambda \, d\lambda$.

In order for an object to be visible, there must be *contrast* between it and its surroundings. If B represents the brightness of the object and B_0 the brightness of the background, then the contrast of the object against its surroundings is defined by

$$C \equiv (B - B_0)/B_0 \tag{7.37}$$

It is clear that this definition is insufficient to distinguish between two objects with different colors but the same brightness. However, this definition is adequate so long as only the contrast between "black" and "white" is considered.

In order to determine visual range, a black object may be chosen which is large enough that the angle under which it appears can easily be resolved by eye. This requires at least 1 minute of arc. At short range B is very small and the contrast is close to minus one. At a greater distance the black object appears to have a certain brightness because light is scattered from all directions into the direction from object to observer. The apparent brightness increases with the distance to the object; the contrast becomes so small that the human eye is no longer able to perceive it at the visual range. For the normal eye the threshold contrast has a value of ± 0.02.

In general, the radiance of the background varies along the line from object to observer as a result of the sum of attenuation and enhancement due to scattering of radiation from other directions. Under fairly uniform conditions the gains and losses are equal. However, the apparent brightness of a black object increases with distance and, according to Beer's law, may be expressed by

$$B_\lambda = B_{s\lambda}(1 - e^{-\sigma_\lambda x})$$

where $B_{s\lambda}$ is the monochromatic brightness of the sky. The total apparent brightness of the black object is given by

$$B = \int_0^\infty B_{s\lambda}(1 - e^{-\sigma_\lambda x}) \, d\lambda$$

and the contrast is given by Eq. (7.37) in the form

$$C = -\int_0^\infty B_{s\lambda} e^{-\sigma_\lambda x} \, d\lambda / B_s$$

Assuming σ_λ to be independent of λ, x may be expressed by

$$x = -(1/\sigma) \ln|C|$$

And taking $C = -0.02$, the visual range becomes

$$x_v = 3.912/\sigma \tag{7.38}$$

Equation (7.38) states an explicit relation which permits calculating the extinction coefficient if the visual range is known. The assumption that σ_λ is independent of wavelength is not quite valid because, as pointed out in Section 7.8, air molecules scatter the blue light more strongly than the red; for this reason distant mountains appear to be blue and distant snowfields yellow. When the scattering is caused by larger particles, dust and small droplets with a size of the order of the wavelength of light, the assumption is much better. So in most cases with poor visibility the visual range is independent of wavelength.

At night it is possible to use artificial light sources for the determination of the extinction coefficient, and this value used in Eq. (7.38) leads to a prediction of the visual range for daytime. The concept of contrast loses its usefulness at night because the light source is many times brighter than the background sky, and the contrast becomes a very large but uncertain quantity. However, the extinction coefficient may be found by applying Beer's law to two identical light sources at two different distances x_1 and x_2. The ratio of the irradiances is then

$$\frac{E_1}{E_2} = \frac{x_2^2}{x_1^2} \exp[\sigma(x_2 - x_1)]$$

or

$$\sigma = \frac{1}{x_2 - x_1} \left(\ln \frac{x_1^2}{x_2^2} + \ln \frac{E_1}{E_2} \right)$$

An exercise is given in Problem 3.

7.12   Airglow

In addition to the photochemical processes discussed in Chapter V which result in absorption of solar radiation, other photochemical reactions occurring in the upper atmosphere result in emission of electromagnetic energy. The following reactions are among a large number of light emitting reactions which result in a faint luminescence of the sky, called the *airglow*.

$$2O \rightarrow O_2 + h\nu$$

$$O + NO \rightarrow NO_2 + h\nu$$

The relatively constant airglow should be distinguished from the highly variable radiation of the aurora which is emitted at somewhat greater heights and is associated with the influx of solar particles in the upper atmosphere. However, ionization and excitation of neutral particles by charged particles contributes also to airglow.

Airglow occurs both during the day and during the night when it can be detected from the earth's surface by the naked eye. Airglow at night is also called *nightglow*; it contributes on a moonless night between 40% and 50% of the total illumination by the night sky, slightly more than the radiation received from the stars. Spectroscopic observations reveal many emission lines and a continuum which extends from below 0.4 μm well into the infrared. *Dayglow* is observed by instruments carried by satellites, rockets, or high flying aircraft.

The constituents responsible for the lines and bands of the airglow spectrum include molecular and atomic oxygen, nitrogen, hydroxyl, and possibly hydrogen. Nightglow originates from ion–electron recombination reactions as well as from photochemical reactions such as those described earlier in this section. In addition to the sources mentioned, dayglow originates also from resonance scattering from sodium, atomic oxygen, nitrogen and nitric oxide.

Quantitative observations of airglow are made from the earth's surface and from rockets and satellites in order to determine composition and properties of the atmosphere and the height at which the airglow originates. Methods of observation are discussed later in this chapter in Part IV.

7.13   Refraction of Light by Air

Electromagnetic waves often are refracted in passing through the atmosphere. To discuss refraction quantitatively we may recognize from Eq. (7.28) that the index of refraction depends on the number of scattering particles per unit volume. The electric dipoles created from atmospheric gases have natural frequencies (ν_0) lying in the ultraviolet, that is, at frequencies higher than the frequencies of visible light (ν). Therefore, upon recognizing that $\kappa^2 \ll n^2$ and neglecting ν^2 compared to ν_0^2, Eq. (7.28) may be expressed approximately for a particular constituent of the atmosphere by

$$n - 1 = Ne^2/8\pi^2 m\varepsilon_0 v_0^2 \tag{7.39}$$

The index of refraction for air can be expressed as the sum of the index for dry air, a uniform and constant mixture of gases, and for water vapor in the form

$$n - 1 = a\rho + b\rho_w/T \tag{7.40}$$

where a and b are characteristic of dry air and water vapor, respectively, and their values depend on wavelength. The temperature enters the second term on the right because the water molecule is polar, and therefore the contribution of water vapor to the index of refraction is temperature dependent. For low temperature the permanent dipoles are readily aligned by an incident electric field, and the contribution to polarization and to index of refraction is consequently large, whereas at high temperatures the probability of a molecule being polarized is small.

For visible light the index of refraction at standard pressure and temperature is 1.000293; the corresponding values of a and b are, respectively, $0.227 \times 10^{-3} \, \text{m}^3 \, \text{kg}^{-1}$ and zero. For a wavelength of 2 cm a again is approximately $0.234 \times 10^{-3} \, \text{m}^3 \, \text{kg}^{-1}$ and b is about $1.75 \, \text{m}^3 \, \text{kg}^{-1} \, \text{K}$. For infrared wavelengths the natural rotational frequency of the water vapor molecule is comparable to the incident frequency, and the index of refraction cannot be represented by Eq. (7.39). This is the region in which anomalous dispersion and absorption occur.

Because for visible light the second term in Eq. (7.40) is negligible, the

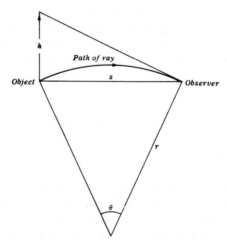

FIG. 7.14.   Geometry of light ray in air in which density decreases upward.

equation shows that for the normal case of decrease of density with height the upper portion of a horizontally propagating wave moves faster than the lower portion, and refraction toward the earth occurs, as shown in Fig. 7.14. Therefore, distant objects appear to be elevated somewhat above their true height, and the sun, the moon, and stars may be visible to an observer on earth before they have risen above the geometrical horizon.

Measurement of the radius of curvature of a horizontal light ray provides a measure of the vertical gradient of air density. To express the radius of curvature, note that the ray shown in Fig. 7.14 is everywhere normal to the wave phase which travels from the object to the observer in the time interval expressed by s/c. This time must be independent of r. Thus $\theta r/c$ is constant, and

$$\frac{1}{c}\frac{dc}{dr} = \frac{1}{r} \tag{7.41}$$

Now $\cos \theta$ is expanded in a Taylor series and only the first two terms are retained with the result

$$\cos \theta = \frac{r}{r+h} = 1 - \frac{s^2}{2r^2}$$

or, approximately

$$\frac{1}{r} = \frac{2h}{s^2} \tag{7.42}$$

Notice that r is positive for refraction toward the earth and negative for refraction away from the earth.

For a nearly horizontal ray the r and z directions are nearly the same, so that Eqs. (7.41) and (7.42) may be combined to give

$$\frac{dc}{dz} = \frac{2hc}{s^2} \tag{7.43}$$

Equation (7.43) shows that h, the difference between the true and apparent height, is proportional to the rate of change with height of the speed of light in the atmosphere. Equations (7.40) and (7.20) together yield

$$\frac{dc}{dz} = -\frac{c(c_0 - c)}{c_0 \rho}\frac{d\rho}{dz}$$

Therefore, upon substituting the equation of state and the hydrostatic equation

$$\frac{dc}{dz} = \frac{c(c_0 - c)}{c_0 T}\left(\frac{g}{R_m} + \frac{\partial T}{\partial z}\right) \tag{7.44}$$

Equations (7.44) and (7.41) may be combined, giving

$$\frac{1}{r} = \frac{n-1}{nT}\left(\frac{g}{R_m} + \frac{\partial T}{\partial z}\right) \tag{7.45}$$

Thus the curvature of the ray depends directly on temperature lapse rate. An application is given in Problem 4.

Mirages

Large vertical temperature gradients occur just above warm horizontal surfaces such as deserts or airport runways or above cold surfaces such as ice or cold bodies of water. In some cases spectacular mirages are observed. For example, above a warm surface, temperature decreases with height as indicated in Fig. 7.15. Curvature is greatest for rays which pass through the air just above the surface and is less for more elevated rays. As shown on the left side of Fig. 7.15, an object above the surface will appear to be below the surface and will be inverted. This is referred to as an *inferior mirage*. This phenomenon may be observed almost any sunny day in driving on a highway when the sky is seen refracted by the air close to the highway. This evidently accounts for the unhappy experience of a pelican found on an asphalt highway near Wichita, Kansas.[†] "The miserable bird had obviously been flying, maybe for hours, across dry wheat stubble and had suddenly spotted what he thought was a long black river, thin but wet, right in the midst of the prairie. He had put down for a cooling swim and knocked himself unconscious." He was discovered by a local farmer, who was kind enough to deliver him to the duck pond in the Wichita zoo.

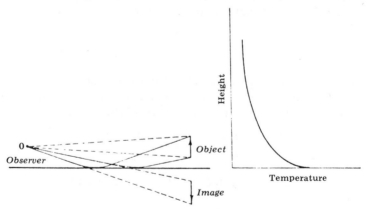

FIG. 7.15. Paths of light rays which form an inverted inferior mirage over a warm surface and typical temperature profile accompanying it.

[†] C. H. Goodrum, *The New Yorker* **38** (8), 115, 1962.

When there is a strong temperature inversion layer at a height of 10 m or so above a layer which is roughly isothermal, the *superior mirage* may be observed. Light rays from distant objects near the surface are refracted downward, so that the object appears elevated. In this case trees, buildings, or banks along the shore may appear greatly extended in the vertical. There are many occasions under which complex mirages may occur, and it often is a challenge to explain the phenomena one sees.

Astronomical Refraction

The refraction of light from outside the atmosphere is illustrated in Fig. 7.16. The angular deviation (δ) of the light ray in passing through the atmosphere is found by recalling from Snell's law that $(\sin i)/c$ is constant along the ray. Thus

$$\sin i/c = \sin Z/c'$$

The problem may be solved quite generally, but the trigonometry is somewhat tedious, so only the approximate solution which neglects the curvature of the earth, and therefore of atmospheric density surfaces, will be discussed. Under these conditions the radii in Fig. 7.16 may be considered parallel, and $i = Z + \delta$. Therefore

$$\frac{\sin Z \cos \delta + \cos Z \sin \delta}{\sin Z} = \frac{c}{c'}$$

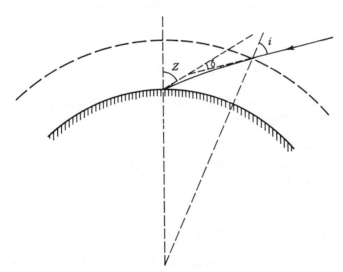

FIG. 7.16. Path through the atmosphere of a light ray from a star or a satellite.

Since δ is very small, $\cos \delta \approx 1$ and

$$\sin \delta = (n - 1) \tan Z \qquad (7.46)$$

Therefore, upon recalling that $n \approx 1.000293$

$$\delta \approx 2.93 \times 10^{-4} \tan Z$$

For zenith angle of $80°$ this gives a δ of $5'48''$ which is within a half-minute of arc or about 10% of the correct value. At larger zenith angles the curvature of the atmosphere must be considered. As a result of atmospheric refraction, the disks of the sun and moon are visible when they are entirely below the geometric horizon, and the disks may appear to be flattened.

Refraction depends on the wavelength, so that the atmosphere disperses light into its component colors in the same way as does a glass or quartz prism. For this reason, the stars may appear near the horizon as vertical lines containing the usual spectral distribution of color with blue and green at the top and red at the bottom. The *green flash* which sometimes is seen just as the limb of the sun vanishes below the horizon is due also to dispersion of light by the atmosphere.

Shimmer and Twinkle of Stars

When the air is turbulent and eddy density gradients are large, the light path from an object to the observer fluctuates rapidly, giving the object the appearance of continuous motion. When air flows over a much warmer or very much colder surface, this phenomenon is marked and is called *shimmer*. Less pronounced density fluctuations occurring along the line of sight are responsible for twinkling of stars.

7.14   Refraction by Ice Crystals

It has been shown in Chapter III that naturally occurring ice crystals are of several different forms with hexagonal basic structure. The great variety of shapes and orientations of the ice particles gives rise to a large number of refraction and reflection phenomena sometimes arresting and at other times hardly noticeable. Only a few of the most common phenomena will be discussed.

Haloes of 22° and 46°

Refraction occurring through hexagonal prisms, which are typical of cirrus clouds, is illustrated in Fig. 7.17a. If the crystal in Fig. 7.17a is rotated about its axis and the direction of the incident ray is kept constant, the angle of deviation D varies in a fashion as sketched in Fig. 7.17b. It is

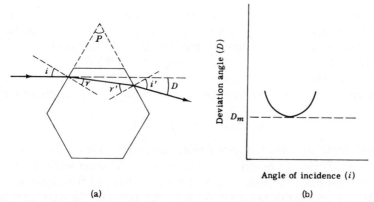

(a)   (b)

FIG. 7.17.   (a) Path of a light ray refracted by a hexagonal crystal. (b) Deviation angle as a function of the angle of incidence.

seen that as the angle of minimum deviation D_m is approached, an increment in angle of incidence brings about only a small change in angle of deviation. Therefore, a group of randomly oriented identical crystals transmit maximum radiance in the direction of minimum deviation. Now imagine an observer looking at the sun or moon through a cloud sheet of hexagonal needles with random orientation. Light reaches him by refraction at many different angles, but the radiance is greatest at the angle of minimum deviation. Thus he sees a ring of bright light (a *halo*) of angular radius equal to the angle of minimum deviation, surrounding a region which is darker than the region outside the halo.

From Fig. 7.17a it follows that the deviation angle may be expressed by

$$D = (i - r) + (i' - r') = i + i' - P$$

For minimum deviation

$$\frac{dD}{di} = 0 = 1 + \frac{di'}{di}$$

Because $r + r' = P$, also

$$\frac{dr'}{dr} + 1 = 0$$

The last two equations are satisfied when $i = i'$ and $r = r'$. Upon applying Snell's law and eliminating i and r

$$\sin \tfrac{1}{2}(D_m + P) = n \sin \tfrac{1}{2}P \qquad (7.47)$$

where n is the index of refraction for ice. An exercise is given in Problem 5.

Since the index of refraction varies with wavelength, white light is dispersed into its component colors with red refracted least and blue refracted most. Notice that this order is the reverse of that observed in the corona, providing a means of distinction between ice crystals and water drops.

The most common halo has angular radius of 22° indicating refraction by hexagonal prisms. This halo can easily be identified because the angle of 22° is just the span of a hand at arm's length. The halo of 46°, which is produced by refraction by rectangular prisms, is less frequently observed.

Sundogs and Tangent Arcs

Horizontally oriented hexagonal plates are responsible for the appearance of *sundogs*. These are bright spots at the same elevation angle as the sun at an angular distance slightly greater than 22° left and right from the sun. The greater angular distance results from the fact that the light rays from sun to observer are not perpendicular to the vertical axes of the prisms, and the angle of minimum deviation is somewhat larger than 22°. The vertical orientation of the prisms prevents the formation of a uniform ring halo and causes bright spots to be visible on either side of the sun.

Tangent arcs on the 22° halo may appear at the highest and lowest points of the circle when sunlight is refracted by hexagonal needles with horizontal orientation. These and many related phenomena caused by refraction and also by reflection by ice crystals are discussed in detail by Humphreys (see references). Once the geometry of the problem is understood the computation of various points of the arcs is merely tedious.

7.15   Refraction by Water Drops

The angle of minimum deviation for light passage through a spherical water drop can be determined in a manner similar to that outlined in the previous section. However, the spherical shape of the droplets makes the

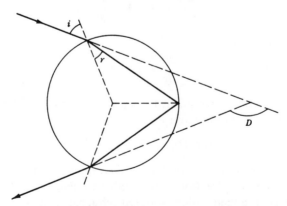

FIG. 7.18.   The geometry of a light ray which contributes to the first order (42°) rainbow.

geometry of the refraction phenomena somewhat different. In Fig. 7.18 the geometry of refraction and reflection of a light ray by a spherical drop is illustrated. From this figure it may be deduced that

$$D = 2(i - r) + \pi - 2r = \pi + 2i - 4r$$

This equation, together with Snell's law and the condition for minimum deviation $(dD/di = 0)$, provides three equations for the three unknowns D_{m}, i, and r. Upon solving for the angles i and r

$$i = \arccos\left(\frac{n^2 - 1}{3}\right)^{1/2} \qquad r = \arccos\frac{2}{n}\left(\frac{n^2 - 1}{3}\right)^{1/2}$$

The corresponding solution for D_{m} yields for the primary rainbow an angular radius of about 42°. It is comparatively simple to extend this development to two or more internal reflections yielding second and higher order rainbows, as required in Problem 6.

The appearance of the rainbow depends rather strongly on the drop size. Large drops (2 mm diameter) are responsible for very clear and colorful

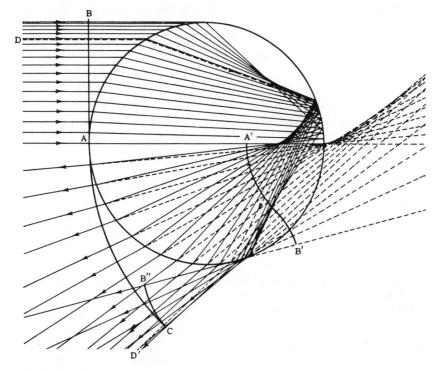

FIG. 7.19. Light rays representing the refraction and reflection of a plane wave surface AB by a spherical water drop (after W. J. Humphreys, *Physics of the Air*, p. 481, McGraw–Hill, New York, 1940).

rainbows with distinct separation between the colors. Small drops (0.2–0.3 mm) produce less colorful rainbows, and very small drops (0.05 mm) produce a white primary bow (mist bow).

When the rainbow is very pronounced, supernumerary bows often become visible. This is a diffraction phenomenon accompanying the refraction of the light in the drop which may be understood by following a wave which strikes the drop as shown in Fig. 7.19. The incident wave phase AB appears when emerging as ACB'', just as if it had come from the virtual phase $A'B'$. The section CB'' trails somewhat behind CA, and this gives rise to interference patterns. When the distance is a half-wavelength, the two sections counteract each other, and when it is a full wavelength they reinforce each other, giving rise to maxima and minima inside the primary bow. These maxima correspond to the supernumerary bows.

7.16   Naturally Occurring Atmospheric Radio Waves

Lightning discharge generates a wide spectrum of electromagnetic waves extending from the ultraviolet to very low-frequency radio waves. Near thunderstorms radio broadcasts may be seriously "jammed" by "sferics" produced in this way. At great distances from the thunderstorm, a remarkable phenomenon may be observed. Radio waves of audio frequency emitted by a lightning discharge may penetrate the ionosphere, travel approximately along the lines of force of the earth's magnetic field and return to the earth's surface at the opposite geomagnetic latitude. The higher frequencies travel faster than the lower frequencies because of dispersive propagation through the upper atmosphere. The result is that an observer with a detector for these low frequencies (e.g., a very long wire and an earphone) hears a whistling tone of steadily falling pitch, called a "whistler."

Whistlers were first observed in 1888 by Pernter and Trabert in Austria using a telephone wire 22 km long as the antenna. The German scientist Barkhausen rediscovered the phenomenon while eavesdropping on Allied military telephone conversations during the First World War. Barkhausen also suggested that whistlers might arise from propagation of sferics over long dispersive paths, thus linking the phenomenon to lightning discharges, but he could not explain how radio waves of such low frequencies could possibly penetrate the ionosphere.

Understanding of whistlers depends upon the magneto-ionic theory, the theory of interaction between electromagnetic waves and charged particles in the earth's magnetic field. This theory shows that the effect of an electron which spins about a magnetic line of force is similar in certain respects to that of a bound charge which oscillates in response to an incident polarized wave. Because the orientation of the electron orbit is determined by the geomagnetic field, the index of refraction has different values for radiation perpendicular and parallel to the magnetic lines of force. The plasma in the magnetic field is said to be *doubly refracting*, and an incident ray is split into two parts, the "ordinary" and "extra-ordinary" rays. The gyrofrequency of the electron plays a role analogous to the resonance frequency of the oscillating dipole. Therefore, the index of refraction may be expected to be expressed by an equation similar to Eq. (7.28). The complete theory shows that the whistlers are caused by the extraordinary mode traveling approximately along the magnetic lines of force and that the index of refraction for this case is given by

$$n^2 = 1 + \frac{Ne^2}{4\pi^2\varepsilon_0 mv(v_G - v)} \tag{7.48}$$

where v_G is the gyrofrequency. By equating the magnetic and centrifugal forces, the gyrofrequency is found to be

$$v_G = \frac{Be}{2\pi m}$$

As has been pointed out in Section 7.6, the index of refraction is the ratio of the speed of light in vacuum to the phase velocity. The signal, however, travels with the group velocity which is given by Eq. (7.19). By substituting Eq. (7.48) into (7.19) and assuming that $Ne^2/4\pi^2\varepsilon_0 m \gg v_G v$ and $v_G > v$, the group velocity may be expressed by

$$c_g = \frac{2c_0 v^{1/2}(v_G - v)^{3/2}}{v_p v_G} \tag{7.49}$$

where $v_p^2 \equiv Ne^2/4\pi^2\varepsilon_0 m$. This equation has a maximum for $v = v_G/4$. For smaller values of v the group velocity decreases with decreasing frequency, giving rise to the dispersion observed in whistlers. The rays tend to follow the lines of geomagnetic force, so that electromagnetic waves generated by a lightning flash at geomagnetic latitude $-\phi$ are observed as a whistler at geomagnetic latitude ϕ in the opposite hemisphere. Whistlers may be reflected from the earth's surface and travel back and forth several times between the Northern and Southern Hemispheres producing the typical form of dispersion illustrated in Fig. 7.20.

In addition to whistlers a wide variety of very low frequency (VLF) phenomena not associated with lightning flashes have been observed. Some of these phenomena have been given the suggestive name of "dawn chorus," describing a fuzzy sound of rapidly increasing pitch. Other fairly frequently observed phenomena are the "hooks" (falling pitch followed by a sharp

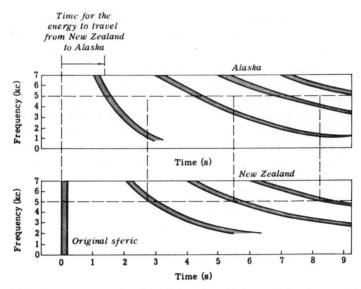

FIG. 7.20.   Frequency as a function of time for whistlers traveling between Unalaska, Alaska, and New Zealand (after M. G. Morgan, *Ann. IGY* **3**, 315, 1957).

rise in pitch) and "risers" (continuous rising pitch). These phenomena are thought to be related to clouds of charged solar particles entering the magnetosphere. Helliwell (see Bibliography) provides further discussion of the effects of the ionosphere and magnetosphere on electromagnetic waves.

7.17   Refraction of Sound Waves

Snell's law, together with Eq. (7.4), shows that refraction of sound waves must occur where temperature gradient exists in the direction normal to the wave propagation. The path of a sound "ray" may be calculated as follows. Imagine the atmosphere to be composed of a large number of iso-thermal horizontal layers each differing in temperature from the underlying layer. Acoustic waves experience refraction at the boundaries between layers as shown in Fig. 7.21. From Snell's law for any two interfaces the following relation must hold

$$\frac{\sin i_1}{c_1} = \frac{\sin r_1}{c_2} = \frac{\sin i_2}{c_2}, \text{ etc.} \tag{7.50}$$

Thus the ratio of the sine of the angle of incidence to speed is constant all

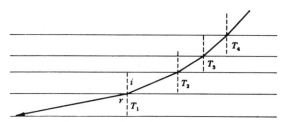

FIG. 7.21. Path of an acoustic ray in passing obliquely through a series of isothermal air layers for the case in which temperature (T) decreases with height.

along the ray, and this holds for continuous as well as for discontinuous variation of temperature. Substitution of Eq. (7.50) into Eq. (7.4) gives

$$\sin i/(T)^{1/2} = \text{const}$$

so long as $(c_p/c_v)R_m$ is constant along the ray. To evaluate the constant, T and i must be known at some point along the ray. For a ray which becomes horizontal at some point, this equation yields

$$\sin i = (T/T')^{1/2}$$

where T' represents the temperature at the point of horizontal incidence. It follows from the Pythagorean theorem that

$$\tan i = [T/(T' - T)]^{1/2}$$

If it is assumed that the temperature is a linear function of height expressed by $T = T' - \gamma z$, where T' represents the temperature at a reference height (usually the ground), and if $\tan i$ is replaced by dx/dz

$$dx = (T/\gamma z)^{1/2}\, dz \qquad (7.51)$$

Equation (7.51) may be integrated (usually numerically) to give the path of the ray. An application is given in Problem 7.

Detection of the Stratosphere

Equation (7.51) provided the basis for the original detection of the stratospheric temperature inversion. It was observed in the early decades of the twentieth century that when loud noises are produced (e.g., artillery fire), several approximately concentric circles could be drawn separating alternate regions of audibility and inaudibility. As shown in Fig. 7.22, within the central region, the "ground wave" was audible, but at greater radii attenuation and upward refraction made the ground wave inaudible. At still greater radii, waves were apparently refracted toward the earth by a layer of warm

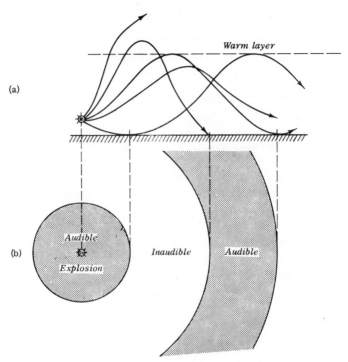

Fig. 7.22.   (a) Propagation of sound waves from an explosion. (b) Distribution of audibility zones.

air at considerable height above the ground. Measurement of the elapsed time between the explosion and the arrival of the sound waves permits an estimate of the height and temperature of the refracting layer. These calculations indicated that the temperature at 50 km above the earth in middle latitudes may be as high as 325 K during daytime. Usually the effect of refraction by wind shear has also to be taken into account, which complicates the interpretation of observations.

From time to time, often on ships at sea, mysterious booming sounds have been heard. These sounds, sometimes called "Barisal guns" or "mistpoefers," have stimulated many imaginings throughout history. Although not all reports of such sounds have been satisfactorily explained, many of them must be the result of distant explosions and sonic booms refracted by the stratosphere.†

† Some of the lore and related physics have been reviewed in the following papers: A. C. Keller, *Studies in the Romance Languages* **72**, Univ. N. Carolina, 151, 1968. J. A. Businger, *Weather* **23**, 497, 1968. D. Shapley, *Science* **199**, 1416, 1978. R. A. Kerr, *Science* **203**, 256, 1979.

The Audibility of Thunder

Sound waves produced by lightning above the earth's surface are refracted away from the earth under conditions of normal atmospheric lapse rate. For any lapse condition, there is a "critical" ray which reaches the earth at grazing incidence as illustrated in Fig. 7.23; at greater distances the only sound reaching the ground results from scattering by atmospheric inhomogeneities and is usually very weak. To determine the maximum range of the critical ray in the case of constant lapse rate, Eq. (7.51) may be integrated simply if T is treated as a constant in the numerator. The path of the sound ray is then given by

$$x - x_0 \approx 2(Tz/\gamma)^{1/2}$$

which is the equation of a parabola whose properties depend on the temperature T and the lapse rate γ. For a lapse rate of $7.5°$ C km^{-1}, temperature of 300 K, and height of 4 km, thunder should be inaudible beyond about 25 km. Lightning which is visible at distances greater than the maximum range of audibility frequently is referred to as *heat* or *sheet* lightning.

Temperature Inversions near the Ground

Within the surface layer temperature inversion, sound rays are refracted downward and may be partially focused with the result that sounds are heard with remarkable clarity. The phenomenon is particularly marked on mornings following clear calm nights and during warm-frontal rain storms,

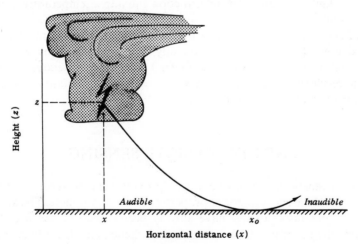

FIG. 7.23. The path of a sound ray emitted by a lightning flash at height z.

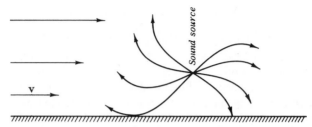

FIG. 7.24.   Sound rays from an elevated source refracted by vertical wind shear.

for these conditions frequently are accompanied by strong temperature inversions.

The noise of jet aircraft or rockets taking off, of ground level explosions, and of other high-intensity sounds are heard at very great distances under inversion conditions. The sound of a plane rising through such an inversion may drop very suddenly as heard by an observer on the ground.

Effect of Wind Shear

Under normal conditions of increasing wind speed with height, acoustic waves moving in the direction of the wind are refracted downward, whereas waves moving against the wind are refracted upward as illustrated in Fig. 7.24. The usual effect of refraction by wind shear near the ground is that audibility in the direction of the wind is increased. For this reason, coastal foghorns are audible at sea at great distances during offshore winds but may be inaudible at short distances during onshore winds. A vertical shear of $6 \text{ m s}^{-1} \text{ km}^{-1}$ has an effect on refraction equal to a temperature gradient of $10° \text{ C km}^{-1}$.

Wind shear near the ground in the vicinity of thunderstorms usually is directed toward the storm. Consequently, thunder originating within the storm is refracted upward, due both to wind shear and temperature gradient, with the result that audibility is markedly reduced. The countryman speaks of "the calm before the storm."

PART IV: REMOTE SENSING

Electromagnetic and acoustic waves provide very powerful means for observing and measuring certain properties of the atmosphere. The simplest and most obvious examples have already been discussed, the use of our eyes and ears in seeing and hearing natural phenomena. Quantitative measurements which greatly extend direct sensory observations are made by radar, by satellite sensors, and many other instruments. Typically, the use of these remote sensing techniques produces enormously greater numbers of data than the use of direct immersion instrumentation; therefore, complex and sophisticated data processing systems often are necessary in conjunction with remote sensing instruments.

Remote sensing techniques are often classified as *passive* if the natural signal is measured, and as *active* if directed electromagnetic or acoustic energy interacts with a target part of the atmosphere and then is measured by an instrument. Observations provide information concerning the absorption, scattering, refraction, polarization, and emission of radiation. There is an obvious advantage in simplicity if an instrument is sensitive to only one of these processes; however, in some cases the received signal is affected by two or more of these processes.

Table 7.1 summarizes some of the remote sensing techniques and instruments which have been used in observing or measuring atmospheric properties. Table 7.1 indicates the versatility of these methods. Remote sensing can be used to measure most of the atmospheric variables of interest, and for some variables there are a variety of techniques and instruments which can be used. Included in this discussion are ground-based and satellite instruments, and also instruments, such as the sonic anemometer–thermometer, in which the air being measured is situated within the structure of the instrument and so is "remote" only in the sense that it is the properties of the air and not the properties of a probe that are measured. Instruments of this latter group are not usually considered as remote sensing instruments, but this useage is convenient here.

Remote sensing has the following inherent advantages over immersion measurements:

(a) Remote sensing provides a line, area, or volume integration of the variable being measured and thus is often more representative of a region than are point measurements.

(b) In many cases the instrument does not have to be transported to the region of observation.

(c) The observation does not disturb or modify the medium appreciably.

(d) Observations often can be made in three space dimensions and time.

(e) High resolution in space and time often is attainable.

(f) The instrumentation often can be made automatic, and data can be processed automatically and continuously.

In recent years remote sensing has greatly extended the ability to observe atmospheric properties both from the earth's surface and from satellites.

TABLE 7.1
SUMMARY OF REMOTE SENSING CAPABILITIES

	Passive	Active
Ultraviolet		Absorption water vapor (Lyman-α hygrometer)
Visible	Scattering visual range cloud imagery Absorption trace gas identification H_2O profiles O_3 and other trace gases Refraction vertical temperature distribution Emission composition of upper atmosphere lightning	Scintillation path averaged velocity path averaged turbulence drop size distribution rainfall rate Absorption H_2O profiles (limb scan) H_2O profiles (lidar) O_3 and other trace gases (limb scan) Scattering Aerosol profiles (lidar) Aerosol distribution (3-D) (polarized lidar) Cloud particle distribution (3-D) (polarized lidar)
Infrared	Absorption water vapor Emission-absorption temperature and H_2O profiles (radiometer) sea surface temperature (radiometer)	Scattering wind velocity (Doppler radar) height, boundary layer (Doppler radar) wave height (Doppler radar) wind velocity (Doppler radar) H_2O profiles (Raman lidar)
Radio	Emission-absorption temperature profile (O_2 radiometer)	Scattering liquid water: 3-D (radar) clouds: 3-D (radar)
Microwave	H_2O profile (H_2O radiometer) integrated liquid water (H_2O radiometer)	ice: 3-D (radar) wind stress on ocean (radar) wind velocity (Doppler radar) turbulence (Doppler radar) wind velocity at ocean surface (Doppler radar)
UHF-VHF		wind velocity (Doppler radar) turbulence (Doppler radar)
HF	Triangulation location of electrical discharges	Refraction magnetic and electrical properties ionosphere height

TABLE 7.1 (*Continued*)

	Passive	Active

Acoustic	Refraction
	  vertical temperature
	    distribution
	  explosion characteristics
		Scattering
		  wind velocity in boundary
		    layer (Doppler echosonde)
		  temperature in boundary
		    layer (Doppler echosonde)
		Velocity of sound
		  temperature
		  wind velocity

The possible applications have not been exhausted, so that further developments are likely. In the following sections a few of the remote sensing techniques and instruments are described.† The examples illustrate the wide variety of techniques and instruments, but do not begin to represent all that have been or could be developed.

7.18   Absorption Techniques

Beer's law describing the reduction in radiance due to absorption provides the basis for several remote sensing techniques and instruments. In some cases absorption is measured over long path lengths in the free atmosphere, and in some cases absorption is measured over short path lengths in the laboratory or within a field instrument.

Long Path Absorption

In principle the simplest example of absorption techniques is the determination of trace gas constituents of the atmosphere by observing absorption lines in the ultraviolet, visible, or infrared solar or stellar spectra. These observations provide measurements of the total mass of absorbing gas along the line of sight; they can be used to study changes in atmospheric composition over extended time periods and geographical differences in long term averages of concentrations. Systematic measurements of total ozone in a vertical column have been made since about 1930 from observations of solar radiation using the Dobson spectrophotometer. Absorption measurements also have been made using long paths within the atmosphere.

† A comprehensive review of the subject up to the early 1970s is provided in *Remote Sensing of the Troposphere*, edited by V. E. Derr.

Limb Scanning

Absorption measurements have been made from satellites along lines of sight which pass through the atmosphere just above the earth's limb. By scanning vertically through this region vertical profiles of water vapor and trace gases such as carbon monoxide and the oxides of nitrogen can be measured using appropriate spectral filters and observing a known light source such as the sun, moon, or another satellite as it passes the limb of the earth. Limb scanning yields good vertical resolution in the stratosphere, and observations can be made rapidly over extensive geographical regions. The technique is not useful in the troposphere because observations are usually obscured by clouds.

The British Meteorological Office has used a limb scanning instrument of this type, flown on a rocket over Australia, to measure the vertical profile of water vapor from the lower stratosphere to a height of 50 km.[†] An absorbing channel of 15 cm^{-1} bandwidth centered at 3830 cm^{-1} ($2.61 \mu\text{m}$), was compared with a channel of no absorption. Average humidity of about 3 parts per million was obtained for the region scanned.

The Lyman-α Hygrometer

A variety of radiometric instruments have been developed for measuring water vapor concentration by absorption of the Lyman-α line of the hydrogen spectrum (1215.6×10^{-10} m). Near this wavelength there is a narrow "window" in the oxygen absorption spectrum, and at the precise Lyman-α wavelength the absorption coefficient for water vapor is 387 cm^{-1} while that for oxygen is only 0.3 cm^{-1}.[‡] On either side of this window oxygen absorbs more strongly than water vapor. Absorption by other atmospheric gases is negligible. Therefore, by passing Lyman-α radiation through a sample of air and measuring the incident and transmitted radiances, the humidity of the air can be calculated. Path lengths may be a few centimeters or less. The instrument is basically simple and small and has almost instantaneous time response.

The Infrared Hygrometer

Successful infrared hygrometers have been developed using radiation of two different wavelength bands for which the absorptivities of water vapor are very different. In one such instrument a single radiation source is used,

† Reported by H. W. Yates, *Remote Sensing of the Troposphere*, op. cit., Chapter 26.

‡ D. L. Randall et al., *Humidity and Moisture: Measurement and Control in Science and Industry*, ed. Arnold Wexler, V. 1, *Principles and Methods of Measuring Humidity in Gases*, ed. R. E. Ruskin, Reinhold, New York, 444–454, 1965.

and wavelength bands centered at 1.37 and 1.24 μm are observed at a fixed path length x. Water vapor absorbs strongly at 1.37 μm but is nearly transparent at 1.24 μm.[†] Beer's law then yields for the specific humidity

$$q = \frac{1}{\rho x(k_2 - k_1)}\left(\ln\frac{L_1}{L_2} - \ln\frac{L_{10}}{L_{20}}\right)$$

where k_1 and k_2 are the absorption coefficients at wavelengths λ_1 and λ_2, respectively, and the 0 subscripts indicate radiances without attenuation by water vapor. However, due to the fact that the spectral bands in practice contain numerous closely spaced lines of different radiances, the relation between humidity and radiance integrated over the bandwidth is not that given by Beer's law, and the relationship must be determined experimentally for each instrument. Because this method depends on comparing radiances for the two absorption bands, it can be used in the presence of haze or fog, assuming that the haze or fog has similar effects on the two measurements. It is particularly useful in measurement of very low humidities. Path lengths of 1 m and 0.305 m have been used.

7.19 Scattering Techniques

As Table 7.1 indicates, there are a large number of remote sensing techniques which depend upon scattering by atmospheric particles or by gradients of air density. The most familiar is the use of visual range to determine properties of light scattering particles. Another familiar example is the use of satellite photography to determine distribution, velocity, and other properties of clouds. Other important examples include the use of radar, the acoustic sounder, and the sonic anemometer–thermometer, which are described in this section.

Conventional or Noncoherent Radar[‡]

Radar waves are emitted as pulses of electromagnetic energy of about a microsecond duration followed by "off periods" of about a millisecond. During the off interval the preceding pulse travels to the target and is scattered back to the receiver. The emitted pulse may contain great power, whereas the signal received after scattering from the target may be many orders of magnitude weaker; therefore, the transmitter must be turned off when the receiver is on and the receiver must be off when the transmitter

[†] W. F. Staats *et al.*, *Humidity and Moisture*, op. cit., 465–480.
[‡] The name "radar" originated as an acronym for "radio detecting and ranging."

is on. The maximum range is limited to the product of the speed of light and half the period between pulses. Radar waves are emitted in beams usually several degrees in width; for some purposes wider beams may be used and for others very narrow beams may be used. The time interval between emission of a pulse and the reflected signal determines the distance or range of the target, while the orientation of the antenna determines the direction.

The electromagnetic energy scattered back to the transmitter by particles small compared to wavelength is expressed by the first term in proportionality (7.34). This shows that the back scattered energy depends strongly on size of scattering particle and on wavelength. Suppose that wavelength is held constant at roughly the centimeter to decimeter range, and imagine the size of scattering particle to change. For very small particles the energy scattered is extremely small, but as size increases scattered energy increases markedly. For millimeter particles (falling rain or ice crystals), enough energy is scattered that sensitive radars may be used to detect and study water distribution in clouds. Clouds which do not contain precipitation-size droplets are nearly transparent to radar wavelengths of 5–10 cm, so that attenuation of the radar beam is important only when strongly scattering (large)

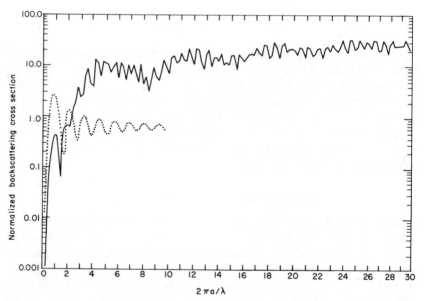

FIG. 7.25. Normalized backscattering cross sections for water (dotted line) and ice (solid line) for a radar wavelength of 3.21 cm calculated from Mie theory. The indices of refraction for ice and water are, respectively, $n' = 1.78 - 0.0024i$ and $n' = 7.14 - 2.89i$ (after B. M. Herman and L. J. Battan, *Quart. J. Roy. Met. Soc.* **87**, 223, 1961).

particles are intercepted. The ratio of the backscattered energy to the energy intercepted by the geometrical cross sectional area of the particles (*normalized backscattering cross section* or scattering efficiency) calculated from Mie theory for water and ice at a radar wavelength of 3.21 cm are shown in Fig. 7.25. Rayleigh scattering is represented on the far left of this figure, where $2\pi a/\lambda$ is less than about 0.6. The Mie series expansions are necessary for $2\pi a/\lambda$ from 0.6 to about 10. At larger values of $2\pi a/\lambda$ scattering efficiency is nearly independent of particle size, and here the methods of geometrical optics are appropriate. The oscillations of the curves shown in Fig. 7.25 have been explained as consequences of alternate constructive and destructive interference between the beams scattered from the front and back faces of the scattering spheres.

Figure 7.25 reveals that at values of a/λ up to about 0.2 water drops are more effective than ice crystals of the same size and mass in scattering radar waves. This results from the fact that for wavelengths of the order of centimeters the index of refraction for liquid water drops is larger than the index of refraction for ice particles; as a result, the first term in proportionality (7.34) for water is about five times the same term for ice. The larger index of refraction for water at radar frequencies results from the fact that each water molecule possesses a dipole moment and that these dipoles can be lined up by electric fields, thereby adding to the polarization of the drop as a whole. This parelectric contribution to polarization is important only for frequencies low enough that the dipoles can be lined up by the electric field before reversal of the field occurs. Liquid and gas molecules are free to move, so that this lining up of dipoles can occur readily, whereas crystal molecules are fixed in position. Therefore, ice has a lower index of refraction than water. At the high frequency of light waves, the molecular dipoles do not respond to the electric field, so at these frequencies the optical indices of refraction of ice and water are practically identical.

As a consequence of greater scattering from water drops than from ice particles, drops of millimeter size are seen more clearly by radar than are ice crystals of comparable mass and size. When ice crystals fall into air warmer than $0°$ C, melting occurs and radar detects the melting as a fairly sharp bright band. The band is brighter than the region below the zero degree isotherm because the raindrops fall faster than the original ice crystals and are therefore farther apart. Coalescence and change in radar cross section with change in shape of the particles may enhance the conspicuousness of the bright band.

For larger particles, such that a/λ exceeds about 0.3, Mie theory indicates that scattering from ice is greater than from water drops, and when a/λ exceeds about 0.7 scattering from ice is about ten times that from water.

Herman and Battan[†] report that this result is associated with the fact that the imaginary part of the index of refraction for ice (absorption) is small compared to the corresponding term for water. When particles are composed of both water and ice, additional effects occur, presumably due to interference between the additional water ice interfaces within the particles. Particle shape and orientation also appear to be important factors.

Radar is used at various ranges out to 200–300 km to delineate the shape and position of clouds containing precipitation-size particles. The reflected signal can provide quantitative measurements of liquid water content, or of rainfall, and can yield changes in these quantities with time. Observations of this kind have been useful in evaluating the importance of the Wegener–Bergeron and the coalescence processes in precipitation and in many other investigations of clouds and cloud systems. Lightning can be detected by radar, and careful evaluations of the observations have provided unique data relating to the mechanism of lightning.

Radar also has been used to measure wind velocity by tracking targets

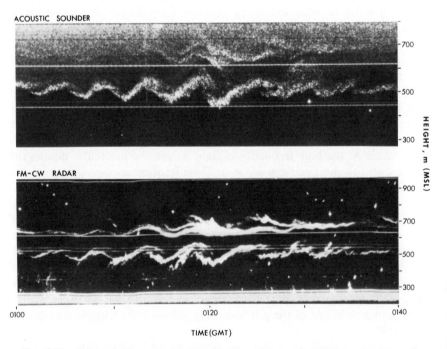

FIG. 7.26. Shear-gravity waves in clear air detected by vertically directed radar and an acoustic sounder (after V. R. Noonkester, *Bull. Am. Meteor. Soc.* **60**, 20–27, 1979).

† B. M. Herman and L. J. Battan, *Quart. J. Roy. Met. Soc.* **87**, 223, 1961.

carried by the wind; reflectors carried by radiosonde balloons are tracked to determine wind velocities as the balloon rises, and in some cases identifiable cloud elements can be assumed to move with the wind. Also, strips of metal foil called "chaff" are sometimes used as radar targets.

Ultrasensitive, high power radars with steerable antennas have been used to observe structures in clear air associated with inhomogeneities in index of refraction (temperature and humidity). The example shown in Fig. 7.26 reveals a series of shear-gravity or Kelvin–Helmholtz waves in the stable boundary layer, observed by both radar and acoustic sounder. Fine details of waves and convection cells, observed by radar, have resulted in improved understanding of mesoscale circulations. Clear air turbulence occurring in the troposphere or lower stratosphere can be detected by radar.

Radar observations made from aircraft† and from the satellite SEASAT, launched in 1978, have been used to infer the wind stress acting on the ocean surface and the wind speed near the surface. In the future such observations made from satellites may provide ocean-wide measurements of surface wind speed.

Coherent or Doppler Radar

The velocity component of the back scattering targets in the direction parallel to the radar beam can be measured by observing the resulting Doppler shift in the carrier frequency. In order to express the relation between the Doppler shift and the target velocity imagine that the target is moving toward the train of waves and is reflecting the train back toward the transmitter. Because frequency is related to speed and wavelength by $v = c/\lambda$, the target intercepts waves with frequency given by

$$v' = (c + v)/\lambda$$

where v represents the speed of the target toward the transmitter. Thus the change in frequency observed by the target is v/λ. An additional and identical change in frequency is observed at the receiver because successive waves are reflected at positions successively closer to the transmitter, so that the total change in frequency is given by

$$\Delta v = 2v/\lambda$$

Although this shift in frequency is very small compared to the carrier frequency, it can be measured with high precision. Obviously, the shift in frequency is negative for target movement away from the transmitter. In Doppler radar observations the radial velocity component is determined

† W. L. Jones et al., IEEE Transactions on Antennas and Propagation AP-25, 1977.

from the shift in carrier frequency at locations in space which are determined from conventional radar data. Vertically pointing Doppler radars can be used to estimate the fall speeds of precipitation and, from knowledge of the drop size and terminal velocity, the vertical air velocity.

Simultaneous Doppler observations from different transmitter locations can be used to determine two components of the velocity field. When the two radars are directed close to the horizontal, this dual Doppler technique can provide a thorough description of the horizontal velocity field. An example is shown in Fig. 7.27. By employing three Doppler radars simultaneously, it is possible to determine the horizontal velocities throughout the precipitation region of a convective storm. The distribution of vertical velocity then can be determined by calculation using the equation of continuity [Eq. (2.31) or Eq. (4.59)].

The Acoustic Sounder

Sound waves emitted from a ground source can be detected after being scattered by inhomogeneities in the boundary layer and therefore can be

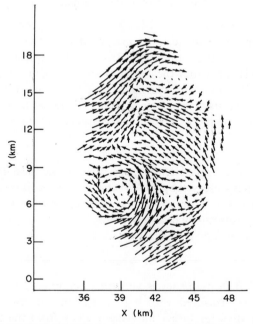

FIG. 7.27.   Horizontal velocity field relative to a moving thunderstorm at a height of 6 km observed by dual Doppler. The observations were made on 28 July 1973 in northeast Colorado as part of the National Hail Research Experiment (after R. A. Kropfli and L. J. Miller, *J. Atmos. Sci.* **33**, 520–529, 1976).

used to detect and track these inhomogeneities. Acoustic sounders or echo-sondes operating on the same principle as radars direct a beam of sound pulses vertically upward and detect the time elapsed between emitted and reflected signal. This provides a measure of the height or heights of reflection and a measure of the energy reflected. Acoustic sounders are used to observe the growth and decay of boundary layer inversions, development of gravity waves on inversions (illustrated in Fig. 7.26), boundary layer convective plumes, and other features of the lower troposphere. Using a hybrid radar–acoustic system temperature profiles in the boundary layer can be determined by using the radar to observe the velocity of the sound wave as a function of height. Doppler echosondes have been used to measure velocities in convective plumes. Figure 7.28 shows an example of an acoustic sounder record of the growth of a convectively mixed layer together with a simultaneous lidar record of scattering from aerosol particles in the boundary layer.

The Sonic Anemometer–Thermometer

Sound waves travel at characteristic speed relative to the air in which they move. If the speed of sound waves is measured in opposite directions with respect to a fixed system, then the difference between the two speeds is twice the wind speed in the direction of measurement. The sum of the two speeds is equal to twice the speed of sound and therefore provides a measure of the temperature. Figure 7.29 illustrates the design of the sonic anemometer-thermometer. An array of two sound pulse emitters E_1 and E_2 and two receivers R_1 and R_2 are placed so that E_1 opposes R_1 and E_2 opposes R_2. The distance between E_1 and R_1 and between E_2 and R_2 is L, whereas the distance between E_1 and R_2 and between E_2 and R_1 is very small. In order for a sound pulse emitted by E_1 to arrive at R_1 it has to be emitted in the direction of $\mathbf{c}t_1$ because the wind carries the pulse in the time t_1 to R_1 along $\mathbf{v}t_1$. Solving for t_1 yields

$$t_1 = \frac{L}{c \cos \beta + v_L}$$

Similarly, the transit time of a pulse from E_2 to R_2 may be expressed by

$$t_2 = \frac{L}{c \cos \beta - v_L}$$

Therefore, after realizing that $c^2 \cos^2 \beta = c^2 - v_n^2$ and $v_n^2 + v_L^2 = v^2$

$$t_2 - t_1 = 2Lv_L/(c^2 - v^2)$$

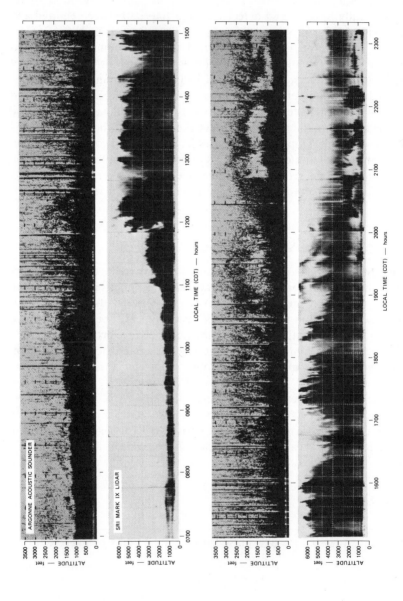

Fig. 7.28. Development of convectively mixed layer as observed by acoustic sounder and by lidar at St. Louis, MO, on 14 August 1972 (courtesy of Edward E. Uthe, SRI International).

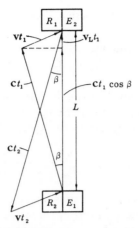

FIG. 7.29. The essential geometry of the sound rays emitted by a sonic anemometer–thermometer. The emitters are indicated by E and the receivers by R.

For $v \ll c$, the component of wind speed in the L direction is approximately

$$v_L = \frac{(t_2 - t_1)c^2}{2L} \tag{7.52}$$

Similarly

$$t_1 + t_2 = \frac{2L(c^2 - v_n^2)^{1/2}}{c^2 - v^2} \approx \frac{2L}{c}$$

Upon combining this equation with (7.4)

$$T = \frac{4c_v L^2}{c_p R_m (t_1 + t_2)^2} \tag{7.53}$$

Equations (7.52) and (7.53) are accurate for dry air to within 1% for wind speeds up to 30 m s^{-1}. However, it is difficult to achieve this level of accuracy in practice. The temperature determination using Eq. (7.53) is strictly valid only for dry air, and the extension necessary for humid air is considered in Problem 8.

The three velocity components can be measured by the three-dimensional sonic anemometer, an instrument consisting of three mutually perpendicular arrays of the type illustrated in Fig. 7.29.

7.20   Refraction Techniques

Although refraction is fundamentally a special aspect of scattering, it is useful to discuss examples of remote sensing by refraction as a separate group.

The discussion of Section 7.13 can be extended to provide an optical method for measuring temperature lapse rate. By combining Eqs. (7.42) and (7.45) the temperature lapse rate may be expressed by

$$-\frac{\partial T}{\partial z} \equiv \gamma = \frac{g}{R_{\mathrm{m}}} - \frac{2nTh}{s^2(n-1)} \tag{7.54}$$

Thus the temperature lapse rate depends directly on the difference between the actual height of an object and its apparent height.

If two telescopes separated by a convenient vertical distance, say 1 m, are focused on targets separated by the same distance, the difference in lapse rates between the two light rays is easily measured. An artillery range finder is a convenient instrument for this purpose. The instrument may be moved vertically in steps to provide a vertical profile of the lapse rate near a uniform surface. Detailed temperature profiles, determined over relatively cold water and over hot land by this method, are shown in Fig. 7.30. These profiles show the very slight anomalies at heights of about 10 cm, probably resulting from radiation exchange with the surfaces as discussed in Section 6.11.

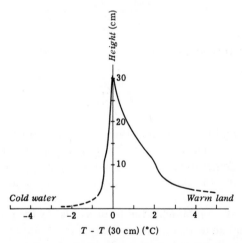

FIG. 7.30.   Vertical profiles of temperature determined by optical measurements over cold water and over warm land (after R. G. Fleagle, *J. Meteor.* **13**, 160, 1956; *Geophys. Research Papers No. 59*, Vol. II, 128, 1959).

Refraction of Radar Waves

To determine the radius of curvature of a horizontal radar ray, express Eq. (7.41) in the form

$$\frac{1}{n}\frac{dn}{dr} = -\frac{1}{r}$$

Then, upon differentiating Eq. (7.40), introducing the equation of state and the hydrostatic equation, and replacing ρ_w by ρq

$$\frac{1}{r} = \frac{n-1}{nT}\left(\frac{g}{R_m} + \frac{\partial T}{\partial z}\right) + \frac{b\rho q}{nT}\left(\frac{1}{T}\frac{\partial T}{\partial z} - \frac{1}{q}\frac{\partial q}{\partial z}\right) \tag{7.55}$$

For radar wavelengths, where b has the value $1.75 \text{ m}^3 \text{ kg}^{-1}$ K, the second term may be as large or larger than the first. Where specific humidity decreases with height, the effect is to refract rays toward the earth. An exercise is given in Problem 9. Under normal conditions of decrease of temperature and humidity with height the radius of the ray is about 4/3 the radius of the earth; consequently, radar usually sees beyond the horizon, but its range is limited.

For rather modest values of humidity lapse or temperature inversion, a horizontal ray may be bent with curvature equal to the curvature of the earth, and the range of the radar is the limited only by its power and sensitivity. Under stronger humidity lapse or temperature inversion even rays emitted with an upward component may be refracted back toward the earth; they are said to be trapped in a radar *duct*, as illustrated in Fig. 7.31. Ducts may develop just above the earth, for instance, when warm dry air moves over

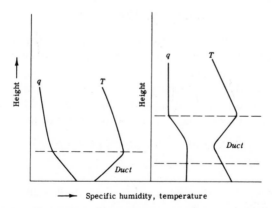

FIG. 7.31. Vertical distributions of temperature (T) and specific humidity (q) which act as radar ducts.

cool water or when moist ground cools by nocturnal radiation. Or elevated ducts may form under dry air which is subsiding or when turbulence creates a lapse layer near the surface in warm air moving over cold water. Usually, ducts at heights of a kilometer or more are only important in interpreting radar observations made from aircraft because signals emitted from the ground are incident on the elevated ducts at angles such that refraction is ineffective.

7.21   Emission–Absorption Techniques

Table 7.1 lists a number of emission and emission–absorption techniques for remote sensing. A particularly direct and simple example is the use of satellite instruments to measure brilliant light flashes, designed for detection of nuclear explosions, but used also for detection, location, and measurement of lightning discharges.

Two important emission–absorption applications are described in this section: determination of upper atmosphere composition and processes from airglow observations, and determination of temperature and water vapor profiles from radiometric measurements.

Airglow

Spectral lines and bands of the airglow emanate from the upper atmosphere, as described briefly in Section 7.12. The heights at which specific lines or bands are emitted can be determined using the van Rhijn method as follows. The radiance of the line or band is measured at the earth's surface as a function of zenith angle, as illustrated in Fig. 7.32. This diagram shows that the radiance emitted in the direction θ by the layer dh is given by

$$L_\lambda(\theta) = L_\lambda(0)\sec(\theta - \phi) \tag{7.56}$$

where $L_\lambda(0)$ is the radiance emitted in the vertical direction. If no absorption takes place between the layer and the earth's surface, this radiance is also measured at the surface.

For a given zenith angle, the angle ϕ is a function of height of the layer. When h is increased by dh, the angle ϕ is increased by $d\phi$. From the small triangle in the layer dh of Fig. 7.32, it can be deduced that

$$dh^2 \sec^2(\theta - \phi) = dh^2 + (R + h)^2\, d\phi^2$$

or

$$\frac{d\phi}{dh} = \frac{[\sec^2(\theta - \phi) - 1]^{1/2}}{R + h} = \frac{\sin(\theta - \phi)}{(R + h)\cos(\theta - \phi)}$$

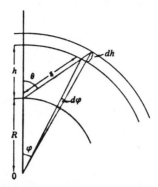

FIG. 7.32. Cross section through the earth and the atmosphere showing the contribution to the radiation received at the surface from a horizontal layer in the upper atmosphere.

where R is the radius of the earth. This equation can be integrated to

$$\frac{R + h}{R} = \frac{\sin \theta}{\sin(\theta - \phi)}$$

or

$$\sec(\theta - \phi) = \left\{ 1 - \left(\frac{R \sin \theta}{R + h} \right)^2 \right\}^{-1/2}$$

Substitution of this equation into (7.56) yields

$$L_\lambda = L_\lambda(0) \left\{ 1 - \left(\frac{R \sin \theta}{R + h} \right)^2 \right\}^{-1/2} \tag{7.57}$$

For $h \ll R$ the right-hand side of Eq. (7.57) approaches $L_\lambda(0) \sec \theta$. In Fig. 7.33, $L_\lambda/L_\lambda(0)$ is plotted for a few values of h/R as a function of θ. Because this height is a small fraction of the earth's radius, very accurate measurements of the radiance are required. Unfortunately, the radiation is weak and variable in space and time, so that accurate observations are very difficult to achieve.

In case the emitted radiation is partly absorbed on the way down, the radiance is reduced according to Beer's law over the path s, and Eq. (7.57) is modified to

$$L_\lambda = L_\lambda(0) \left\{ 1 - \left(\frac{R \sin \theta}{R + h} \right)^2 \right\}^{-1/2} \exp\left[-k_\lambda \int_0^s \rho \, ds \right]$$

where k_λ is the absorption coefficient and s is a function of h for constant θ

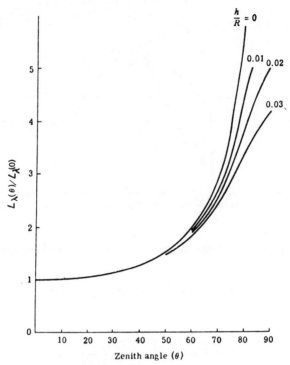

FIG. 7.33.   Ratio of radiance from a thin layer at zenith angle θ to radiance at zero zenith angle as calculated from Eq. (7.57).

given by

$$s = R \cos\theta\left\{\left(1 + \frac{2Rh + h^2}{R^2 \cos^2\theta}\right)^{1/2} - 1\right\}$$

The heights found by various investigators using this method have varied between 70 and 300 km, indicating that most of the radiation is emitted by the upper mesosphere and lower thermosphere. Direct measurements of airglow emission using rocket and satellite photometers indicate that most of the airglow originates between heights of 80 and 120 km.

Radiometric Sounding

In Chapter V we considered the problem of calculating the radiation emitted and absorbed by the atmosphere given the distributions of temperature, water vapor, and other absorbing and radiating gases. In this section we consider the inverse problem, namely, to determine the distributions of temperature and of radiating gases given radiances measured at a number of wavelengths at the top or bottom of the atmosphere. To determine the distribution of temperature, it is assumed that a gas such as carbon dioxide or oxygen is uniformly mixed throughout the vertical column; it is then possible to interpret differences in measured radiances of spectral lines emitted by the gas so as to determine the vertical distribution of temperature. Carbon dioxide lines in the infrared part of the spectrum are used for sounding in clear areas, and oxygen lines in the microwave spectrum may be used in areas of clouds and precipitation. The method is outlined below. After determining the temperature distribution, it is then possible to select lines in the water vapor spectrum and from radiances measured for these lines to determine the vertical distribution of water vapor.

To determine the vertical temperature distribution consider the infrared radiance as observed from a satellite by a vertically directed instrument. For a particular wavelength in the carbon dioxide spectrum, Schwarzshild's equation (5.10) can be integrated, after multiplying by e^u, to yield

$$e^u L_\lambda \Big|_0^{u_t} = \int_0^{u_t} L_\lambda^* e^u \, du$$

where the optical thickness u represents $\int k_\lambda \rho q_c \, dz$, q_c is the carbon dioxide mixing ratio, and k_λ is the absorption coefficient. The integration proceeds upward in the direction of the radiation. Therefore,

$$L_\lambda = L_{\lambda 0} e^{-u_t} + \int_0^{u_t} L_\lambda^* e^{-(u_t - u)} \, du$$

Upon defining the transmissivity by $\tau_\lambda(u_t - u) \equiv e^{-(u_t - u)}$ and $\tau_{\lambda 0} \equiv e^{-u_t}$, the equation of radiative transfer becomes

$$L_\lambda = L_{\lambda 0} \tau_{\lambda 0} + \int_0^{u_t} L_\lambda^* \frac{d\tau_\lambda(u_t - u)}{du} \, du \tag{7.58}$$

Equation (7.58) states that the monochromatic radiance measured by the satellite instrument represents the sum of the upward radiance at the earth's surface times the transmissivity of the atmosphere plus the integral over the vertical column of the black-body radiance times the differential of transmissivity. A similar equation expresses the measured radiance for each wavelength observed by the satellite instrument. For downward directed radiation observed at the earth's surface a similar equation may be written, but in this case the second term alone represents the radiance since there is negligible downward radiance at the top of the atmosphere. For the remainder of this discussion we will consider that wavelengths are chosen which do not penetrate all the way through the atmosphere, so that the first term can be omitted for upward radiation.

If transmissivity is assumed independent of temperature and if n closely spaced wavelengths are chosen, Eq. (7.58) may be written after transforming from u as the vertical coordinate to the geopotential Φ.

$$L_i = \int_0^\infty L^*(\bar{\lambda}, \Phi)K_i(\Phi)\,d\Phi, \qquad i = 1, 2, \dots, n \qquad (7.59)$$

where $\bar{\lambda}$ is an average of the closely spaced wavelengths and $K_i \equiv d\tau_i/d\Phi$. Equation (7.59) states that each measured radiance is a weighted mean of the Planck function profile L^*, which is the unknown, with K_i representing the weighting function for each chosen wavelength.

The weighting function is analogous to the vertical distribution of the rate of absorption of ultraviolet radiation which was discussed in Section 5.20. The weighting function can be expressed as

$$K_i = \frac{d\tau_i}{d\Phi} = \frac{k_i q_c p_0}{g R_m T} \exp\left[-\frac{\Phi}{R_m T} - \frac{k_i q_c p_0}{g} \exp\left(\frac{\Phi}{R_m T}\right) \right] \qquad (7.60)$$

The shapes of these weighting functions are determined by the product of a decreasing exponential of height and a function which increases with height to a finite value at the top of the atmosphere. The height of maximum value of the weighting function can be found by setting the derivative of Eq. (7.60) equal to zero. It can be expressed approximately as

$$\Phi_m = gH \ln(k_i q_c \rho_0 H) \qquad (7.61)$$

where $H \equiv \overline{R_m T}/g$. Problem 10 provides an exercise which shows that for each wavelength maximum radiance is emitted from the height corresponding to a transmissivity of $1/e$.

In practice, since the instrument receives and senses a finite bandwidth rather than a single wavelength, the transmissivity, and therefore the weighting function, is not represented exactly by Eq. (7.60). Figure 7.34a shows the actual weighting functions determined experimentally for the eight channel Satellite Infra Red Spectrometer (SIRS) flown on Nimbus 3. Each of the chosen wavelength bands emanates from a different layer of the atmosphere. The channel at $669\ \text{cm}^{-1}$ senses radiation from the center of a strongly absorbing and emitting band; the instrument therefore sees only a relatively short distance downward into the atmosphere. The other channels were chosen to select radiation from successively lower layers of the atmosphere, and the final channel was chosen in a nearly transparent region of the spectrum in order to measure the temperature of the earth's surface.

However, a finite number of observations cannot describe uniquely the continuous profile (L^*). There could be a variety of temperature profiles which would result in the same measured radiance for one or more of the chosen wavelengths. Therefore, methods have been devised to approximate the temperature distribution given the measured radiances. These methods, which are too elaborate to be discussed here, yield temperature soundings in clear air which have a root-mean-square deviation from a radiosonde sounding of 1°C to 2°C. An example is shown in Fig. 7.35. In areas which are partially cloudy, measurements from regions between clouds often can be used. In areas of complete cloud cover, it is possible in principle to use a microwave radiometer which uses wavelengths from the O_2 spectrum for which clouds are transparent. Since the emitted radiances are very small at microwave frequencies, this requires very high sensitivity.

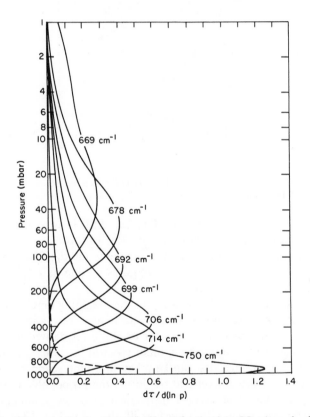

FIG. 7.34a.   The weighting functions ($d\tau/d(\ln p)$) for the eight CO_2 channels of the Satellite Infra Red Spectrometer (SIRS-A) flown on Nimbus 3 (after W. L. Smith, *Applied Optics* **9**, 1993–1999, 1970).

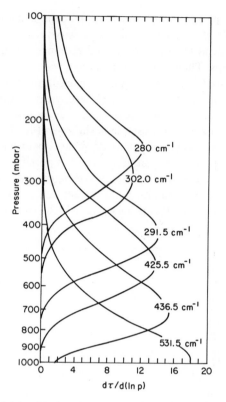

FIG. 7.34b. The weighting functions for the six H_2O channels of SIRS-B (after W. L. Smith, *Applied Optics* **9**, 1993–1999, 1970).

To determine the water vapor profile, Eq. (7.58) is applied to radiances observed at selected water vapor wavelengths. In this case τ_λ becomes the unknown. The water vapor weighting functions for the SIRS experiment are shown in Fig. 7.34b. Problem 11 requires an alternate formulation of the radiative transfer equation.

The spectral intervals corresponding to the weighting functions shown in Figs. 7.34a and b were chosen to provide temperature and water vapor soundings in the troposphere. If greater resolution is desired in the stratosphere and less in the troposphere, other spectral intervals can be chosen.

Radiometric soundings also can be taken in a limb scanning mode, and this provides much greater vertical resolution than is possible in the downward vertical mode. The vertical instrument observes a radiance whose source can be associated with a layer as illustrated in Fig. 7.34a. The limb scanning beam which observes this same layer intercepts a considerably

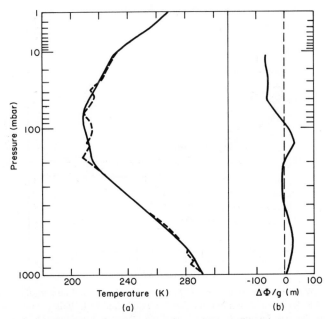

FIG. 7.35. (a) Temperature soundings determined by the SIRS-A instrument on Nimbus 3 (*solid line*) and by radiosonde (*dotted line*) in clear air near Little Rock, Ark., 25 April 1969. (b) Differences in height of pressure surfaces ($\Delta\Phi/g$) calculated for the soundings on the left side (after D. Q. Wark, *Applied Optics* **9,** 1761–1766, 1970).

greater mass of air than does the vertical beam, so that the measured radiance is greater. The height at which most of the radiation is emitted is determined from the orientation of the instrument. Vertical resolution is achieved by rotating the instrument in a vertical plane, so that it samples the atmosphere at closely spaced vertical intervals. The weighting functions shown in Fig. 7.36 for the instrument flown on Nimbus 6 show clearly that considerably greater vertical resolution can be achieved by limb scanning than from vertical radiance measurements. However, because of the presence of clouds in the troposphere, the method has been limited to the stratosphere and above.

Soundings also have been made using radiometers on the ground looking upward either vertically or along slant paths. In these cases the corresponding weighting functions decrease monotonically upward because both transmissivity and air density decrease upward. Since the ground-based radiometer is best suited to providing low level data, while the satellite radiometer is best suited to providing high level data, a combined system has special promise. Such a system has not yet been developed.

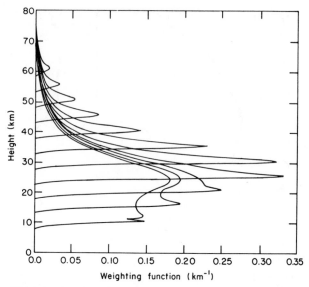

FIG. 7.36.   Weighting functions for a CO_2 absorption band for Limb Radiance Inversion Radiometer (LRIR) flown on Nimbus 6 (after J. C. Gille and P. L. Bailey, *Inversion Methods in Atmospheric Remote Sensing*, A. Deepak, ed., 195–216, Academic Press, New York, 1977).

List of Symbols

		First used in Section
a	Radius of scattering element	7.7
A	Amplitude, area	7.1, 7.3
B	Brightness, scalar magnetic induction	7.11, 7.16
$\mathbf{B}$	Magnetic induction	7.3
a, b	Constants	7.13
c	Phase speed	7.1
c_g	Group velocity	7.4
c_0	Speed of light in vacuum	7.6
c_p, c_v	Specific heat at constant pressure and constant volume, respectively	7.2
C	Contrast	7.11
D	Displacement, angle of deviation	7.3, 7.14
$\mathbf{D}$	Vector displacement	7.3
e	Elementary electric charge	7.6
E	Electric field strength (scalar), irradiance	7.3, 7.11
$\mathbf{E}$	Electric field strength	7.3
F	Force	7.2
$\mathbf{F_M}$	Magnetic force	7.3
g	Force of gravity per unit mass	7.2

		First used in Section
h	Depth of water column, difference between true and apparent height, height above earth's surface	7.2, 7.13, 7.21
H	Magnetic field strength (scalar), scale height	7.3, 7.21
$\mathbf{H}$	Magnetic field strength	7.3
i	Angle of incidence	7.10
I	Electrical current	7.3
$\mathbf{J}$	Electrical current density	7.3
$\mathbf{J_D}$	Displacement current density	7.3
k	Wave number, absorption coefficient	7.4, 7.6
K	Weighting function	7.21
L	Radiance, distance between emitter and receiver	7.11, 7.18
m	Mass per unit length, mass of electrical charge	7.2, 7.6
n	Real index of refraction, integer	7.6, 7.9
n'	Complex index of refraction	7.6
N	Number of scattering particles per unit volume	7.6
p	Pressure	7.2
P	Prism angle	7.14
$\mathbf{P}$	Polarization	7.3
q	Electrical charge, specific humidity	7.3, 7.18
r	Radius, distance, angle of refraction	7.2, 7.7, 7.10
R	Radius of earth	7.21
R_m	Specific gas constant for air	7.2
s	Path length, displacement	7.2, 7.6
t	Time	7.1
T	Tension, temperature	7.2, 7.13
u	Optical thickness	7.21
v	Speed	7.18
$\mathbf{v}_q$	Velocity of charge	7.3
x_v	Visual range	7.11
x, y, z	Cartesian coordinates	7.1
Z	Apparent zenith angle	7.13
α	Specific volume, polarizability, angular radius of diffraction band	7.2, 7.6, 7.9
α'	Half-width of absorption line	7.6
β	Angle between sound ray and direct path	7.18
γ	Damping coefficient, temperature lapse rate	7.6, 7.17
δ	Angular deviation in astronomical refraction	7.13
ε	Permittivity	7.3
θ	Angle between dipole axis and direction of wave propagation, angular radius of diffraction ring, angle, zenith angle	7.7, 7.9, 7.13, 7.21
κ	Absorption index	7.6
λ	Wavelength	7.1
μ	Permeability	7.3
ν	Frequency	7.6
ν_G	Gyrofrequency	7.16
ρ	Density	7.2
ρ_w	Density of water vapor	7.13
σ	Surface area, extinction coefficient	7.7, 7.11
τ	Transmissivity	7.21

		First used in Section
ϕ	Longitude angle in spherical coordinates, angle defined in Fig. 7.30	7.7, 7.21
Φ	Geopotential	7.21
ψ_λ	Sensitivity of the human eye	7.11
χ	Susceptibility	7.3
ω	Angular frequency	7.4

Subscripts

0	Value in vacuum, resonance
m	Maximum value
ν	Frequency
λ	Wavelength
L	Component along array
n	Component normal to array
t	Total
c	Carbon dioxide

Problems

1. Show that the differential equation which governs a wave on a stretched string may be written

$$\frac{\partial^2 z}{\partial t^2} = \frac{T}{m}\frac{\partial^2 z}{\partial y^2}$$

where T represents the tension in the string and m the mass per unit length.

2. Find the average diameter of cloud droplets which produce a white corona of 4° angular radius. What may be concluded about the range of sizes of the cloud droplets?

3. For a particular object, contrast at short range appears to be -0.5. What is the maximum distance that this object can be seen when the visual range is 10 km?

4. Find the temperature lapse rate required for light to travel around the earth at sea level. Assume the air temperature is 20° C. What value of g would result in curvature around a planet whose size is the same as the earth and whose atmosphere is isothermal if the index of refraction and the gas constant are the same as that in our atmosphere?

5. Find the difference in angular radius of the red and violet rings of the haloes formed by prism angles of 60° and 90° if

$\lambda(\mu m)$	n
0.656 (red)	1.307
0.405 (violet)	1.317

6. Derive the angle of minimum deviation for the secondary bow. Describe the color sequence of the primary and secondary bow using the following indices of refraction and wavelengths.

$\lambda(\mu m)$	n
0.656	1.332
0.405	1.344

7. Find the critical temperature lapse rate just sufficient to prevent the sound of a volcanic eruption at the top of a mountain from reaching an observer by a direct path for the following case: The mountain is 100 km from the observer, and the top is 4×10^3 m higher than the observer. Neglect the curvature of the earth and assume that the average temperature is 290 K.

8. If the specific humidity is 10 g kg^{-1}, find the error in temperature made by using Eq. (7.53) for dry air.

9. (a) Find the increase of specific humidity with height which would result in straight line propagation of radar waves at sea level if the specific humidity is 5 g kg^{-1} and the temperature is 290 K and is constant with height.

(b) Find under the same conditions the change of specific humidity with height which results in curvature of a horizontal beam equal to the curvature of the earth.

10. Show that the height given by Eq. (7.61) corresponds to an optical path length $(u_t - u)$ of unity.

11. Show that the spectral radiance observed above the radiating atmosphere can be expressed by

$$L_\lambda = L_\lambda^*(u_t) + \int_{L_\lambda^*(u_t)}^{L_\lambda^*_0} \tau_\lambda \, dL_\lambda^*$$

where it is assumed that the surface of the earth radiates as a black body. Discuss the two terms in the cases of an isothermal atmosphere and an atmosphere with uniform lapse rate.

Solutions

1. As shown in Section 7.2, the force in the downward direction experienced by a segment of string ΔS displaced vertically upward is expressed by

$$F = T \, \Delta S / r$$

where r is the radius of curvature of the string. From Newton's second law this force must equal the mass of the segment $m \, \Delta S$ times the vertical acceleration $\partial^2 z/\partial t^2$. Upon recognizing that for infinitesimal displacement the curvature of the string $(1/r)$ is $\partial^2 z/\partial y^2$, where y is distance along the string, we have

$$m \, \Delta S \frac{\partial^2 z}{\partial t^2} = T \, \Delta S \frac{\partial^2 z}{\partial y^2}$$

and

$$\frac{\partial^2 z}{\partial t^2} = \frac{T}{m} \frac{\partial^2 z}{\partial y^2}$$

2. Average droplet diameter: $\sim 14 \, \mu\text{m}$, Range: ~ 11 to $16 \, \mu\text{m}$.

3. Contrast of -0.5 indicates that the object at close range is half as bright as the background. Contrast decreases with distance according to $C(x) = C_0 e^{-\sigma x}$. Therefore, the maximum distance the object can be seen is

$$x_m = -\frac{1}{\sigma} \ln \left| \frac{0.02}{0.5} \right|$$

The scattering coefficient σ is found from Eq. (7.38) to be $3.912 \times 10^{-4} \text{ m}^{-1}$. Therefore, $x_m = 8.2$ km.

4. $\gamma = -0.123°\,C\,m^{-1}$, $g = 45\,m\,s^{-2}$

5. For $P = 60°$, $\delta D_m = 45'$. For $P = 90°$, $\delta D_m = 2°12'$.

6. The equations for i, r, and $\pi - D_m$ for the primary rainbow which are developed in Section 7.14, yield the following values

n	i	r	$\pi - D_m$
1.332 (red)	59.47°	40.29°	42.22°
1.344 (violet)	58.77°	39.51°	40.51°

Therefore, the red appears on the outside, violet on the inside.

For two internal reflections the angle of deviation is

$$D = 2(i - r) + 2(\pi - 2r) = 2i - 6r + 2\pi$$

Upon differentiating and setting the derivative with respect to i equal to zero

$$di = 3\,dr$$

From Snell's law, $\sin i = n \sin r$, and upon differentiating, $3 \cos i = n \cos r$. And after squaring and adding

$$\cos i = [(n^2 - 1)/8]^{1/2} \qquad \cos r = 3/n[(n^2 - 1)/8]^{1/2}$$

From these two equations for i and r corresponding to minimum deviation and the original equation for D, we may calculate the three angles for the given values of n. This gives for the secondary bow

n	i	r	$D_m - \pi$
1.332 (red)	71.87°	45.52°	50.63°
1.344 (violet)	71.49°	44.87°	53.73°

Therefore, the secondary bow is seen outside the primary bow and violet appears on the outside, red on the inside, the reverse of the order of the primary bow. The reversal of order occurs because in the secondary bow $D_m > \pi$, that is, the returning ray crosses the incident ray.

7. Equation (7.50) can be applied directly. Upon integrating

$$\gamma = \frac{4Tz_1}{x_1^2}$$

where the subscripts represent the values for the top of the mountain. Substitution gives

$$\gamma = 0.46°\,C\,km^{-1}$$

that is, decrease of temperature with height at a greater rate will result in the sound not reaching the observer directly.

8. Humidity affects c_p/c_v and R_m. Therefore, the effect of these factors on the temperature calculated from Eq. (7.53) can be expressed by

$$\Delta T = T\left\{\frac{\Delta(c_v/c_p)}{c_v/c_p} - \frac{\Delta R_m}{R_m}\right\}$$

For specific humidity of $10 \, \text{g kg}^{-1}$ the following values can be computed from Table 2.3.

$$\frac{\Delta(c_v/c_p)}{c_v/c_p} = 5.8 + 10^{-4}$$

$$\frac{\Delta R_m}{R_m} = 0.61 \times 10^{-2}$$

Therefore, $\Delta T/T = -55 \times 10^{-4} = -0.55\%$ and $\Delta T \approx -1.65°$ C (for $T = 300$ K). Therefore, use of Eq. (7.53) for dry air will yield temperatures too high by about $1.65°$ C.

9. (a) For the conditions specified, Eq. (7.55) yields

$$\frac{\partial q}{\partial z} = \frac{n-1}{b\rho}\frac{g}{R_m} = \left(\frac{a}{b} + \frac{q}{T}\right)\frac{g}{R_m} = 0.051 \times 10^{-4} \, \text{m}^{-1} = 0.51 \, \text{g kg}^{-1}(100 \, \text{m})^{-1}$$

(b) Upon substituting the radius of the earth for r, Eq. (7.55) yields

$$\frac{\partial q}{\partial z} = \frac{n-1}{b\rho}\frac{g}{R_m} - \frac{nT}{b\rho r_E} = (0.051 - 0.208) \times 10^{-4} \, \text{m}^{-1} = -1.57 \, \text{g kg}^{-1}(100 \, \text{m})^{-1}$$

10. The definition of optical thickness gives

$$u_t - u = \int_z^\infty k_\lambda \rho q_c \, dz \approx k_\lambda \rho_0 q_c \int_z^\infty e^{-gz/\overline{R_m T}} \, dz$$

Integrating and setting $\overline{R_m T}/g \equiv H$ gives

$$u_t - u = k_\lambda \rho_0 q_c H e^{-z/H}$$

Substitution of this relation in Eq. (7.61) results in

$$u_t - u = 1$$

showing that the height given by Eq. (7.61) corresponds to an optical thickness of unity. From the definition of transmissivity, it follows that where $u_t - u = 1$, $\tau = e^{-1}$.

11. Equation (7.58) may be integrated by parts to yield

$$L_\lambda = L_{\lambda 0}\tau_{\lambda 0} + L_\lambda^* \tau_\lambda \Big|_0^{u_t} - \int_{L_{\lambda 0}^*}^{L_\lambda^*(u_t)} \tau_\lambda \, dL_\lambda^*$$

$$L_\lambda = L_\lambda^*(u_t) + \int_{L_\lambda^*(u_t)}^{L_{\lambda 0}^*} \tau_\lambda \, dL_\lambda^*$$

The first term on the right is the Planck radiance at the top of the radiating atmosphere, that is where the transmissivity is unity. The second term represents the integral through the vertical column of the product of transmissivity and differential of Planck radiance. For an isothermal atmosphere, the integral vanishes and the total radiance is the Planck radiance for the top of the atmosphere. For a uniform lapse rate (temperature increasing downward), the integral contributes positively. Each layer contributes an amount proportional to the product of transmissivity and difference in temperature between the bottom and top.

General References

Battan, *Radar Observation of the Atmosphere*, gives a coherent, well-organized account of the application of radar to atmospheric observation and measurement.

Derr, *Remote Sensing of the Troposphere*, a compilation of 30 papers, covers background and many specific applications.

Helliwell, *Whistlers and Related Ionospheric Phenomena*, provides a comprehensive account of these topics which are only marginally discussed here.

Humphreys, *Physics of the Air*, presents the physical basis for a great range of atmospheric phenomena, especially those which are distinctly acoustic or optical.

Johnson, *Physical Meteorology*, discusses a wide range of topics in atmospheric physics. His discussion of visibility and other topics related to scattering is particularly useful.

Joos, *Theoretical Physics*, gives an excellent discussion of the physical basis of index of refraction, diffraction, and electromagnetic waves.

Panofsky and Phillips, *Classical Electricity and Magnetism*, gives an excellent treatment on an advanced level of dipole radiation and other fundamental topics concerned with electromagnetic waves.

Sears, *Principles of Physics*, Vol. 3, gives an elementary but correct and clear account of optics including the ray treatment of diffraction and refraction.

Unsöld, *Physik der Sternatmosphären*, gives a comprehensive account on an advanced level of the mechanism of absorption of radiation.

Van de Hulst, *Light Scattering by Small Particles*, provides an advanced account of the problems associated with scattering.

Mathematical Topics

A. Partial Differentiation

The quantities which serve to describe the state of the atmosphere, for example, temperature, specific humidity, and velocity, are called dependent variables. They depend on four independent variables: time and the three space coordinates.

First consider a function represented by $z = f(x)$ which is continuous, has continuous derivatives, and is single valued. The dependent variable is z, and the independent variable is x. The derivative is defined by

$$\frac{dz}{dx} \equiv \lim_{\Delta x \to 0} \frac{f(x + \Delta x) - f(x)}{\Delta x}$$

which describes the slope of the curve of z in the xz plane. Now suppose that the function z represents the height of the surface of a mountain. The height must depend on both x and y, the two horizontal coordinates. It is clear that a derivative of z may be calculated with respect to x along any of an infinite number of directions depending on the direction chosen for x. If x is chosen along the line of steepest ascent, dz/dx is large, but if the direction chosen is along a contour, dz/dx vanishes. To avoid this ambiguity, derivatives of functions of two or more variables are defined holding all but one independent variable constant. Thus for a function of two independent variables

$$\left(\frac{dz}{dx}\right)_y \equiv \frac{\partial z}{\partial x} \equiv \lim_{\Delta x \to 0} \frac{f(x + \Delta x, y) - f(x, y)}{\Delta x}$$

This is called the *partial derivative* of z with respect to x. The derivative with respect to y holding x constant is written

$$\left(\frac{dz}{dy}\right)_x \quad \text{or} \quad \frac{\partial z}{\partial y}$$

For a function of more than two variables, the partial derivative is formed by holding all but one independent variable constant. Higher partial derivatives may be formed in the same way; the order of differentiation may be interchanged without affecting the result. Thus

$$\frac{\partial^2 z}{\partial x\, \partial y} = \frac{\partial^2 z}{\partial y\, \partial x}$$

as may be demonstrated by choosing an elementary function of x and y and carrying out the differentiation in the indicated order.

In the examples, z has been used as a dependent variable. In many problems in geophysics z also is used as an independent variable.

B.   Elementary Vector Operation

A quantity which is described by a direction and a magnitude, force, for example, is called a *vector* quantity; whereas a quantity which is described by magnitude alone, temperature, or pressure, for example, is called a scalar quantity. Vectors may be represented by bold-faced letters, for example. **A**.

Vector Addition

The vectors **A** and **B** may be added by making their lengths proportional to their respective magnitudes and attaching the tail of **B** to the head of **A**. The vector connecting the tail of **A** and the head of **B** is defined as the vector sum; it is written **A** + **B**. Vectors are equivalent if they have the same direction and magnitude. It is clear from Fig. A-1 that the vector **A** may be expressed as the sum of three vectors taken along the three coordinate axes x, y, and z. This sum may be expressed by the vector equation

$$\mathbf{A} = \mathbf{A}_x + \mathbf{A}_y + \mathbf{A}_z$$

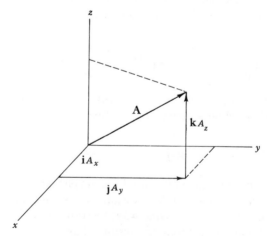

FIG. A-1.   Projections of the vector **A** on the axes x, y, and z.

It is customary to define unit vectors in the x, y, and z directions by the equations

$$A_x/A_x \equiv \mathbf{i} \qquad A_y/A_y \equiv \mathbf{j} \qquad A_z/A_z \equiv \mathbf{k}$$

It is important to realize that a unit vector describes a direction but has no dimensions and has unit magnitude in any system of units. The use of unit vectors permits expanding $\mathbf{A}$ in the form

$$\mathbf{A} = \mathbf{i}A_x + \mathbf{j}A_y + \mathbf{k}A_z$$

The Pythagorean theorem shows that

$$A^2 = A_x^2 + A_y^2 + A_z^2$$

Vector Multiplication

The product of a scalar and a vector is a vector with magnitude equal to the product of the magnitudes and the direction of the original vector. Multiplication by a negative number reverses the direction of the vector.

The product of two vectors is defined in two different ways, reflecting two physical relationships between vectors. First, the *scalar* or *dot* product of two vectors is defined by the equation

$$\mathbf{A} \cdot \mathbf{B} \equiv AB \cos(\mathbf{A}, \mathbf{B})$$

and is read, A dot B. Because the cosine of an angle equals the cosine of the corresponding negative angle, $\mathbf{A} \cdot \mathbf{B} = \mathbf{B} \cdot \mathbf{A}$. An example of the use of the scalar, or dot, product is the definition of work

$$W \equiv \int \mathbf{F} \cdot d\mathbf{s}$$

where $\mathbf{F}$ represents force, $\mathbf{s}$ represents displacement, and the integral is taken along a specified line or contour.

Second, the vector product of vectors $\mathbf{A}$ and $\mathbf{B}$ is defined as a vector normal to the plane of $\mathbf{A}$ and $\mathbf{B}$ in the direction which a right-hand screw would advance if rotated from $\mathbf{A}$ into $\mathbf{B}$ through the smaller angle and with a magnitude given by $AB \sin(\mathbf{A}, \mathbf{B})$. The *vector* or *cross* product is written $\mathbf{A} \times \mathbf{B}$ and is read $\mathbf{A}$ cross $\mathbf{B}$. Because the order of the vectors determines the direction of the vector product, $\mathbf{A} \times \mathbf{B} = -\mathbf{B} \times \mathbf{A}$.

An example of the vector product is the definition of the angular momentum in the form, $m\mathbf{r} \times \mathbf{v}$. From Fig. A-2, it is seen that rotation of $\mathbf{r}$ into $\mathbf{v}$ generates a vector directed out of the paper as a right-hand screw would advance if rotated from $\mathbf{r}$ into $\mathbf{v}$.

From Fig. A-2 it may also be recognized that the angular frequency vector ω may be defined by

$$\omega \equiv \frac{\mathbf{r} \times \mathbf{v}}{r^2}$$

It follows then that $\omega \times \mathbf{r} = \mathbf{v}$ and that the angular momentum may be defined by $m\mathbf{r} \times (\omega \times \mathbf{r})$. These relations hold whether or not the vector $\mathbf{r}$ lies in the plane normal to ω.

Unit Vector Relations

The following vector products of the unit vectors can be readily recognized

$$
\begin{array}{ll}
\mathbf{i} \cdot \mathbf{i} = 1 & \mathbf{j} \times \mathbf{i} = -\mathbf{k} \\
\mathbf{j} \cdot \mathbf{j} = 1 & \mathbf{j} \times \mathbf{k} = \mathbf{i} \\
\mathbf{k} \cdot \mathbf{k} = 1 & \mathbf{k} \times \mathbf{i} = \mathbf{j} \\
\mathbf{i} \cdot \mathbf{j} = 0 & \mathbf{j} \times \mathbf{j} = 0 \\
\mathbf{j} \cdot \mathbf{k} = 0 & \mathbf{k} \times \mathbf{k} = 0 \\
\mathbf{i} \cdot \mathbf{k} = 0 & \mathbf{i} \times \mathbf{i} = 0
\end{array}
$$

The Vector Operator ∇

It is convenient to introduce the *del* operator which is defined in Cartesian coordinates by

$$\nabla \equiv \mathbf{i}\,\frac{\partial}{\partial x} + \mathbf{j}\,\frac{\partial}{\partial y} + \mathbf{k}\,\frac{\partial}{\partial z}$$

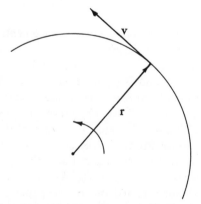

FIG. A-2.   The velocity (**v**) and position vector (**r**) of a point rotation.

It may operate on a scalar field $\phi(x, y, z)$ in which case the *gradient* of ϕ is expressed by

$$\nabla\phi = \mathbf{i}\,\frac{\partial\phi}{\partial x} + \mathbf{j}\,\frac{\partial\phi}{\partial y} + \mathbf{k}\,\frac{\partial\phi}{\partial z}$$

Or ∇ may operate on a vector in either of two ways. The *dot* operation of ∇ into $\mathbf{A}$, called the *divergence* of $\mathbf{A}$, is defined as

$$\nabla\cdot\mathbf{A} \equiv \mathbf{i}\cdot\frac{\partial\mathbf{A}}{\partial x} + \mathbf{j}\cdot\frac{\partial\mathbf{A}}{\partial y} + \mathbf{k}\cdot\frac{\partial\mathbf{A}}{\partial z} = \frac{\partial A_x}{\partial x} + \frac{\partial A_y}{\partial y} + \frac{\partial A_z}{\partial z}$$

The *cross* operation of ∇ into $\mathbf{A}$, called the *curl* of $\mathbf{A}$, is defined as

$$\nabla\times\mathbf{A} \equiv \mathbf{i}\times\frac{\partial\mathbf{A}}{\partial x} + \mathbf{j}\times\frac{\partial\mathbf{A}}{\partial y} + \mathbf{k}\times\frac{\partial\mathbf{A}}{\partial z}$$

$$= \mathbf{i}\left(\frac{\partial A_z}{\partial y} - \frac{\partial A_y}{\partial z}\right) + \mathbf{j}\left(\frac{\partial A_x}{\partial z} - \frac{\partial A_z}{\partial x}\right) + \mathbf{k}\left(\frac{\partial A_y}{\partial x} - \frac{\partial A_x}{\partial y}\right)$$

C. Taylor Series

It is often important to expand a function in a series of terms about a particular point. Examples occur in the solution of differential equations (both by analytic and numerical methods), in the analysis of errors and in the evaluation of indeterminate forms of functions. The Taylor series provides a systematic method for such expansion. It may be illustrated by a geometrical example: Suppose that it is desirable to evaluate a function $f(x)$, as shown in Fig. A-3, by measurements made in the neighborhood of the point x_0. Taylor's theorem states that a function which has continuous

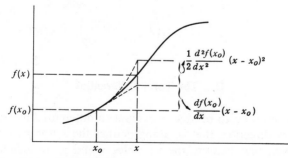

FIG. A-3. Geometrical representation of the expansion of $f(x)$ about the point x_0.

derivatives in a certain interval may be expressed in that interval by an infinite series of the form

$$f(x) = f(x_0) + \frac{df(x_0)}{dx}(x - x_0) + \frac{d^2f}{dx^2}\frac{(x - x_0)^2}{2!} + \cdots$$

The first term is the value of the function at x_0, the second is the slope of the tangent at x_0 times the increment $x - x_0$, the third is proportional to the curvature at x_0 times $(x - x_0)^2$ for small increments. In the illustration the third term overshoots the point $f(x)$, but if the interval $x - x_0$ is not too large, the overshooting is counteracted by the next term which is proportional to the third derivative. It is important that the series converge and that enough terms of the infinite series be used to reduce the remainder of the series to negligible magnitude.

It is clear that because each succeeding term is multiplied by $x - x_0$ to an increasing power, the Taylor series converges more rapidly the smaller the interval $x - x_0$. For differential quantities all differentials of order higher than the first vanish, so that the Taylor series becomes

$$f(x) = f(x_0 + dx) = f(x_0) + \frac{df}{dx}dx$$

which is a particularly useful expansion.

The Taylor series can be extended to a function of two or more independent variables. For a function of two variables for which all partial derivatives exist the expansion is

$$f(x + h, y + k) = f(x, y) + \left(h\frac{\partial}{\partial x} + k\frac{\partial}{\partial y} \right)f(x, y)$$

$$+ \frac{1}{2!}\left(h\frac{\partial}{\partial x} + k\frac{\partial}{\partial y} \right)^2 f(x, y)$$

$$+ \frac{1}{3!}\left(h\frac{\partial}{\partial x} + k\frac{\partial}{\partial y} \right)^3 f(x, y) + \cdots$$

D. The Total Differential

It is often important to be able to evaluate a differential increment taken in an arbitrary direction, that is, along neither the x nor y axis. It is easy to see, however, that because the variables x and y are independent, the increment can be represented by the sum of increments taken along each of the axes. Thus for the function $z = f(x, y)$

$$dz = dz_x + dz_y = \frac{\partial z}{\partial x} dx + \frac{\partial z}{\partial y} dy$$

and the expansion of the differential of temperature as a function of space and time is

$$dT = \frac{\partial T}{\partial t} dt + \frac{\partial T}{\partial x} dx + \frac{\partial T}{\partial y} dy + \frac{\partial T}{\partial z} dz$$

This equation is quite general. It may be applied to many different problems by specifying the displacement in an appropriate manner. Where the increment is taken following the displacement of a particular element, dT/dt is called the *individual derivative* of temperature with respect to time.

When differentials having more than a single geometrical or physical meaning are employed, it is convenient to distinguish them by using different symbols. For example dT usually represents the individual differential of temperature, whereas δT may be employed to represent a space differential of temperature.

E. The Exact Differential

Functions whose value is a unique function of position, like the height of a mountainside, have an important mathematical property. The line integral of the differential of such a function from point A to point B is independent of the path followed from A to B. Having reached B one might return to A by any other path, and then should have found that the line integral around the closed path is zero. A differential which has this property is called an *exact differential*. Examples, in addition to height of a continuous surface, are the geopotential and the temperature distribution in a gas at a fixed time.

F. Gauss' Theorem

This theorem is of great utility in transforming vector functions which are continuous and have continuous partial derivatives. Gauss' theorem (or the divergence theorem) expresses a general relation between surface and volume integrals which may be clearly visualized for the case of fluid flow. The mass of fluid passing a unit area per unit time, the fluid flux, is represented by $\rho\mathbf{V}$, and the component of flux normal to the surface is expressed by $\rho\mathbf{V}\cdot\mathbf{n}$, where $\mathbf{n}$ represents the unit vector directed normally outward from the surface. The rate of outflow from an enclosed volume is then

$$\oiint \rho\mathbf{V}\cdot\mathbf{n}\,d\sigma$$

Now divide the volume into differential rectangular elements as shown in Fig. A-4, and express the flux through each face by expression in a Taylor series about the origin. Then from Fig. A-4, the net outflow through the two xy faces is $(\partial\rho v_y/\partial y)\,dy\,dx\,dz$, and the net outflows through the xy and yz faces are, respectively, $(\partial\rho v_z/\partial z)\,dz\,dx\,dy$ and $(\partial\rho v_x/\partial x)\,dx\,dy\,dz$. If the contributions to net outflow from all the differential volume elements are added, the net outflow from the entire volume is expressed by $\iiint \mathbf{V}\cdot\rho\mathbf{V}\,d\tau$ where $d\tau$ represents the differential volume element. Upon equating the two expressions for net outflow, Gauss' theorem appears in the form

$$\oiint \rho\mathbf{V}\cdot\mathbf{n}\,d\sigma = \iiint \mathbf{V}\cdot\rho\mathbf{V}\,d\tau \tag{F.1}$$

If there is no source of mass within the enclosed volume, the left side of

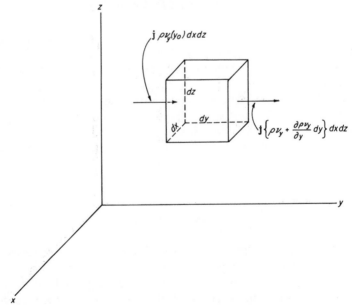

FIG. A-4.   Fluid flux through two faces of a differential rectangular element.

Eq. (F.1) can be expressed by the volume integral $-\iiint (\partial\rho/\partial t)\,d\tau$, and therefore conservation of mass is expressed by

$$-\frac{\partial\rho}{\partial t} = \mathbf{V}\cdot\rho\mathbf{V} \tag{F.2}$$

An equation of this same form can be used for other conservative properties as well. In the case of a nonconservative property q, which is created within a fixed enclosed volume, the appropriate relationship becomes

$$\frac{dq}{dt} = \frac{\partial q}{\partial t} + \mathbf{V}\cdot\mathbf{E}_q \tag{F.3}$$

where $\mathbf{E}_q$ represents the flux of q. This equation is applied in discussing creation of entropy in Section 2.16.

G. Stokes' Theorem

Stokes' theorem relates the area integral of the curl of a vector and the line integral of the tangential component of the vector taken around the area. To develop this theorem, consider the component of $\mathbf{V}\times\mathbf{V}$ perpendicular to the xy plane as shown in Fig. A-5. Evaluation of the line integral around the figure in the positive or counterclockwise sense gives

$$\oint_{dx\,dy} \mathbf{V}\cdot d\mathbf{l} = \left(\frac{\partial v_y}{\partial x} - \frac{\partial v_x}{\partial y}\right) dx\,dy$$

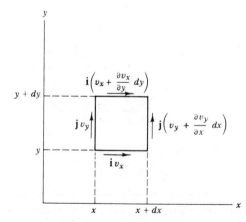

FIG. A-5. Differential element in the xy plane and the components of the vector v along the sides dx and dy.

Now consider a finite area enclosed by a curve as shown in Fig. A-6. Divide the area into differential elements and evaluate the line integral around each element. All contributions to the sum of the line integrals cancel except for the sides of the differential elements which lie along the bounding curve. Therefore, if equations for the enclosed area are summed

$$\oint \mathbf{V} \cdot d\mathbf{l} = \int\int \left(\frac{\partial v_y}{\partial y} - \frac{\partial v_x}{\partial x} \right) dx\, dy$$

which is Stokes' theorem for the plane. Since orientation of the surface is arbitrary, a more general form is

$$\oint \mathbf{V} \cdot d\mathbf{l} = \int\int \mathbf{\nabla} \times \mathbf{V} \cdot \mathbf{n}\, d\sigma \qquad (G.1)$$

where $\mathbf{n}$ represents the outwardly directed unit vector and $d\sigma$ represents the differential element of area.

H.   The Potential Function

Irrotational fields, for which $\mathbf{\nabla} \times \mathbf{V} = 0$, form an important class of vector fields. Such vectors can be represented by the gradient of a scalar "potential" function ϕ, as is evident if one substitutes the components of $\mathbf{\nabla}\phi$ into

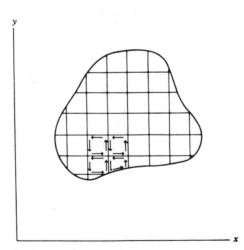

FIG. A-6.   Differential area elements enclosed by a curve and contributions of several elements to the line integral taken around the boundary.

the expansion of $\mathbf{V} \times \mathbf{V}$. The right-hand side of Stokes' theorem now vanishes and the left-hand side becomes

$$\oint \mathbf{V}\phi \cdot d\mathbf{l} = \oint d\phi = 0$$

Evidently, the existence of an exact differential implies that the gradient of the scalar function is irrotational. Also, it follows that a vector which is irrotational may be expressed by the gradient of a scalar function.

I.   Solid Angle

In dealing with several physical problems an understanding of the concept of solid angle is required. Consider a cone with vertex at the origin of a concentric spherical surface as shown in Fig. A-7. The solid angle is defined as the ratio of the area of the sphere intercepted by the cone to the square of the radius. Thus

$$\omega \equiv A/r^2$$

The unit of solid angle is the steradian. The area cut out of a sphere by a steradian is equal to the square of the radius.

The solid angle encompassing all directions at a point is given by the total area of a circumscribed sphere divided by the square of the radius. Therefore, integration over all angles yields

$$\omega = 4\pi \text{ sr}$$

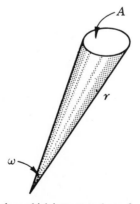

FIG. A-7.   A cone of solid angle ω which intercepts the surface A at distance r from the origin.

Special problems dealing with solid angle are the determination of the number of collisions at a surface area of a wall (Chapter II) and the determination of the radiation flux passing through a unit area (Chapter V). In these problems the orientation of the area is important, and it is convenient to consider the solid angle with respect to the plane dA shown in Fig. A-8. An increment of solid angle can then be represented using spherical coordinates by

$$d\omega_{\theta\phi} = \sin\theta\, d\phi\, d\theta \tag{I.1}$$

In the special case that the vertical axis is an axis of circular symmetry, $d\omega$ may be integrated over ϕ, giving a solid angle increment

$$d\omega_{\theta} = 2\pi \sin\theta\, d\theta \tag{I.2}$$

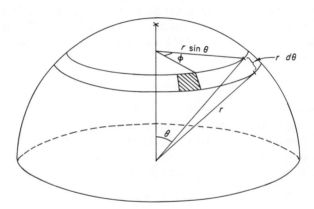

FIG. A-8.   Geometry of solid angle with respect to a fixed plane.

Physical Topics

A. Units and Dimensions

Various internally consistent systems of units have been used in scientific work. The International System of Units (Le Systeme International–SI) is now widely accepted and is used in this book; other units in common use are referred to where particularly relevant.† The SI system utilizes the following base units

Name	Quantity	Symbol
meter	length	m
kilogram	mass	kg
second	time	s
ampere	electric current	A
kelvin	thermodynamic temperature	K
mole	amount of substance	mole
candela	luminous intensity	cd

The *meter* was originally defined as 10^{-7} part of the arc length between the north pole and the equator. It is now defined as 1650763.73 times the wavelength in vacuum of the orange-red line of krypton 86. The *kilogram* was originally defined as 1000 times the mass of a cubic centimeter of distilled water at the temperature of its greatest density, but since 1889 it has been defined by the *standard kilogram*, a platinum–iridium cylinder which is carefully preserved in its original state. The *second*, formerly defined by the rotation rate of the earth, was defined by the 12th General Conference on Weights and Measures in 1967 as the duration of 9,192,631,770 periods of the radiation corresponding to the transition between two hyperfine levels of the ground state of cesium 133.

The quantities of length, mass, time are referred to as the *dimensions* of mechanics and are designated by L, M, and t, respectively. Other mechanical quantities, such as force, momentum, energy, etc., have dimensions which are made up of the three fundamental dimensions. The dimensions of all mechanical quantities may be determined from the definition of the quantity

† The SI system is comprehensively summarized in *The SI Metric Handbook*, John L. Firrer, Scribner, New York, Chas. A. Bennett Co., Peoria, 1977.

or from a physical relation between quantities. Thus the dimensions of momentum and force are, respectively,

$$\text{mass} \times \text{velocity} = MLt^{-1}$$

$$\text{mass} \times \text{acceleration} = MLt^{-2}$$

Electric current (A), thermodynamic temperature (K), and amount of substance (mole) also are regarded as dimensions in the SI system.

B. Significant Figures[†]

It is important in calculation to distinguish between *mathematical* numbers which are known to any accuracy required and *physical* numbers whose accuracy is limited by errors of measurement. It follows that the last digit of a physical number is uncertain. The number of digits in such a number, after any zeros to the left of the first number different from zero, is called the number of significant figures. Thus, if we can read a distance as 98 km, we say that there are two significant figures. We are not permitted to say that this is 98,000 m, for this erroneously gives the impression of a measurement accurate to the order of meters. Instead, we should say that the distance is 98×10^3 m.

Mathematical numbers, which carry any number of significant figures, are treated as though they have at least as many significant figures as the largest number of significant figures in the problem.

Often one more figure may be used than is significant in order to prevent round-off error from influencing the result. The mean of more than 10 and less than 1000 numbers may contain one more significant figure than the individual observations. This may sometimes be used to advantage, as Chapman demonstrated in finding the lunar tide in high latitudes from barometric observations.[‡]

In adding or subtracting, all numbers should have the same number of significant *decimal places*. In order to achieve this, digits beyond the last significant decimal place should be dropped.

In multiplying and dividing the number of significant figures in the result is equal to the number of significant figures in that number which has the smallest number of significant figures.

Special rules may be developed for handling the number of significant

[†] An excellent book on the subject is Yardley Beers, *Introduction to the Theory of Error*, 2nd ed., Addison–Wesley, Cambridge, Massachusetts, 1957.

[‡] S. Chapman, *Quart. J. Roy. Meteorol. Soc.* **44**, 271 (1918).

figures of functions. The basic idea, however, is that the uncertainty of a function $y = f(x)$ is expressed by

$$dy = \frac{df}{dx} dx$$

where dx represents the uncertainty of the error of measurement of x.

C.   Electromagnetic Conversion Table[a]

Quantity	Symbol	SI system	Rationalized CGSe system
Force	F	1 newton (1 N)	$= 10^5$ dyn
Charge	Q	1 coulomb (1 C)	$= 2.997925 \times 10^9$ Fr
Current	I	1 ampere (1 A)	$= 2.997925 \times 10^9$ Fr s^{-1}
Potential	V	1 volt (1 joule C^{-1})	$= 3.335640 \times 10^{-3}$ erg Fr^{-1}
Resistance	R	1 ohm (1 JA^{-2} s^{-1})	$= 1.112649 \times 10^{-12}$ erg s Fr^{-2}
Capacitance	C	1 farad (1 CV^{-1})	$= 8.987556 \times 10^{11}$ Fr^2 erg^{-1}
Electrical field strength	E	1 NC^{-1}	$= 3.335640 \times 10^{-5}$ dyn Fr^{-1}
Magnetic induction[b]	B	1 weber m^{-2} (1 NA^{-1} m^{-1})	$= 3.335640 \times 10^{-7}$ dyn s Fr^{-1} cm^{-1}
Work	W	1 joule (1 J)	$= 10^7$ erg
Power	P	1 watt (1 W = 1 J s^{-1})	$= 10^7$ erg s^{-1}

Permeability (vacuum) $\mu_0 = 4\pi \times 10^{-7}$ weber A^{-1} m$^{-1} = 4\pi \times 1.112649 \times 10^{-21}$ dyn s^2 Fr^{-2}
Permittivity (vacuum) $\varepsilon_0 = 10^7 (4\pi c_0^2)^{-1}$ farad m$^{-1} = \frac{1}{4\pi}$ Fr^2 dyn^{-1} cm^{-2}

[a] Calculated from Nederlandse Natuurkundige Vereniging, *Jaarboek*, p. 131, 1974–5.
[b] The magnetic induction is also expressed in the CGS electromagnetic unit, called the gauss, which is 10^{-4} weber m^{-2}.

D.   Table of Physical Constants

Universal Constants

Velocity of light in vacuum[a]	(c_0)	$(2.99792458 \pm 0.000000012) \times 10^8$ m s^{-1}
Boltzmann constant[a]	(k)	$(1.380662 \pm 0.000044) \times 10^{-23}$ J K^{-1}
Planck constant[a]	(h)	$(6.626176 \pm 0.000036) \times 10^{-34}$ J s
Avogadro's number[a]	(N_0)	$(6.022045 \pm 0.000031) \times 10^{23}$ mole^{-1}
Mass of proton[a]	(m_+)	$(1.6726485 \pm 0.0000086) \times 10^{-27}$ kg
Mass of electron[a]	(m_-)	$(9.109534 \pm 0.000047) \times 10^{-31}$ kg
Charge of electron[a]	(e)	$(1.6021892 \pm 0.0000046) \times 10^{-19}$ C
Stefan–Boltzmann constant[a]	(σ)	$(5.67032 \pm 0.00071) \times 10^{-8}$ W m^{-2} K^{-4}
Standard molar volume of gas[a]	(V_m)	$(2.241383 \pm 0.000070) \times 10^{-2}$ m^3 mole^{-1}
Molar gas constant[a]	(R)	(8.31441 ± 0.00026) J mole^{-1} K^{-1}
Gravitation constant[a]	(G)	$(6.6720 \pm 0.0041) \times 10^{-11}$ N m^2 kg^{-2}
Wien's displacement constant[b]	(α)	(2897.82 ± 0.013) μm K

(*cont.*)

D. Table of Physical Constants (*Continued*)

Atomic mass, carbon-12 nucleus[b]	(C^{12})	12 (exact)
Atomic mass, hydrogen-1 atom[b]	(H^1)	1.007822 ± 0.000003
Atomic mass, oxygen-16 nucleus[b]	(O^{16})	15.994915
Calorie[e]		4.18580 J

Sun

Solar radius[c]		$(6.960 \pm 0.001) \times 10^8$ m
Solar mass[c]		$(1.991 \pm 0.002) \times 10^{30}$ kg
Solar constant[d]	$(\bar{E}_s)$	(1370 ± 1) W m^{-2}

Earth

Earth mass[c]		$(5.977 \pm 0.004) \times 10^{24}$ kg
Mean solar day[b]		1.00273791 sidereal days
Length of year[b]		365.24219878 days
Acceleration of gravity at sea level[e] 45° 32′ 33″	(g_0)	9.80665 m s^{-2}
Effective earth radius[e]		6356.766 km
Angular frequency of earth's rotation	(Ω)	7.29212×10^{-5} s^{-1}

Air

Standard sea level pressure[e]	(p_0)	1.013250×10^2 kPa
Standard sea level temperature[e]	(T_0)	288.15 K
Average molecular mass at sea level[e]	(M_d)	28.9644
Thermal conductivity at sea level, 15°C[e]	(λ)	2.5326×10^{-3} W m^{-1} K^{-1}
Viscosity at sea level, 15°C[e]	(μ)	1.7894×10^{-5} kg m^{-1} s^{-1}
Diffusivity of water vapor in air at sea level, 15°C[f]	(D)	2.49×10^{-4} m^2 s^{-1}

Water

Molecular mass[g]	(M_w)	18.0160
Ice point[c]	$(0°C)$	(273.155 ± 0.015) K

[a] International Council of Scientific Unions, *CODATA Bulletin No. 11*, Dec. 1973.
[b] Nederlandse Natuurkundige Vereniging, *Jaarboek*, 1962.
[c] C. W. Allen, *Astrophysical Quantities*, Univ. London, p. 11, 1955.
[d] R. C. Wilson, J. R. Hickey, *The Solar Output and its Variation*, O. R. White, ed., Univ. Colorado, 1977.
[e] *U.S. Standard Atmosphere*, 1976, NOAA, NASA, USAF, Washington, D.C., 1976.
[f] R. B. Montgomery, *J. Meteor.*, **4**, 193, 1947.
[g] W. E. Forsythe, *Smithsonian Physical Tables*, 9th rev. ed., Smithsonian Institution, Washington, D.C., 1959.

E.   U.S. Standard Atmosphere 1976[a] and COSPAR International Reference Atmosphere (CIRA)[b]

Height[a] (km)	Temp.[a] (K)	Pressure[a] (kPa)	Density[a] (kg m^{-3})	Temp.[b] (K)	Pressure[b] (kPa)	Density[b] (kg m^{-3})
0	288.15	101.330	1.225			
1	281.65	89.880	1.112			
2	275.15	79.500	1.007			
3	268.66	70.120	9.093×10^{-1}			
4	262.17	61.660	8.194×10^{-1}			
5	255.68	54.050	7.364×10^{-1}			
6	249.19	47.220	6.601×10^{-1}			
8	236.22	35.650	5.258×10^{-1}			
10	223.25	26.500	4.135×10^{-1}			
12	216.65	19.400	3.119×10^{-1}			
15	216.65	12.110	1.948×10^{-1}			
20	216.65	5.529	8.891×10^{-2}			
25	221.55	2.549	4.008×10^{-2}	221.7	2.483	3.899×10^{-2}
30	226.51	1.197	1.841×10^{-2}	230.7	1.175	1.774×10^{-2}
35	236.51	5.746×10^{-1}	8.463×10^{-3}	241.5	5.741×10^{-1}	8.279×10^{-3}
40	250.35	2.871×10^{-1}	3.996×10^{-3}	255.3	2.911×10^{-1}	3.972×10^{-3}
45	264.16	1.491×10^{-1}	1.966×10^{-3}	267.7	1.525×10^{-1}	1.995×10^{-3}
50	270.65	7.978×10^{-2}	1.027×10^{-3}	271.6	8.241×10^{-2}	1.057×10^{-3}
55	260.77	4.253×10^{-2}	5.681×10^{-4}	263.9	4.406×10^{-2}	5.821×10^{-3}
60	247.02	2.196×10^{-2}	3.097×10^{-4}	249.3	2.296×10^{-2}	3.206×10^{-3}
65	233.29	1.093×10^{-2}	1.632×10^{-4}	232.7	1.146×10^{-2}	1.718×10^{-3}
70	219.59	5.221×10^{-3}	8.283×10^{-5}	216.2	5.445×10^{-3}	8.770×10^{-5}
75	208.40	2.388×10^{-3}	3.992×10^{-5}	205.0	2.460×10^{-3}	4.178×10^{-5}
80	198.64	1.052×10^{-3}	1.846×10^{-5}	195.0	1.067×10^{-3}	1.905×10^{-5}
85	188.89	4.957×10^{-4}	8.220×10^{-6}	185.1	4.426×10^{-4}	8.337×10^{-6}
90	186.87	1.836×10^{-4}	3.416×10^{-6}	183.8	1.795×10^{-4}	3.396×10^{-6}
95	188.42	7.597×10^{-5}	1.393×10^{-6}	190.3	7.345×10^{-5}	1.343×10^{-6}
100	195.08	3.201×10^{-5}	5.604×10^{-7}	203.5	3.090×10^{-5}	5.297×10^{-7}
105	208.84	1.448×10^{-5}	2.325×10^{-7}	228.0	1.422×10^{-5}	2.173×10^{-7}
110	240.00	7.104×10^{-6}	9.708×10^{-8}	265.5	7.362×10^{-6}	9.661×10^{-8}
115	300.00	4.010×10^{-6}	4.289×10^{-8}	317.1	4.236×10^{-6}	4.645×10^{-8}
120	360.00	2.538×10^{-6}	2.222×10^{-8}	380.6	2.667×10^{-6}	2.438×10^{-8}

[a] *U.S. Standard Atmosphere*, 1976, NOAA, NASA, USAF, Washington, D.C., 1976.

[b] *COSPAR International Reference Atmosphere*, 1972, Akademie Verlag, Berlin, 1972.

Bibliography

Allis, W. P., and M. A. Herlin, "Thermodynamics and Statistical Mechanics." McGraw–Hill, New York, 1952. (Chapter 2.)

Banks, P. M., and G. Kockarts, "Aeronomy." Academic Press, New York, 1973. (Chapter 2.)

Battan, L. J., "Radar Observation of the Atmosphere." Univ. of Chicago Press, Chicago, Illinois, 1973. (Chapter 7.)

Berry, F. A., et al. (eds.), "Handbook of Meteorology," p. 283: J. Charney, Radiation. McGraw–Hill, New York, 1945. (Chapter 5.)

Brunt, D., "Physical and Dynamical Meteorology," 2nd ed. Cambridge Univ. Press, London and New York, 1941. (Chapters 4 and 6.)

Budyko, M. I., "The Heat Balance of the Earth's Surface." U.S. Weather Bureau, Washington, D.C., 1956 (translated by Nina A. Stepanova, 1958). (Chapter 6.)

Chalmers, J. A., "Atmospheric Electricity," 2nd ed. Pergamon Press, New York, 1967. (Chapter 3.)

Chamberlain, J. W., "Physics of the Aurora and Airglow." Academic Press, New York, 1961. (Chapter 5.)

Chandrasekhar, S., "Radiative Transfer." Oxford Univ. Press, London and New York, 1950. (Chapter 5.)

Chapman, S., and R. S. Lindzen, "Atmospheric Tides, Thermal and Gravitational." D. Reidel, Dordrecht, Holland, 1970. (Chapter 1.)

Craig, R. A., "The Upper Atmosphere: Meteorology and Physics." Academic Press, New York, 1965. (Chapter 5.)

Derr, V. E., ed., "Remote Sensing of the Troposphere." U.S. Dept. Commerce and Univ. of Colorado, U.S. Govt. Printing Office, Washington, D.C., 1972. (Chapter 7.)

Dutton, J. A., "The Ceaseless Wind. An Introduction to the Theory of Atmospheric Motion." McGraw–Hill, New York, 1976. (Chapter 4.)

Elsasser, W. M. with M. F. Culbertson, Atmospheric radiation tables. *Meteorol. Monographs* **4**, No. 23 (1960). Am. Meteorol. Soc., Boston, Massachusetts. (Chapter 5.)

Fletcher, N. H., "The Physics of Rainclouds." Cambridge Univ. Press, London and New York, 1962. (Chapter 3.)

Geiger, R., "The Climate Near the Ground." (translated from 4th German edition). Harvard Univ. Press, Cambridge, Mass., 1965. (Chapter 6.)

Gish, O. H., Universal aspects of atmospheric electricity. *Compendium Meteorol.*, p. 101 (1951). Am. Meteorol. Soc., Boston, Massachusetts. (Chapter 3.)

Goody, R. M., "Atmospheric Radiation, Vol. 1, Theoretical Basis." Clarendon Press, Oxford, 1964. (Chapter 5.)

Gossard, E. E., and W. H. Hooke, "Waves in the Atmosphere." Elsevier, Amsterdam, 1975. (Chapter 4.)

Gribbin, J., ed., "Climatic Change." Cambridge Univ. Press, London and New York, 1978. (Chapter 6.)

Halliday, D., and R. Resnick, "Physics for Students of Science and Engineering." Wiley, New York, 1960. (Chapters 3 and 4.)

Haltiner, G. J., and F. L. Martin, "Dynamical and Physical Meteorology." McGraw–Hill, New York, 1957. (Chapters 1, 4, and 5.)

Haugen, D. A., ed., "Workshop on Micrometeorology." Am. Meteorol. Soc., Boston, Massachusetts, 1973. (Chapter 6.)

Helliwell, R. A., "Whistlers and Related Ionospheric Phenomena." Stanford Univ. Press, Palo Alto, 1965. (Chapter 7.)

Hess, S., "Introduction to Theoretical Meteorology." Henry Holt, New York, 1959. (Chapter 4.)

Hinze, J. O., "Turbulence," 2nd ed. McGraw–Hill, New York, 1975. (Chapter 6.)

Hirschfelder, J. O., C. F. Curtiss, and R. B. Bird, "Molecular Theory of Gases and Liquids." Wiley, New York, 1954. (Chapters 2 and 3.)

Holmboe, J., G. E. Forsythe, and W. Gustin, "Dynamic Meteorology." Wiley, New York, 1945. (Chapters 2 and 4.)

Holton, J. R., "An Introduction to Dynamic Meteorology." Academic Press, New York, 1972. (Chapter 4.)

Humphreys, W. J., "Physics of the Air," 3rd ed. McGraw–Hill, New York, 1940. (Chapter 7.)

Johnson, J. C., "Physical Meteorology." M.I.T. Press and Wiley, New York, 1954. (Chapter 7.)

Joos, G., "Theoretical Physics," 2nd ed. (translated by Ira M. Freeman). Hafner, New York, 1950. (Chapters 1 and 7.)

Junge, C. E., "Air Chemistry and Radioactivity." Academic Press, New York, 1963. (Chapter 3.)

Kennard, E. H., "Kinetic Theory of Gases." McGraw–Hill, New York, 1938. (Chapter 2.)

Kondratyev, K. Ya., "Radiation in the Atmosphere." Academic Press, New York, 1969. (Chapter 5.)

Kourganoff, V., "Basic Methods in Transfer Problems." Oxford Univ. Press, London and New York, 1953. (Chapter 5.)

Lettau, H., and B. Davidson, eds., "Exploring the Atmosphere's First Mile," Vols. 1 and 2. Pergamon, New York, 1957. (Chapter 6.)

Lumley, J. L., and H. A. Panofsky, "The Structure of Atmospheric Turbulence." Wiley (Interscience), New York, 1964. (Chapter 6.)

Mason, B. J., "The Physics of Clouds," 2nd ed. Oxford Univ. Press, London and New York, 1971. (Chapter 3.)

Mason, B. J., "Clouds, Rain and Rainmaking," 2nd ed. Cambridge Univ. Press, London and New York, 1975. (Chapter 3.)

Möller, F., Strahlung in der Unteren Atmosphäre in "Handbuch der Physik" (J. Bartels, ed.), Vol. 48, p. 155. Springer, Berlin, 1957. (Chapter 5.)

Morel, P., ed., "Dynamic Meteorology." D. Reidel, Dordrecht, Holland, 1973. (Chapter 4.)

Morse, P. M., "Thermal Physics," 2nd ed. W. A. Benjamin, New York, 1969. (Chapter 2.)

Panofsky, W. K. H., and M. Phillips, "Classical Electricity and Magnetism." Addison–Wesley, Reading, Massachusetts, 1955. (Chapter 7.)

Pasquill, F., "Atmospheric Diffusion." D. Van Nostrand, London, New York, Princeton, Toronto, 1962. (Chapter 6.)

Pedlosky, J., "Geophysical Fluid Dynamics." Springer-Verlag, New York, 1979 (Chapter 4).

Priestley, C. H. B., "Turbulent Transfer in the Lower Atmosphere." Univ. of Chicago Press, Chicago, Illinois, 1959. (Chapter 6.)

Pruppacher, H. R., and J. D. Klett, "Microphysics of Clouds and Precipitation." D. Reidel, Dordrecht, Holland, 1978. (Chapter 3.)

Ridenour, L. N., ed., "Modern Physics for the Engineer," p. 330: L. B. Loeb, Thunderstorms and lightning strokes. McGraw–Hill, New York, 1954. (Chapter 3.)

Sears, F. W., "Principles of physics," Vol. 1: Mechanics, Heat, and Sound. Addison–Wesley, Reading, Massachusetts, 1947. (Chapter 1.)

Sears, F. W., "Principles of Physics," Vol. 2: Electricity and Magnetism. Addison–Wesley, Reading, Massachusetts, 1947. (Chapters 3 and 7.)

Sears, F. W., "Principles of Physics," Vol. 3: Optics. Addison–Wesley, Reading, Massachusetts, 1947. (Chapter 7.)

Sears, F. W., "An Introduction to Thermodynamics, the Kinetic Theory of Gases and Statistical Mechanics." Addison–Wesley, Reading, Massachusetts, 1953. (Chapter 2.)

Sellers, W. D., "Physical Climatology." Univ. of Chicago Press, Chicago, Illinois, 1965. (Chapter 6.)

Sommerfeld, A., "Thermodynamics and Statistical Mechanics." Academic Press, New York, 1956. (Chapter 2.)

Tennekes, H., and J. L. Lumley, "A First Course in Turbulence." M.I.T. Press, Cambridge, Massachusetts, 1972. (Chapter 6.)

Unsöld, A., "Physik der Sternatmosphären." Springer, Berlin, 1955. (Chapter 7.)

Van de Hulst, H. C., "Light Scattering by Small Particles." Wiley, New York, 1957. (Chapter 7.)

Wallace, J. M., and P. V. Hobbs, "Atmospheric Sciences: An Introductory Survey." Academic Press, New York, 1977. (Chapters 2 and 3.)

Index

International Geophysics Series

EDITED BY

J. VAN MIEGHEM
(July 1959–July 1976)

ANTON L. HALES
(January 1972–December 1979)

WILLIAM L. DONN
Lamont-Doherty Geological Observatory
Columbia University
Palisades, New York